Social Mendelism

Who was the scientific progenitor of eugenic thought? Amir Teicher challenges the preoccupation with Darwin's eugenic legacy by uncovering the extent to which Gregor Mendel's theory of heredity became crucial in the formation – and radicalization – of eugenic ideas. Through a compelling analysis of the entrenchment of genetic thinking in the social and political policies in Germany between 1900 and 1948, Teicher exposes how Mendelian heredity became saturated with cultural meaning, fed racial anxieties, reshaped the ideal of the purification of the German national body and ultimately defined eugenic programs. Drawing on scientific manuscripts and memoirs, bureaucratic correspondence, court records, school notebooks and Hitler's table talk as well as popular plays and films, *Social Mendelism* presents a new paradigm for understanding links between genetics and racism, and between biological and social thought.

Amir Teicher is Assistant Professor in the Department of History, Tel Aviv University.

Social Mendelism

Genetics and the Politics of Race in Germany, 1900–1948

Amir Teicher

Tel Aviv University

CAMBRIDGE
UNIVERSITY PRESS

CAMBRIDGE
UNIVERSITY PRESS

University Printing House, Cambridge CB2 8BS, United Kingdom

One Liberty Plaza, 20th Floor, New York, NY 10006, USA

477 Williamstown Road, Port Melbourne, VIC 3207, Australia

314–321, 3rd Floor, Plot 3, Splendor Forum, Jasola District Centre,
New Delhi – 110025, India

79 Anson Road, #06–04/06, Singapore 079906

Cambridge University Press is part of the University of Cambridge.

It furthers the University's mission by disseminating knowledge in the pursuit of
education, learning, and research at the highest international levels of excellence.

www.cambridge.org
Information on this title: www.cambridge.org/9781108499491
DOI: 10.1017/9781108583190

First published 2020

Printed in the United Kingdom by TJ International Ltd. Padstow Cornwall

A catalogue record for this publication is available from the British Library.

ISBN 978-1-108-49949-1 Hardback

Cambridge University Press has no responsibility for the persistence or accuracy
of URLs for external or third-party internet websites referred to in this publication
and does not guarantee that any content on such websites is, or will remain,
accurate or appropriate.

For my grandparents

The hereditary mass of an individual is a mosaic from numerous hereditary units, each having a 50% chance to be combined [during sexual reproduction] with the hereditary units of the other partner. Each one of these genes (hereditary units) maintains its integrity in the course of generations. If one inter-crosses two different races, none of these individual hereditary units can ever fully disappear, but each hereditary trait will appear among the offspring following ratios that can be computed according to Mendel's hereditary laws. So, for example, the probability that a certain hereditary disposition from a particular grandparent would appear by a grandchild is one fourth.

Certain hereditary factors, when they come together with certain others, penetrate through, that is, they leave their mark on the hybrid (dominant). The hereditary factor that is thereby repressed (recessive) is however not lost, but will become noticeable again in further crossings, if the joining of hereditary factors would result in a combination that contains only the recessive dispositions.

With respect to the application of the first Mendelian law, each hereditary unit needs to be considered on its own (it mendels independently).

– The opening sentences of a memorandum titled "What demands must the hereditary biologist raise regarding the solution of the Jewish question?"prepared by the German Interior Ministry expert on national health, Arthur Gütt, for a discussion regarding the legal definition of Jews and Jewish *Mischlinge*, ten days after the proclamation of the Nuremberg Laws, 1935)

"I wanted to out-mendel myself (Ich wollte mich herausmendeln)!"
– Nuclear physicist Fritz Houtermans, explaining to his wife why he decided to identify himself as Jewish, despite having only one Jewish grandparent, 1935

Contents

Figures

Every effort has been made to contact the relevant copyright holders for the images reproduced in this book. In the event of any error, the publisher will be pleased to make corrections in any reprints or future editions.

Acknowledgments

This book grew out of a doctoral dissertation, written at Tel Aviv University under the caring supervision of two distinguished scholars, Shulamit Volkov and Eva Jablonka. Volkov provided me uncompromising guidance and support since my days as an undergraduate student; much of what I know – and love – about historical research comes from her. Jablonka, my second "Doktormutter," saved me from many pitfalls and warmly encouraged me throughout the entire project. I owe her a great deal.

Snait Gissis has been my academic guardian angel from very early on. Her unconditional backing and unbelievable modesty are a rare diamond in our world.

During the first four years of my doctoral studies, my research was supported by a President of the University and Rector's Scholarship for Excellent PhD Students (2008–2012). Three research visits to Germany were funded by the Minerva Foundation (short-term research grants). In Germany, I was hosted twice by Veronika Lipphardt's group in the Max-Planck Institute for the History of Science (MPIWG) as a pre-doctoral fellow. In my fifth year as a doctoral student, I had the privilege of becoming a fellow at the Edmund J. Safra Center for Ethics and benefited from both its financial support as well as its stimulating intellectual environment. At Tel Aviv University, the Zvi Yavetz School of Historical Studies, the Minerva Institute for German History, and the Stephen Roth Institute for the Study of Contemporary Antisemitism and Racism all supported my work, financially and otherwise. At the final stages, a grant from the Israeli Science Foundation (ISF grant 145/18) proved crucial for bringing this project to the finish line.

Many scholars have commented on parts of this work at one point or another, and their critique and advice were of immense importance. Among them I should mention Gadi Algazi, Batya Amir, Mitchell G. Ash, Jose Bruner, Eric Engstrom, Raphael Falk, Michael Hagner, Shai Lavi, Veronika Lipphardt, Pauline Mazumdar, Billie Melman, Amos Morris-Reich, Staffan Müller-Wille, Boaz Neumann, Diana B. Paul,

Theodore M. Porter, Oded Rabinovitch, Iris Rachmimov, David Sabean, Galili Schachar, Falko Schmieder, Stefanie Schüller-Springorum, Sagie Shaeffer, Michal Shapira, Simon Teuscher, Scott Ury, Paul Weindling, Yossi Yovel and Michael Zakim. Among the archivists who helped me in my work I would like to thank in particular Britta Leise (MPIP), Annett Krefft and Bettina Reimers (DIPF/BBF), Johann Zilien (HHStAW), Klaus-Dieter Rack (HStAD), the staff of the MPIWG library and the staff of the Wiener Library (TAU).

Finally, my English may have been only partially comprehensible were it not for the proofreading and corrections of Sarah Mandel; and I would probably not have been able to complete this project without the immense help of my two diligent research assistants, Lea Herzig and Clara Hestermann. Thank you.

Abbreviations

ARGB	Archiv für Rassen- und Gesellschaftsbiologie
BFL	Erwin Baur, Eugen Fischer & Fritz Lenz, *Menschliche Erblichkeitslehre und Rassenhygiene* (3rd ed.), Munich: J. F. Lehmanns, 1927.
FB	Familiengeschichtliche Blätter
Oxford Handbook	Alison Bashford & Philippa Levine (eds.), *The Oxford Handbook of the History of Eugenics*, New York, NY: Oxford University Press, 2010.
ZgNP	Zeitschrift für die gesamte Neurologie und Psychiatrie
ZMA	Zeitschrift für Morphologie und Anthropologie

Archival Material

BArch	Bundesarchiv Berlin-Lichterfelde, Berlin, Germany
BayHStA	Bayerisches Hauptstaatsarchiv
DIPF/BBF/Archiv	Deutsches Institut für internationale pädagogische Forschung, Bibliothek für bildungsgeschichtliche Forschung
HStAD	Hessisches Staatsarchiv, Darmstadt
HHStAW	Hessisches Hauptstaatsarchiv, Wiesbaden
IfZ Munich	Institut für Zeitgeschichte, München
LAB	Landesarchiv Berlin
MPG Archive	Max-Planck-Gesellschaft Archiv
MPIP	Historisches Archiv des Max-Planck-Instituts für Psychiatrie, Munich, Germany
Staatsarchiv Munich OLG	Staatsarchiv München Oberlandesgericht
YVA	Yad-Vashem Archives, Jerusalem, Israel

Introduction

In 1865, at two consecutive meetings of the Natural Science Society in the Moravian city of Brünn (Brno), an Augustinian friar by the name of Gregor Mendel outlined the results of fertilization experiments he conducted on peas. Carefully planned and meticulously analyzed, Mendel's investigations suggested that the pattern of inheritance of individual pea traits, such as height, color or shape, was governed by a simple mechanism, wherein each trait was determined by an independent hereditary factor. Not before the dawn of the twentieth century, thirty-five years later, did Mendel's results garner their due appreciation. Recognized today as the cornerstone of modern genetics, Mendel's laws of heredity transformed forever our understanding of the process of heredity in plants, animals and humans.

The importance of Mendel's theory for the growth of modern genetics is uncontested. But how did the rediscovery of Mendel's theorem, and the research practices that it enshrined, impact the broader domain of racial and eugenic thought throughout the first half of the previous century? In the case of Germany, what changes did Mendel's laws promote in the scientific perceptions of anthropologists, psychiatrists and genealogists, and how did these changes affect the social programs initiated, supported and finally implemented by those same scientists, before and after the Nazis came to power? How did Mendelian ideas influence the way Germans of all ranks understood the nature of the hereditary threat that racial-aliens and the hereditarily damaged posed to their well-being and to their future, and how did Mendelian reasoning inform the measures chosen in order to face those dangers?

The importance of these questions goes beyond the specific case of Nazi Germany's eugenic policies. Eugenics and its related sciences serve as a prototype for how scientific and social thought can become intertwined, with potentially horrendous results. Studying eugenics provides an opportunity to explore the dynamics between scientific theories, cultural perceptions, political agendas and social policies in the modern world. The links between science, culture and society have fascinated scholars and the public during the course of the past several decades. While the popular image of science is all too often that of an autonomous kingdom of the mind – progressing from one achievement to the next through the materialization of genius ideas and cognitive breakthroughs of great, white-bearded men – for scholars studying science, and probably for most practicing scientists themselves, it is obvious that science and scientific thought, while having their own unique dynamics of development, are also embedded in certain cultural environments, shaped by social institutions and constrained by material considerations. In the flourishing field of Science Studies, there are many open questions – and plenty of intriguing answers – regarding the intricate dynamics of the interaction between scientific practices and perceptions and their social or ideological pretext, or impact.

These issues are of special interest in the case of Nazi bio-political thought. Infamously, the Nazis prided themselves on grounding their world-view in biological principles. Consequently, historical accounts on the intellectual origins of National Socialism often invoke the name of Charles Darwin or of his German disciple, Ernst Haeckel, whose evolutionary theories, so it is argued, supplied racial theorists and politicians with an overall framework for constructing narratives on racial struggles between peoples and nations, struggles in which those who survived were necessarily superior in physical, intellectual or moral qualities. For racial theorists, the argument continues, the survival of the fittest was elevated from a characterization of evolutionary dynamics to the sphere of normative judgment. The science of eugenics – a term that was coined by Darwin's half cousin, the British polymath Francis Galton – set out to redirect human evolutionary processes and to combat physical, mental and social degeneration by regulating human reproduction. The rise of eugenic thought in Germany and its eventual manifestation in Nazism are therefore characterized as the epitome of "Social Darwinism," or of the application of Darwin's ideas to society in an attempt to eliminate parasitic elements and alien races.[1]

---

[1] Analysis of the relations between Darwinian thinking and racism informs many works on German racial and eugenic thought, from Hanna Arendt, *Elements of Totalitarianism* (New York, NY: Harcourt Brace Jovanovich, 1951), 178–179, Daniel Gasman, *The Scientific Origins of National Socialism* (London: Macdonald & Co., 1971), and Peter Weingart, Jürgen Kroll and Kurt Bayertz, *Rasse, Blut und Gene: Geschichte der Eugenik und Rassenhygiene in Deutschland* (Frankfurt a. M.: Suhrkamp, 1988), to Jerry Bergman, *Hitler and the Nazi Darwinian Worldview: How the Nazi Eugenic Crusade for a Superior Race Caused the Greatest Holocaust in World History* (Kitchener, ON: Joshua Press, 2012). The relationship between Darwin's thought and Nazism is outlined most explicitly in Richard Weikart, *From Darwin to Hitler: Evolutionary Ethics, Eugenics, and Racism in*

How does Mendel – or rather, his theory, as understood in the early twentieth century – fit into this story? Current historiography provides three possible answers to this question. The first identifies Mendelism with the notion of "hard heredity," or the idea that genes are immutable units of heredity, indifferent to environmental influences. Mendel was in reality not the progenitor of this idea – August Weismann is probably the more natural candidate for that role – but it is nevertheless true that many early twentieth-century individuals associated Mendel with it, or saw Mendelian theory as providing the strongest counterargument to the opposite, environmentalist, "Lamarckian" view. The main problem with equating Mendel with hard heredity is not its historical inaccuracy, but its narrowness, as it fails to capture the complexity and richness of Mendelian thinking and therefore also misses its wide-ranging impact on various domains of social and scientific thinking.[2] Mendelian theory, as this book shows, had much more to offer than the mere notion of hard heredity.

Another answer provided by existing historiography acknowledges other elements of Mendelian theory but reduces them to mere technicalities, with few implications of any real significance. Michael Burleigh and Wolfgang Wippermann's authoritative *The Racial State* is a case in point. After describing Darwin as the "involuntary progenitor of racist ideology," Burleigh and Wippermann move on to discuss Hitler's *Mein Kampf*

Germany (New York, NY: Palgrave Macmillan, 2004), a book that was received with reservations by many in the academic community. See for instance the criticism offered in Robert J. Richards, *Was Hitler a Darwinian? Disputed Questions in the History of Evolutionary Theory* (Chicago, IL: University of Chicago Press, 2013), 192–242. Weikart's response may be found at www.evolutionnews.org/2014/01/was_hitler_a_da080761.html. As these latter examples demonstrate, in American public discourse the purported links between Darwin's theory and Nazism have been mobilized by creationists to discredit evolutionary thought as supposedly having proto-Nazi ingredients. The scholarly debate on this theme thus became overburdened with political implications, which didn't always work in its favor. For more balanced assessments on the meaning of Social Darwinism in Germany, cf. Richard J. Evans, "In Search of German Social Darwinism. The History and Historiography of a Concept," in Manfred Berg and Geoffrey Cocks (eds.), *Medicine and Modernity: Public Health and Medical Care in 19th- and 20th-Century Germany* (Washington, DC: German Historical Institute, 1997), 55–79, and Diane B. Paul, "Darwin, Social Darwinism and Eugenics," in Jonathan Hodge and Gregory Radick (eds.), *The Cambridge Companion to Darwin*, 2nd ed. (Cambridge: Cambridge University Press 2009), 219–245.

[2] In the words of Staffan Müller-Wille and Hans-Jörg Rheinberger: "Most historical studies of eugenics do not scrutinize the concepts of heredity that underwrote these movements. ... If any effort is made to characterize notions of heredity, the well-worn dichotomy of 'hard' versus 'soft' is usually employed." Staffan Müller-Wille and Hans-Jörg Rheinberger, *A Cultural History of Heredity* (Chicago, IL: University of Chicago Press, 2012), 99. For a discussion, see Maurizio Meloni, *Political Biology* (Basingstoke: Palgrave Macmillan, 2016).

and state that "Hitler eschewed technical scientific terms like Weismann's 'germ plasm' or Mendelian 'hereditary properties' in favor of calls for the 'maintenance of the purity of the blood.'" They fail to recognize that Hitler's perception of such blood purification had Mendelian concepts already absorbed into it, concepts that were used by Hitler and other Nazi leaders on several different occasions.[3] Another example for the marginalization of Mendelism can be found in Saul Friedländer's masterful account, *Nazi Germany and the Jews*. Friedländer does cite Hitler's statement that in "the seventh, eighth, or ninth generation the Jewish part would be 'out-Mendeled'" but is quick to clarify: "ausgemendelt – a pun on the name of the Czech monk, Gregor Mendel, who discovered the laws of heredity."[4] The present study argues that references to Mendel, by Hitler as well as by others, were much more than "a pun," and that they often testified to the internalization of a particular way of thinking that was enshrined by Mendelian theory and that needs to be acknowledged, explored and analyzed, not dismissed.

Finally, some writers on racism and its relation to genetics portray Mendelian theory as antiracist and benevolent. Mendelism may have been abused and oversimplified by its early proponents, such as the American eugenicist Charles B. Davenport, but ultimately it led to the dissolution of racial dogmas and to the emergence of more progressive, liberally minded biology.[5] This perception of Mendelism, as the final section of this book shows, is in itself an outgrowth of post–World War II

---

[3] Examples of Nazi leaders' usage of Mendelian concepts will be provided in Chapter 5. Burleigh and Wippermann further claim, "In contrast to racial-hygienicists, Hitler expected no immediate results from these [racial-hygienic] measures," because he prophesied that the process of regenerating the nation and weeding out the physically degenerate and mentally ill would require six hundred years. Such views seem (and are presented as) senseless. An acknowledgment of their underlying Mendelian propositions makes them intelligible and, in contrast to Burleigh and Wippermann's assertion, fully compatible with the views of other racial-hygienicists. See Michael Burleigh and Wolfgang Wippermann, *The Racial State: Germany, 1933–1945* (Cambridge: Cambridge University Press, 1991), 28, 40. For a recent critical assessment of their work, see Devin O. Pendas, Mark Roseman and Richard F. Wetzell (eds.), *Beyond the Racial State: Rethinking Nazi Germany* (Cambridge: Cambridge University Press, 2017).

[4] Saul Friedländer, *Nazi Germany and the Jews*, vol. 2, *The Years of Extermination* (New York, NY: HarperCollins, 2007), 278.

[5] For examples, consider the description provided in the Human Genome Project website: www.genome.gov/25019887/online-education-kit-timeline-from-darwin-and-mendel-to-the-human-genome-project/, as well as the Wikipedia entry: https://en.wikipedia.org/wiki/History_of_genetics. See also Jonathan Harwood and Michael Banton, *The Race Concept* (Newton Abbot: David & Charles, 1975), 47–49, where Mendel's laws (specifically, the recognition that during sexual reproduction, genes for different traits are "shuffled like a pack of playing cards," p. 49) are presented as striking a definite blow to essentialist racial thinking, and Lucius Outlaw, "Toward a Critical Theory of 'Race,'" in Bernard Boxill (ed.), *Race and Racism* (Oxford: Oxford University Press, 2001), 58–82.

developments. While it is true that Mendelian thinking posed certain challenges to racial thinking, it also provided solutions to other pressing disciplinary problems and was eagerly embraced by racial thinkers, including top-ranking Nazis. Hence, while presenting Mendelism as essentially antiracist may be reassuring, it would be a gross anachronism.

Contrary to all of these descriptions, the present book argues that Mendel's theory of heredity had a far-reaching impact on how Germans and Nazis thought about society, purity, national renewal and medical dangers. The extension of Mendelian thinking to the human domain – an extension that Mendel himself could neither foresee nor imagine – is what I term Social Mendelism. If the principles of Darwinian thought gradually converged around the struggle over limited resources, nature's selection powers and the ultimately benevolent outcome of free competition, Mendelian thought offered its own independent toolbox for thinking about the sociobiological dynamics leading from the past through the present to the future. Significantly, the notions and axioms Mendelian theory provided were distinct from those associated with Darwin's work, and most of them did not necessitate an underlying evolutionary narrative. Naturally, however, in certain areas – most notably racial-hygiene – Mendelian principles and Darwinian reasoning were meshed together. Disentangling the threads of Mendelian thinking and identifying their multifarious manifestations is necessary for reconstructing the Nazi mind-set, and helps us appreciate the unique relations between the biological and the cultural sides of Nazi policies regarding both Jews and the mentally ill.

Toward the end of this introduction I will provide a succinct description of the essential components of Social Mendelian thinking. Before doing so, however, I will highlight a significant transformation in the role of Mendelian theory for scholars interested in human heredity, which took place during the mid-1910s, and present an overview of the content of the book. I will then turn to the question of the scientific/pseudoscientific nature of Mendelian theory, eugenics and Nazi science in general. At the close of this introduction, I include a short glossary of terms to help orient readers who feel less comfortable with the biological lexicon. This brief introduction should provide the uninitiated with all the necessary preliminary knowledge for reading this book.

Mendelism and the Study of Human Heredity

As odd as it may sound, until the mid-1950s, humans did not yet have DNA. In the nineteenth century, what humans did have were hereditary transmissions, hereditary substances, hereditary influences and hereditary burdens. Were these running through the blood stream or residing within the germ cells? Indifferent or sensitive to environmental signals? Tending to blend into each other or gradually accumulating? Containing messages from remote ancestors or sensitive to the sex of their carriers? Inheritance was an undeniable phenomenon with occurrences in everyday life; but the nature of inherited properties was an enigma, and the patterns of generational transmission were fertile ground for competing hypotheses.

The new century brought with it the rediscovery of Mendel's work, which offered a solution to some of these mysteries by defining a simple mechanism governing the transmission of hereditary factors, later to be called genes. Throughout the first decade of the twentieth century, Mendelian mathematical and experimental methods were being defined, as the theory gained increasing popularity in various scientific communities. Mendelism was adopted enthusiastically in the United States, where during the 1910s, and as a result of its obvious success, its experimental ramifications became ever more sophisticated. Acknowledging the role of chromosomes in the hereditary process, the American "Lord of the Flies," Thomas Hunt Morgan, performed detailed studies on fruit flies that transcended the original framework of Mendelian thought, applying new experimental methods and mobilizing new interpretative techniques. Concepts such as chromosomal crossover and chromosomal maps became established; later, in the 1920s, novel technologies were also piloted, such as inducing mutations via radiation. At the same time, alternative ways of studying different facets of the hereditary process took shape worldwide. A decade and a half after the rediscovery of Mendel's work, for both supporters and opponents of Mendelism, studying heredity came to require increasing engagement with cellular mechanisms and developmental processes, way beyond the initial mathematical and experimental framework that Mendel had laid out.[6]

This was true in Germany, too. From the mid-1910s, most German geneticists dealing with heredity in animals and plants were either anti-Mendelian or simply non-Mendelian. Jonathan Harwood's 1993 book on the German genetics community and later works by Bernd Gausemeier demonstrated that many experimental geneticists and plant breeders distanced themselves from the American-led (Mendelian) focus

---

[6] For a general overview, see Ernst Mayr, *The Growth of Biological Thought* (Cambridge, MA: Belknap Press of Harvard University Press, 1982); Peter J. Bowler, *The Mendelian Revolution: The Emergence of Hereditarian Concepts in Modern Science and Society* (London: Athlone Press, 1989); Hans-Jörg Rheinberger and Staffan Müller-Wille, *The Gene: From Genetics to Post-Genomics* (Chicago, IL: University of Chicago Press, 2017).

on transmission genetics and preferred instead to study developmental and embryological issues, as well as cytoplasmic inheritance.[7] Anthropologists, in turn, far from confining themselves to pedigree analyses that would emulate Mendelian experimentation, gradually moved to direct their energies toward alternative research techniques, such as twin-studies.[8] In the field of psychiatry, Pauline Mazumdar differentiated between a Mendelian and a non-Mendelian style of inquiry and showed that German psychiatrists chose the latter path, a fact corroborated by other works on German psychiatry.[9] When taken together, historical scholarship gives the impression that in the professional communities of geneticists, anthropologists and psychiatrists, Mendelism was overshadowed by competing or complementary strategies of research and interpretation.

This analysis is valid when one looks primarily at the methods of inquiry and at experimental procedures (what is commonly referred to as "the context of discovery"), but less so when one examines the overall conceptual framework and the regime of legitimization ("the context of justification"). It overlooks the fact that from 1913 to 1933, Mendelism changed its role: during this period its ability to inspire experimental work or offer direct solutions to professional challenges indeed waned, while at the same time its status as a legitimizer of results – obtained in whatever way – became more established. As a result, Mendelism remained prominent in the study of human traits, even if and when the methods for analyzing heritability transgressed the Mendelian framework. This new role for Mendelian theory would later transform it into an important political and propaganda weapon with lasting results. In a sense, it is the transformation of Mendelism from a research framework

---

[7] Jonathan Harwood, *Styles of Scientific Thought: The German Genetics Community, 1900–1933* (Chicago, IL/London: University of Chicago Press, 1993); Bernd Gausemeier, *Natürliche Ordnungen und politische Allianzen: Biologische und biochemische Forschung an Kaiser-Wilhelm-Instituten, 1933–1945* (Göttingen: Wallstein, 2005); Gausemeier, "Genetics as a Modernization Program: Biological Research at the Kaiser Wilhelm Institutes and the Political Economy of the Nazi State," *Historical Studies in the Natural Sciences* 40, no. 4 (2010): 429–456.

[8] Karl Heinz Roth, "Schöner neuer Mensch. Der Paradigmenwechsel der klassischen Genetik und seine Auswirkungen auf die Bevölkerungsbiologie des 'Dritten Reichs,'" in Heidrun Kaupen-Haas (ed.), *Der Griff nach der Bevölkerung* (Nördlingen: F. Greno, 1986), 11–63; Hans-Walter Schmuhl, *The Kaiser Wilhelm Institute for Anthropology, Human Heredity, and Eugenics, 1927–1945: Crossing Boundaries*, Boston Studies in the Philosophy of Science, vol. 259 (Dordrecht: Springer, 2008).

[9] Pauline Mazumdar, "Two Models for Human Genetics: Blood Grouping and Psychiatry in Germany between the World Wars," *Bulletin of the History of Medicine* 70 (1996): 609–657; Volker Roelcke, "Programm und Praxis der psychiatrischen Genetik an der deutschen Forschungsanstalt für Psychiatrie unter Ernst Rüdin: Zum Verhältnis von Wissenschaft, Politik und Rasse-Begriff vor und nach 1933," *Medizinhistorisches Journal* 37 (2002): 21–55.

to an interpretative scheme that set the stage for its next metamorphosis – into an overall framework for pursuing human improvement, a social theory.

To an extent, this shift was intrinsic to the attempt – or pretension – to apply Mendelian thinking to the human sphere, and not just to the plant and animal domains. Among biologists, botanists and zoologists, studying heredity and searching for Mendelian patterns required carefully designed breeding experiments. These experiments constituted the first part of any research project; the analysis of the results comprised the second part. On the face of it, practical obstacles should have hindered the possibility of similarly identifying Mendelian patterns among humans, since humans cannot be hybridized like plants and animals, let alone self-pollinated. Nevertheless, attempts to apply the Mendelian framework to the human domain – first, to human pathologies, then to normal human traits – began as early as 1902 and intensified in the decade that followed.[10] To bypass the non-feasibility of experimentation, scholars interested in human heredity studied the mating conducted inadvertently by humans throughout history: namely, they looked at human pedigrees. Family histories, explained one psychiatrist in 1925, functioned as "the protocol of an experiment that man unconsciously performed throughout generations."[11]

In theory, if properly selected and adequately analyzed, pedigrees could substitute directed hybridizations, but only on the condition that their formal characteristics made them comparable with Mendelian experimental conditions. Mendel crossed pure, clearly distinct and constant strains. Was it not also possible to find human populations that were clearly distinct – namely, different races – and examine the traits among those who were racially crossed? Mendel counted the ratios of traits among the progeny of his cross-pollinated plants. Couldn't one similarly count the prevalence of certain traits in human families? True, human families did not have progenies in numbers approaching those of the plant and animal kingdoms. But couldn't this problem be overcome with the aid of statistical techniques and large-scale sampling?

The search for methodological substitutes to the controlled laboratory experiment yielded various solutions, and each of these found favor in a different scientific community. Studying the mixture of distinct

---

[10] The first study to identify Mendelian transmission in humans was Archibald E. Garrod, "The Incidence of Alkaptonuria: A Study in Chemical Individuality," *Lancet* 160, no. 4137 (1902): 1616–1620.

[11] Friedrich Meggendorfer, "Erblichkeitsforschung und Psychiatrie," *Zeitschrift der Zentralstelle für niedersächsische Familiengeschichte* 7, no. 10 (1925): 225–229 (quote from p. 229).

human populations with an eye on Mendelian reasoning seemed like a promising direction to follow in anthropological research. Encouragingly, the dynamics of racial mixture appeared to corroborate the validity of Mendel's laws for human crossings. This opened new paths of research into the Mendelian characterization of separate traits and into the implications of Mendelian theory for anthropological research in general. Tracking the manifestation of traits in individual families, irrespective of racial affiliation, could have appealed to genealogists, who were eager to improve the scientific status of their field by incorporating biological notions into their familial studies. Finally, substituting experiments with mass familial statistics became the domain of psychiatrists. On the basis of information gathered from clinics and asylums, they computed the prevalence of diseases among patients and their relatives and inferred from them the patterns of inheritance of mental disorders.

The beginning seemed promising. Revealingly, the two scientists who attempted most rigorously to introduce Mendelian concepts into their respective fields in the early 1910s – Eugen Fischer (in anthropology) and Ernst Rüdin (in psychiatry) – would later become among the most prominent scientists in the Nazi academic world. These issues were in fact related: Fischer and Rüdin's wish to "mendelize" their fields indicated that they were both more than capable of adapting scientific work at the frontline of science while adopting novel techniques and abandoning outdated concepts. They thus reshaped their fields, which won them repute in Germany and abroad, and both came to head prestigious research institutes. When the Nazis seized power, both were in a position to exert academic influence and play an active role in shaping policies within the Nazi administration. Their work has justly received great scholarly attention and is analyzed here as well.

As it soon turned out, however, there was a limit to what Mendelian teaching could offer to each of these three disciplines. In the case of genealogy, the Mendelian framework demanded that genealogists give up too much of their traditional methods and their professional identity. They therefore adopted very little of it. Among psychiatrists, there were vocal opponents to Mendelian inquiries, but the impact of Mendelism was nevertheless deep and lasting. The anthropological discipline fully embraced and adapted itself to Mendelian ideas, though simultaneously recognized the limitations of Mendelian analysis. Developments in these three scholarly communities are explored in Chapters 1 and 2, which also show that at the end of the 1920s Mendelism was still a ruling paradigm both in anthropology and psychiatry, even if it no longer supplied actual tools for scientific inquiry.

But Mendelism did not only reign high among scientists interested in processes of human heredity; as Chapter 3 demonstrates, it also became integral to the refashioning of concepts with larger social, political and cultural implications. Mendelian theory offered a new way for thinking about the meaning of purity and (racial) hybridity – or, as the Germans preferred to call it, "bastardization" – and it suggested a new concept that quickly became a focal point for scientific, medical and social anxieties: recessive traits. The new toolbox that Mendelism offered impinged directly on the emerging visions for racial and national regeneration. In particular, Mendelian teaching became part of the discussions centered on the pathological nature of Jews, on the one hand, and of the cultural exaltation of the peasanthood as the fountain of racial renewal, on the other hand. Intriguingly, it was the same Mendelian mechanisms, only differentially applied, that could account for the negative evaluation of the Jews and the positive character of the peasants.

All of these ideas were transformed from theoretical speculations into social realities after Hitler became chancellor. Chapter 4 examines how Mendelian logic informed the attitude of the Nazis toward the mentally ill and how it shaped the Nazi sterilization policy. It begins by showing that Mendelian reasoning led to the inclusion of certain disease categories in the Nazi Sterilization Law of July 1933, an inclusion that later helped the Nazis to argue that their sterilization campaign was based on Mendelian teaching. It then moves to analyze the way Mendelian theory and the sterilization policy were explained to high school students and exposes the multiple functions that Mendelian reasoning performed in the Nazi classroom. Finally, an examination of the implementation of the sterilization law and the proceedings in different Hereditary Courts reveals that, although the law was implemented without explicit dependence on Mendelian theory, it was still informed by and imbued with Mendelian suppositions.

With respect to antisemitic racial policy, Chapter 5 shows that Mendelian thinking left a clear mark on the legislation of the 1935 Nuremberg Laws, the most important anti-Jewish legislation during the Third Reich. Degrading the status of Jews to second-class citizens, the Nuremberg Laws also redefined who was to be considered a Jew, half-Jew, quarter-Jew or German-blooded. These definitions were not only informed by Mendelian reasoning; they were also propagated as such. This was true in the public domain, and was certainly true in German high schools, where Mendelism and racial theory were intimately intertwined. After revealing the (at times surprising) interconnections between racial and Mendelian teaching, the chapter moves to examine the praxis of racial diagnosis, exposing the Mendelian assumptions that underlay it. It ends by highlighting the way Mendelian anxieties of the recoupling of recessive Jewish traits shaped Nazi policies toward the Jewish *Mischlinge*. These anxieties, shared by Nazi leaders up to Hitler himself, successfully merged with prior and even competing modes of anti-Jewish hatred.

Were these social trajectories of Mendel's teaching unique to Nazi Germany? The epilogue argues to the contrary. While Social Mendelism certainly gained its most explicit meaning in Germany under the Nazi rule, many of its underlying assumptions were shared by scholars and social reformers elsewhere. Moreover, the legacy of Social Mendelism did not suddenly disappear with the collapse of the Third Reich. The efforts made by German eugenicists to continue the sterilization campaign in the immediate postwar years attest to the persistence of Mendelism as a legitimizing framework. Moreover, they show that a racial-antisemitic worldview continued to inform eugenic efforts, under the guise of nonideological Mendelian thinking. The story, then, does not sit comfortably within the boundaries of the 1900–45 timeline. Neither is it a purely German story. There were great national differences in the way Mendel's theory was received, adopted and applied, and in some nations its influence was marginal. But in others, like the United States and Britain, it had great social and cultural impact, whose exact nature would require further research. What is already clear, however, is that the aforementioned image of Mendelism as a nonideological, possibly even antiracist theory, is no more than a mirage. The book ends by examining the political implications of Mendelian thinking and the way its nonpolitical image was construed after World War II, to yield our contemporary, cleansed view of Mendelian theory.

Eugenics, Pseudo-science and Nazism

One year after the end of World War II, when analyzing the contribution of German doctors to the promotion of Nazi goals, the Russian-Jewish linguist Max Weinreich stated, "There were in the history of mankind Jenghiz Khans and Eugen Fischers but never before had a Jenghiz Khan joined hands with an Eugen Fischer. For this reason, the blow was deadly efficient."[12] Since that time, and especially (in fact, exponentially) since the 1980s, historical studies exposed the intimate links that existed between physicians, psychiatrists, anthropologists and the Nazi state. There is now little doubt that medical professionals were deeply involved in defining Nazi aspirations and translating them into dreadful

---

[12] Max Weinreich, *Hitler's Professors: The Part of Scholarship in Germany's Crimes against the Jewish People* (New York, NY: Yiddish Scientific Institute, 1946), 240.

realities.[13] The metaphor of boundary-crossing – both ethical and moral, but also between the scientific and the political domains – is a dominant one in the writing on German/Nazi sciences, along with the idea of a "Faustian bargain," which scientists made.[14]

These metaphors are useful as long as we don't take them to mean that the sciences involved in the preparation and realization of Nazi programs were inherently different from, or inferior to, other branches of science, or to the sciences of our own days. The work of German scholars in the fields of biology and human heredity was highly regarded by most international scholars, national particularities notwithstanding. The expulsion of Jewish scientists and the corruption of the Nazi dictatorship may have harmed the quality of research, and the need to align one's rhetoric with Nazi goals surely did not promote free intellectual exchange among scholars, but these were not sufficient to turn entire fields of scholarship into nonscientific ones. Scientific work rarely operates in an environment free from external pressures – whether financial, political, social or cultural. The Nazified academy was undoubtedly an extreme environment to work in, and the conditions of total war radicalized many of the dynamics of scientific work and sometimes introduced new limitations and pressures. Still, scholarly fields did not dissolve in an instant, and most scholars retained their scientific identity, with many (though not all) of its internal codes of conduct.[15]

[13] There is a rich literature on the relations between the medical profession, racial-anthropology and the horrors that took place in Nazi asylums, concentration camps and extermination camps. See Ernst Klee, *"Euthanasie" im NS-Staat: Die "Vernichtung Lebensunwerten Lebens,"* 2nd ed. (Frankfurt a. M.: S. Fischer, 1983); Robert J. Lifton, *The Nazi Doctors: Medical Killing and the Psychology of Genocide* (New York, NY: Basic Books, 1986); Robert Proctor, *Racial Hygiene: Medicine under the Nazis* (Cambridge, MA: Harvard University Press, 1988); Michael Burleigh, *Death and Deliverance: "Euthanasia" in Germany, 1900–1945* (Cambridge: Cambridge University Press, 1994); Götz Aly, *Cleansing the Fatherland: Nazi Medicine and Racial Hygiene* (Baltimore, MD: Johns Hopkins University Press, 1994); Benno Müller-Hill, *Murderous Science: Elimination by Scientific Selection of Jews, Gypsies, and Others in Germany, 1933–1945,* trans. George R. Fraser (Oxford: Oxford University Press, 1988); Susan Bachrach and Dieter Kuntz (eds.), *Deadly Medicine: Creating the Master Race* (Washington, DC: United States Holocaust Memorial Museum, 2004); Paul Weindling, *Nazi Medicine and the Nuremberg Trials: From Medical War Crimes to Informed Consent* (New York, NY: Palgrave Macmillan, 2004); Weindling (ed.), *From Clinic to Concentration Camp: Reassessing Nazi Medical and Racial Research, 1933–1945* (London/New York, NY: Routledge, 2017).
[14] Schmuhl, *Crossing Boundaries*; Sheila F. Weiss, *The Nazi Symbiosis* (Chicago, IL: University of Chicago Press, 2010).
[15] Weiss, *The Nazi Symbiosis*; Alan Beyerchen, "What We Now Know About Nazism and Science," *Social Research* 59 (1992): 615–641; Richard J. Evans, *The Third Reich at War* (London/New York, NY: Allen Lane, 2008), 593–613.

These issues need to be recognized in the face of the understandable temptation to portray the scholarly work associated with, contributing to, or leading toward Nazism as faulty science. Such a characterization may be true for certain, extremely politicized research projects or institutes, but is invalid as a general characterization of the German sciences from 1933 to 1945.[16] Underlying the association of Nazism with "bad science" is an understandable wish to correlate rationality with morality and to contrast the smart and good with the evil and ignorant. Few contemporary scholars would dare to publicly defend such pairings, which, when stated explicitly, carry the irritating odor of Western elitism, cultural arrogance and even racism. At the same time, these pairings are flourishing in the public discourse, and for academics, whose professional lives revolve around research and teaching, they are almost inevitable. The skills and actions required of those pursuing a scientific career – devising and conducting experiments, paying attention to small details, analyzing complex sets of data, choosing between competing hypotheses, constructing solid and consistent arguments, widening the intellectual horizons, learning about other eras, cultures or worlds – should ideally also be of value outside the laboratory or library. It is commonly assumed that people with scientific training, broadly defined, will be able to think better (more critically, more profoundly, more widely, more thoroughly) also on issues with moral and social consequences.

These hopes, whose cultural roots go back to Greek philosophy and the European Enlightenment, are severely frustrated by the case of Nazi Germany. Nazi perpetrators could not have achieved their goals without the men of letters and numbers, the "architects of annihilation" – the bureaucrats, clerks and scientists who provided them with efficient tools, technology and legitimization.[17] The sciences were neither abused nor distorted; in fact, from the perspective of the scholars involved, the Nazi regime was initially seen as harboring a great promise. Physicians in particular supported Hitler even before he became the leader of the German state, and saw the Nazi movement as the first whose political plans were built on scientific principles.[18] Some of these scientists were

---

[16] For instance, see Heather Pringle, *The Master Plan: Himmler's Scholars and the Holocaust* (New York, NY: Hyperion, 2006).

[17] The expression is taken from Götz Aly and Susanne Heim, *Architects of Annihilation: Auschwitz and the Logic of Destruction*, trans. A. G. Blunden (London: Weidenfeld and Nicolson, 2002).

[18] Ute Deichmann, *Biologists under Hitler* (Cambridge, MA: Harvard University Press, 1999); Michael H. Kater, *Doctors under Hitler* (Chapel Hill: University of North Carolina Press, 1989); Proctor, *Racial Hygiene*. Proctor stresses that "Nazi racial theory and practice were not the product of a tiny band of marginal and psychotic individuals. Nazi racial hygienists were among the top professionals in their fields. ...

men of great reputation, known at home and abroad for their intellectual achievements and scientific breakthroughs. It is certainly true that, as Mitchell Ash put it, science and politics served as mutual resources, with individuals in each domain making the most of their relation to the other – exploiting financial opportunities, drawing legitimation, supplying information, mobilizing expertise.[19] And yet "Nazi" scientific work – not free of its own share of mistakes and biases – cannot simply be dismissed as "pseudo-science."

In general, the term "pseudo-science" is recognized today as primarily a "political battle concept, which is meant to discredit the reliability of a particular science and of those who apply it, in order to conversely claim for the purity, independence and in-contaminable nature of the sciences."[20] In 2003, the *Oxford Companion to the History of Modern Science* summarized the entry on pseudo-science by arguing, "there is more to be lost than gained historically by seeking retrospectively to draw sharp distinctions between the 'real' and the 'pseudo' in science."[21] I share this view. This does not imply that the mistakes, biases, forgeries and dogmatism of past scientists should not be critically examined and, when necessary, condemned – quite the contrary. It does mean, however, that

---

Racial hygienists like Lenz, Fischer, and Verschuer were not men whose scientific or medical credentials could be questioned" (p. 284) and, "One could well argue that the Nazis were not, properly speaking, abusing the results of science but rather were merely putting into practice what doctors and scientists had themselves already initiated. Nazi racial science in this sense was not an abuse of eugenics but rather an attempt to bring to practical fruition trends already implicit in the structure of this branch of science" (p. 296).

[19] Mitchell G. Ash, "Wissenschaft und Politik als Ressourcen füreinander," in Rüdiger vom Bruch (ed.), *Wissenschaften und Wissenschaftspolitik – Bestandaufnahmen zu Formationen, Brüchen und Kontinuitäten im Deutschland des 20. Jahrhunderts* (Stuttgart: Franz Steiner, 2002), 32–51.

[20] Michael Hagner, "Bye-Bye Science, Welcome Pseudoscience? Reflexionen über einen beschädigten Status," in Dirk Rupnow et al. (eds.), *Pseudowissenschaft: Konzeptionen von Nichtwissenschaftlichkeit in der Wissenschaftsgeschichte* (Frankfurt a. M.: Suhrkamp, 2008), 21–50, esp. 22, 26. See also the insightful analysis offered in Michael D. Gordin, *The Pseudo-Science Wars: Immanuel Velikovsky and the Birth of the Modern Fringe* (Chicago, IL/ London: University of Chicago Press, 2012). With regard to German racial science and its status as real/pseudo-science, see esp. Veronika Lipphardt, "Das 'schwarze Schaf' der Biowissenschaftler. Ausgrenzungen und Rehabilitierungen der Rassenforschung im 20. Jahrhundert," in Lipphardt (ed.), *Pseudowissenschaft: Konzeptionen von Nicht-/ Wissenschaftlichkeit in der Wissenschaftsgeschichte.* (Frankfurt a. M.: Suhrkamp, 2008), 223–250; Lipphardt, "Isolates and Crosses in Human Population Genetics; or, A Contextualization of German Race Science," *Current Anthropology* 53, no. S5, special issue: "The Biological Anthropology of Living Human Populations: World Histories, National Styles, and International Networks," (2012): S69–S82.

[21] Roger Cooter, "Pseudo-science and Quackery," in John L. Heilborn (ed.), *The Oxford Companion to the History of Modern Science* (New York, NY: Oxford University Press, 2003), 683–684.

the marking of entire fields (e.g. "racial science") as nonscientific distances us both from the historical understanding of agents' motivations and actions and from the ability to note the mistakes, biases, forgeries and dogmatism that still exist, and will probably always exist, in other sciences, in past times as well as in the present and future.

One exemplary scientific simplification, of particular relevance to the present study, is worth pointing out here. In Mendelian analyses of hereditary patterns, the most common assertion regarding human traits was that they were either "dominant" or "recessive." Practically, in many works, the "dominant/recessive trait" functioned as substitute for an earlier, already considered outdated notion of "continuous/discontinuous heredity." Thus, whenever heredity was continuous (that is, the same trait manifested itself in parents and children), that trait was referred to as dominant; when heredity was discontinuous (skipping a generation or two), the trait was considered recessive. Strictly speaking, such a substitution is both untenable and misleading. It is untenable because recessive traits could also display continuous heredity. It is misleading because while "continuous/discontinuous heredity" described observable phenomena, their substitution with Mendelian terminology implied that the underlying genotypic level was understood as well, and furthermore that the traits conformed to other parts of Mendel's theory – for example, that they segregated in certain ratios. Such a terminological replacement therefore suggested the applicability of genetic models that often could not be truly sustained and the existence of concrete scientific knowledge where no such foundation existed. In addition, an identification of dominance with continuous heredity and recessivity with discontinuous heredity precluded a more nuanced understanding, which was available already from the 1910s, of the structure and character of the genetic substrate. The language of dominance/recessivity ignored ideas such as epistasis (the dependence of the expression of one gene on the presence of other "modifier genes") or inhibiting factors (the suppression of certain genes by others). It reduced the complexity of interactions between genes and between genes and environmental influences to a single term, or mechanism, that revolved solely around the relative strength of one trait (or gene) with respect to another. Finally, regardless of the intentions of the speaker, naming a trait "dominant" or "recessive" gave the impression that the trait was determined by a single hereditary factor; it therefore inadvertently undermined the idea of multi-allelic inheritance, that is, that several genes worked together to determine a trait. Despite the fact that German scholars were clearly aware of the various advancements in heredity science since the days of early Mendelism, these methodological deficiencies were quite abundant in the professional literature of the time.[22]

For those who are acquainted with contemporary genetic discourse, these remarks may raise an unintended smile, sounding all too familiar; many of the problems that they describe exist to this very day. This alone does not disqualify contemporary genetic science, nor does it distract from its sophistication. But neither should these faults lead to the repudiation of early eugenics as groundless or ridiculous. Indeed, if we rule out eugenics on the basis of several piquant examples showing the dogmatism of scholars such as Charles B. Davenport,[23] we run the risk of similarly discrediting the entire Human Genome Project

---

[22] See, for instance, the chapters of Fischer and Lenz in Erwin Baur, Eugen Fischer and Fritz Lenz, *Menschliche Erblichkeitslehre und Rassenhygiene*, 3rd ed. (Munich: J. F. Lehmanns, 1927) [hereafter *BFL* (1927)]. Many of these methodological deficiencies were noted in 1922 by the Finnish scholar Harry Federley. After analyzing these fallacies in detail, Federley suggested that the tendency to mark each and every disease as "dominant" or "recessive" may have been an outgrowth of the early years of Mendelian research, when the "Prevalence Rule" was considered one of the cornerstones of Mendelian theory (see Chapter 1 for a detailed explanation). Later it was realized that such prevalence, or dominance, was not all-pervasive, and that the most fundamental law of Mendelian theory was actually the Segregation Rule. The classification of phenomena into "dominant" and "recessive" was therefore a remnant of the historical development of the field of genetics, which became disciplinarily fixated and misled researchers who then defined complex features too simplistically. Harry Federley, "Zur Methodik des Mendelismus in Bezug auf den Menschen," *Acta Medica Scandinavica* 56, no. 1 (1922): 393–410.

[23] The two favorites among historians are the following: (1) Davenport asserted that Black-White racial miscegenation leads to the birth of children who would be "characterized by the long legs of the Negro and the short arms of the white, which would put them at a disadvantage in picking up things from the ground." This claim was already countered at the time when William Castle observed that the reported average difference in the length of the limbs was such that, at worst, it would force the hybrids in question to stretch an extra three-eighths of an inch. Interestingly and quite revealingly, Davenport did not include this explicit reference to the alleged difficulty in "picking up things from the ground" in his official scientific report, but only in the popular account of it. Compare Charles B. Davenport and Morris Steggerda, *Race Crossing in Jamaica*, (Washington, DC: Carnegie Institution, 1929), 471, with Charles B. Davenport, "Race Crossing in Jamaica," *Scientific Monthly* 27 (1928): 225–238 (238); see Castle's criticism in William E. Castle, "Race Mixture and Physical Disharmonies," *Science* 71 (1930): 604–605. (2) Another often-cited example for the dogmatism of early proponents of Mendelism is that of the Mendelian gene for nomadism that Davenport thought he had discovered. Charles B. Davenport, *Nomadism, or the Wandering Impulse, with Special Reference to Heredity*, The Feebly Inhibited II (Washington DC: Carnegie Institute, 1915). Examples from German scholarship are less easy to find in the literature on eugenics in general; perhaps it seems trivial, and therefore requires no proof, that German work on human heredity was unfounded, oversimplistic or overridden by ideological considerations. The methodological divergence from the strict Mendelian path in most fields during the 1920s also seems to make these examples of less immediate relevance for what followed under the Nazis.

because of a popular article announcing that the gene for criminality has been found.[24]

More importantly, it is doubtful if terminological ambiguities were a hindrance to scientific progress. To take two closely related comparative cases, the concept of "gene" in biology and the concept of "paradigm" in the sociology of science have both been defined, redefined, altered and mobilized for various and sometimes conflicting purposes in their respective scientific communities, as well as in the popular sphere. Conceptual obscurity and terminological ambiguity undoubtedly caused some confusion, but, given enough continuities, they also enabled the development of further ideas and made room for competing hypotheses and methodological innovations. A certain amount of terminological fuzziness is not necessarily detrimental; sometimes it can become scientifically productive, allowing for flexibility and innovation at the temporal expense of rigidness and philosophical rigor.[25]

In the case of Mendelian thinking, some of these deficiencies can be seen retrospectively, as a precondition for the ability to apply Mendelian thinking to the human domain to begin with. The usage of Mendelian terms did encourage a simplistic understanding of hereditary processes; but at the same time, it enabled the replacement of older concepts with new ones, led to a refinement of the tools of research and eventually to their substitution by other, non-Mendelian tools. In the public domain, the use of Mendelian concepts certainly lagged far behind contemporary scientific advances, but that is probably a general feature of structures of thought, which migrate from one domain to another, and especially of the popularization of scientific theories. For Mendelism to become

---

[24] www.telegraph.co.uk/news/2016/04/06/sex-offending-is-written-in-dna-of-some-men-oxford-university-fi/; www.telegraph.co.uk/news/science/11526272/Can-our-DNA-turn-us-into-criminals.html.

[25] Margaret Masterman, "The Nature of a Paradigm," in Imre Lakatos and Alan Musgrave (eds.), *Criticism and the Growth of Knowledge: Proceedings of the 1965 International Colloquium in the Philosophy of Science*, vol. 4, 3rd ed. (Cambridge: Cambridge University Press, 1970 [1965]), 59–90; Petter Portin and Adam S. Wilkins, "The Evolving Definition of the Term 'Gene,'" *Genetics* 205, no. 4 (2017): 1353–1364; Rheinberger and Müller-Wille, *The Gene*. For similar reasons I prefer not to define too precisely two terms that I use frequently in this study – "Mendelism" (or "Mendelian thinking") and "German." Thus, "Mendelism" generally refers here to the variety of epistemological suppositions, ontological principles, experimental procedures and terms that became associated with Mendel's name by historical actors in Germany during the examined period. "German" generally refers to the state of Germany, but can also refer to scholarship written in German or published in German journals by Austrian, Danish or Swiss authors. That said, non-German (by nationality) scholars who either wrote in German or had substantial influence on German scholarship will be dealt with here only to the extent that their work influenced events or discussions in Germany itself.

Social Mendelism, its content had to be reduced to a limited number of basic principles.

What, then, were the main components of Social Mendelian thinking? First, foreshadowing Richard Dawkins' "Selfish Gene," it was not variable humans that were struggling against each other, but fixed, non-altering genes that were competing for quantitative prominence. The basic mechanisms governing this competition were laid out by Mendel; in essence, they showed that nature favored continuity, not change. Indeed, if Darwinian thinking associated change and variability with progress, Mendelian thinking saw genetic change (i.e., mutation) as inherently detrimental; improvement therefore had to be sought in reestablishing past glories, not in creating novel forms.[26] Second, contrary to the Darwinian emphasis on struggle and competition, the most important dynamics governing genetic relations were those of concealing and unmasking. Some genes cloaked the presence of others; it was the task of humans to discover and sometimes annihilate those hidden genes. The emphasis, however, was not on managing reproduction for redirecting evolution, but first and foremost on identifying those genes, which seemed to be playing a sinister game of hide-and-seek with humans. Third, human genetic detection was essential because it was precisely those undercover genes that posed the greatest risk to human society, being both pathological in nature and elusive in character. Individuals or social groups carrying those genes needed to come under scrutiny; strikingly, those individuals or groups often displayed devious behaviors that were similar in kind to those of their deluding genes. To these principles we may add the atomization of humans into discrete traits and the need to prevent at any cost malignant elements from recoupling. Importantly, these Mendelian ideas were rarely treated as abstract concepts; on the contrary, they took a concrete form in discussions on Nordic or Jewish traits, dominant and recessive inheritance, and the uncovering and "out-mendeling" of deviant or malignant factors.

This book is devoted to the analysis of the multiple meanings that Mendelian theory acquired in the German scientific and social domains. It does so, first of all, by looking closely at the practices of

---

[26] There were certainly geneticists who looked favorably on mutations; among them was one of the "rediscoverers" of Mendel's work, Dutch botanist Hugo de Vries, who was also the father of the mutation theory. Nevertheless, as will be shown throughout the following chapters, the predominant view among those concerned with human heredity in the examined period was that mutations were usually harmful. Therefore, any aspiration to improve humans biologically needed to rely on selecting from, and combining successfully, existing genetic varieties, not on inducing genetic changes.

research employed by scholars who studied human heredity from 1900 to 1945. Significantly, many of these scholars were also active in the public domain, as authors of popular textbooks, educators, lecturers and leading figures in eugenic societies. Both before and after the Nazis seized power, many of them offered their services to the German state. In three different scholarly fields – physical anthropology, genealogy and psychiatry – heredity occupied a central place. Each of these fields, in its turn, helped to shape the principal concepts and practices of eugenics. Without the physical examination of human bodies, the construction of pedigrees and the delineation and diagnosis of mental illnesses, eugenic thought would remain precisely that: a thought. It became a tool of social control and a force that reshaped (or destroyed) German society by transcending idle intellectual contemplation, translating ideas into policies and perceptions into social programs. Methodological questions on how to cluster races, define familial relations and diagnose disease had consequences way beyond the laboratory and university hall. This is why this study goes beyond scientists' desktops and also examines the actions and perceptions of schoolchildren and educators, legislators and bureaucrats, citizens labeled as hereditarily diseased and those categorized as *Mischlinge*, and high-ranking Nazis. In the hands of the latter, perceptions were translated into policies; this book traces the impact of Mendelian thinking on the actual measures implemented by the Nazi state to combat perceived racial and medical degeneration. On the basis of a variety of historical sources – notebooks, letters, scientific publications and governmental correspondences, films and plays, school exams and court decisions – we will reveal the centrality of Mendelian thinking, the set of tools that it provided, and the conceptual shifts that it enshrined, which together shaped social programs in Germany and translated Nazi visions into hideous realities.

Glossary of Terms, as They Were Used and Understood by the Historical Actors

The following terms are commonly used throughout the book. The given definitions follow the way that these terms were understood by Germans during the first half of the twentieth century; they are simplified and generalized and may therefore differ in certain respects from nuanced or contemporary understandings of the same terms. Instead of alphabetically, they are ordered in a way that would ease the uninitiated to read through them continuously and thereby become acquainted with the basic concepts of Mendelian theory.

homozygosity	the state of having two identical copies of a certain hereditary factor, each of them inherited from a different parent. Denoted with two identical letters – 'AA' (if the factor is dominant) or 'aa' (if it is recessive). An organism whose inherited hereditary factors for a particular trait are equal will be called **homozygous** with respect to that trait.
heterozygosity	the state of having two dissimilar copies of a certain hereditary factor, each of them inherited from a different parent. Often denoted with two different letters – 'Aa' or 'DR.' An organism whose inherited hereditary factors for a particular trait are different will be called **heterozygous** with respect to that trait.
dominant	a hereditary factor 'A' is called dominant if it exerts its influence (disregarding potential environmental effects) also in a heterozygous state. **One** copy of a dominant factor is therefore enough for the respective trait to manifest itself. Dominance makes a homozygous 'AA' and a heterozygous 'Aa' externally indistinguishable. If 'A' is the dominant factor for brown eyes, it is enough to inherit it from one parent in order for the offspring to have brown eyes.
recessive	a hereditary factor 'a' is called recessive if it exerts its influence (disregarding potential environmental effects) only in a homozygous state. **Two** copies of a recessive factor are therefore necessary for the respective trait to manifest itself, a condition which requires that the respective factor be inherited from both one's father and one's mother.
gene	a theoretical, computational unit, whose exact chemical/organic nature is still unknown and, for many practical

purposes, insignificant. Genes are inherited units that are highly stable and largely impervious to external influences (with the exception of mutations, which are rare and usually detrimental). They are transmitted from parents to offspring following simple probability rules, which Mendel discovered; and they don't blend into each other. Thus, a heterozygous parent 'Aa' has equal chances to transmit to any of its descendants either the 'A' or the 'a' factor.

allele

a version of a gene. For example, assuming there are genes for eye color, there is an allele for blue eyes ('a') and another allele for brown eyes ('A').

phenotype

whatever is externally expressed by an organism will belong to its phenotype. Traits – for example, blue eyes – are part of the phenotype.

genotype

whatever constitutes the genetic makeup of an organism is designated with the word "genotype." Hereditary factors, or genes ('A', 'a'), are part of the genotype.

hybrids / Bastards / *Mischlinge*

the results of crossing two different races.

P / F1 / F2

when conducting crossing experiments, **P** represents the generation of the parents (as a rule, of pure race); **F1** represents their direct descendants; and **F2** represents the result of crossing these F1 descendants among themselves.

mendelize / mendel-out / *(her)ausmendeln*

a verb used either to connote that certain traits follow Mendel's laws, or that ancestral traits or even entire racial types that have been hidden (recessive) for several generations in a mixed population suddenly reappear.

1   Mendel's Laws and Their Application to Humans, 1865–1913

Between 1856 and 1863, Johann Gregor Mendel, an Augustinian monk from Brno, experimented with the hybridization of different varieties of peas (*Pisum sativum*) and noted stable ratios in the progeny of his hybrid plants. He postulated a hereditary mechanism that could account for these ratios, which included "dominant" and "recessive" factors for separate pea traits, designated with capital and lowercase letters, respectively (A, a). These factors were inherited from both paternal and maternal sides, remained intact in the reproduction cells of the hybridized plant and had equal probability of being passed on to the next generation. Mendel lectured on his findings and published them in the mid-1860s but received little attention from the scientific community. Three and a half decades would have to pass before Mendel's ideas would gain widespread scientific recognition.[1]

The decade that saw Mendel lecturing on his work also saw German anthropology organizing itself as an academic discipline. In 1861 German anthropologists convened in Göttingen for the first time to discuss the coordination of scientific work and the standardization of anthropological measurements. Within ten years the German Society for Anthropology, Ethnology and Prehistory (*Deutsche Gesellschaft für Anthropologie, Ethnologie und Urgeschichte*) was founded. In the interim, two new journals were launched: the *Archive for Anthropology* (*Archiv für Anthropologie*, 1866) and the *Journal for Ethnology* (*Zeitschrift für Ethnologie*, 1869). Two more decades would elapse before the first university chair for anthropology was established in Munich (1886), to be followed, in the next decade, by similar chairs in Leipzig and Berlin. One year before the century came to a close, anatomist Gustav Schwalbe founded

---

[1] Curt Stern and Eva R. Sherwood, *The Origins of Genetics* (San Francisco/London: W. H. Freeman, 1966); Peter J. Bowler, *The Mendelian Revolution: The Emergence of Hereditarian Concepts in Modern Science and Society* (London: Athlone Press, 1989); Vítězslav Orel, *Gregor Mendel: The First Geneticist* (Oxford: Oxford University Press, 1996).

the *Journal for Morphology and Anthropology* (*Zeitschrift für Morphologie und Anthropologie*).[2]

The very name of the German Society for Anthropology, Ethnology and Prehistory is telling. Anthropological research in Germany was perceived as related to, and having common scientific interests with, ethnology (or: cultural anthropology) and prehistory (including paleontology and archeology); but anthropology per se was distinct from both of these fields. Anthropologists in Germany emphasized that their object of study was the physical attribute of humans, not to be equated with human culture or human mentality. Historian Robert Proctor explained this "dominance of the physicalist tradition" in Germany as stemming from the academic growth of German anthropology as a subdiscipline of medicine. The vast majority of German anthropologists were physicians by training, a fact which had clear repercussions for the profession's distinct anatomic color, Proctor claims.[3] Others have related this focus on physical measuring to an overall humanistic, empiricist and anti-Darwinian tendency of nineteenth century German anthropology, one that rejected the attempts to read cultural or evolutionary distinctions into the bodies of racial others.[4]

The political and scientific career of the notable nineteenth-century anatomist Rudolf Virchow demonstrates that anthropological and racial thought were indeed compatible with liberal political ideals and not intrinsically bound up with right-wing agendas. Virchow saw the advocacy of public medicine and political reforms as integral to his scientific vocation; he was both the unrivaled chief of the anthropological

---

[2] Accounts of the history of German Anthropology are offered in Ilse Schwidetzky, "History of Biological Anthropology in Germany," *International Association of Human Biologists: Occasional Papers* 3, no. 4 (1992); Frank Spencer (ed.), *History of Physical Anthropology: An Encyclopedia* (New York, NY/London: Garland, 1997), 423–434; Uwe Hossfeld, *Geschichte der biologischen Anthropologie in Deutschland* (Stuttgart: Franz Steiner, 2005); Dirk Preuss, Uwe Hossfeld and Olaf Breidbach, *Anthropologie nach Haeckel* (Munich: Franz Steiner, 2006).

[3] Robert Proctor, "From 'Anthropologie' to 'Rassenkunde' in the German Anthropological Tradition," in George W. Stocking (ed.), *Bones, Bodies, Behavior. Essays on Biological Anthropology* (Madison: University of Wisconsin Press, 1988), 138–179, esp. 140–142.

[4] See Benoît Massin, "From Virchow to Fischer: Physical Anthropology and 'Modern Race Theories' in Wilhelmine Germany (1890–1914)," in George W. Stocking (ed.), *Volksgeist as Method and Ethic. Essays on Boasian Ethnography and the German Anthropological Tradition* (Madison: University of Wisconsin Press, 1996), 79–154; Woodruff D. Smith, *Politics and the Sciences of Culture in Germany, 1840–1920* (Oxford: Oxford University Press, 1991); H. Glenn Penny and Matti Bunzl (eds.), *Worldly Provincialism: German Anthropology in the Age of Empire* (Ann Arbor: University of Michigan Press, 2003). See also Andrew Zimmerman, *Anthropology and Antihumanism in Imperial Germany* (Chicago, IL/London: University of Chicago Press, 2001) for a different view.

profession and the cofounder of a liberal party, as well as a vocal opponent of Bismarck, who once challenged him to a duel.[5] Such liberal leanings began to give way to a more nationalistic-conservative anthropology around the turn of the century, when political movements started to propagate the virtues of the Nordic race and the need to retain racial purity to prevent physical, mental and cultural deterioration. Ideas on Aryan cultural supremacy and on the hazards of racial bastardization were publicized in the academic domain through the writings of the French novelist Arthur de Gobineau, and they found a receptive audience – albeit somewhat belatedly – also among German anthropologists. This direction gained momentum with the public and in the academic sphere, informing the discussion on races and their character. Throughout these shifts in the ideological tone of the discipline, it was obvious to all those involved that the study of human races was never a purely intellectual pursuit; it was charged with political and social implications and was therefore characterized by public involvement, by word or by deed, of its advocates.

One basic assumption of anthropological study was that relatively distinct human races existed at one time or another, and could be defined, delineated and characterized. Different typologies of such races were already available to German scholars by 1900 in works by the Swede Anders Retzius, the American William Z. Ripley, the French-Russian Joseph Deniker and the Briton-turned-German Houston Stewart Chamberlain. In later years, these typologies were reworked and refined.[6] Europe, it came to be agreed, was composed of between four and six prime races. The Nordic, Alpine, Mediterranean and Dinaric were four such basic categories, which later became popularized through the works of the racial thinker Hans F. K. Günther. There were disagreements on some of the details, the names and the subdivisions, but also a general consensus on their overall evolutionary history, their geographical distributions and characterizations as well as on the idea that present-day Europeans were highly intermixed.

[5] Byron A. Boyd, *Rudolf Virchow: The Scientist as Citizen* (New York, NY: Garland, 1991); Ian Farrell McNeely, *Medicine on a Grand Scale: Rudolf Virchow, Health Politics, and Liberal Social Reform in Nineteenth-Century Germany* (London: The Welcome Trust Centre for the History of Medicine at University College London, 2002); Constantin Goschler, *Rudolf Virchow: Mediziner, Anthropologe, Politiker* (Cologne: Boehlau, 2002).

[6] Houston Stewart Chamberlain, *Die Grundlagen des neunzehnten Jahrhunderts* (Munich: F. Bruckmann, 1899); Joseph Deniker, *Les races et les peuples de la terre* (Paris: Masson, 1900); William Z. Ripley, *The Races of Europe: A Sociological Study* (New York, NY: D. Appleton and Co., 1899). For an analysis, see Richard McMahon, *The Races of Europe: Anthropological Race Classification of Europeans, 1839–1939* (London: Palgrave Macmillan, 2016).

If one of the major tasks of racial-anthropology was to identify and characterize distinct racial types, one of the principal aims of late nineteenth-century psychiatry was to define and delineate, as accurately as possible, the diseases or disorders that subvert human mental capacities. During the middle decades of the nineteenth century, psychiatrists strove to achieve this goal by drawing methods and inspiration from physics and biology. Following the notable neurologist Wilhelm Griesinger, growing numbers of psychiatrists adhered to the idea that "mental diseases were brain diseases" and should therefore be investigated using tools borrowed from the natural sciences. As historian of psychiatry Eric J. Engstrom has shown, the production of knowledge on human mental disturbances gradually shifted from the personal experiences of alienists to the laboratory, the autopsy table and the microscope. The 1870s and 1880s saw the heyday of this neuro-anatomic direction of the psychiatric profession. During the second half of the 1880s and the early 1890s there was a relative stagnation in the accumulation of new knowledge on the structure and function of the brain. Alternative methods and techniques, from hypnosis to experimental psychological tests, came to the fore; insights into the mental field, so it was now believed, necessitated not the dissection of the cerebral tissues of dead patients, but intensive clinical observations of living ones.[7]

Unfortunately, none of these methodological approaches yielded a reliable classification of mental ailments. At the same time, the idea advocated by Griesinger and other notable psychiatrists, that all psychoses emanated from a single cause, gradually became less and less tenable. In particular, the clinical studies of Karl Ludwig Kahlbaum, who focused especially on the course of various illnesses, and of Ewald Hecker, pointed to the existence of many different psychic disturbances.[8] It was eventually Emil Kraepelin, "Freud's psychiatric antipode,"[9] who settled the issue. The sixth edition of Kraepelin's Compendium, which was published in 1896 under the title *Psychiatry: A Textbook for*

---

[7] See Eric J. Engstrom, *Clinical Psychiatry in Imperial Germany* (Ithaca, NY/London: Cornell University Press, 2003), 88–93, 98–102, 123–127.

[8] Matthias M. Weber et al., "Emil Kraepelin (1856–1926): Zwischen klinischen Krankheitsbildern und 'psychischer Volkshygiene,'" *Deutsches Ärzteblatt* 103, no. 41 (2006): 2685–2690; German E. Berrios and Dominic Beer, "The Notion of Unitary Psychosis: A Conceptual History," *History of Psychiatry* 5, no. 17 (1994): 13–36.

[9] See Weber et al., "Zwischen klinischen Krankheitsbildern." See also Eric J. Engstrom, "Emil Kraepelin. Leben und Werk des Psychiaters im Spannungsfeld zwischen positivistischer Wissenschaft und Irrationalitaet," MA thesis, University of Munich, 1990.

Students and Physicians, introduced a new and improved classification system for mental disorders, based less on symptoms and more on the background, course and final state of the diseases.[10] Kraepelin's novel nosological categories and his differentiation between two major disease complexes – dementia praecox (later: schizophrenia) and manic-depressive insanity – became the dominant paradigm of German psychiatry for years to come.[11]

Kraepelin was also an adherent of the degeneration theories of his time.[12] The theory of "dégénérescence," propagated by the French psychiatrists Benedict Auguste (Augustin) Morel and Valentin Magnan, was dominant in European thought during the later decades of the nineteenth century. According to Morel, degeneration was the progressive deterioration of anomalous dispositions from one generation to the next. This deterioration happened in certain families in four consecutive steps: members of the first generation displayed character anomalies such as nervous irritability; their children were neurotic and physically infirm; the next generation would show signs of mental disorders such as psychoses and mental retardation; and the fourth generation would consist of idiots and often "die out."[13] Psychiatrists' keen interest in degeneration and its overall consequences underlines the fact that their efforts were not directed solely at curing individual maladies; familial and societal mental deterioration was high on their list of concerns as well.

The last four decades of the nineteenth century also saw the flourishing of another field of research: genealogy. In 1869, a national genealogical society, "Herold, Society for Heraldry, Genealogy and Related Sciences"

---

[10] Emil Kraepelin, *Psychiatrie: Ein Lehrbuch für Studierende und Aerzte*, vol. VI (Leipzig: Johann Ambrosius Barth, 1899).

[11] Engstrom, "Emil Kraepelin. Leben und Werk"; Volker Roelcke, "Die Entwicklung der Psychiatrie zwischen 1880 und 1932. Theoriebildung, Institutionen, Interaktionen mit zeitgenössischer Wissenschafts- und Sozialpolitik," in Rüdiger vom Bruch and Brigitte Kaderas (eds.), *Wissenschaften und Wissenschaftspolitik: Bestandsaufnahmen zu Formationen, Brüchen und Kontinuitäten im Deutschland des 20. Jahrhunderts* (Stuttgart: Franz Steiner, 2002), 109–124.

[12] See Eric J. Engstrom, "'On the Question of Degeneration' by Emil Kraepelin (1908)," *History of Psychiatry* 18, no. 3 (2007): 389–404, esp. 392; Paul Hoff, "Kraepelin and Degeneration Theory," *European Archives of Psychiatry and Clinical Neuroscience* 258, no. 2 (Suppl.) (2008): 12–17, esp. 13.

[13] For background on degeneration theory, see Daniel Pick, *Faces of Degeneration: A European Disorder, c.1848–c.1918* (Cambridge: Cambridge University Press, 1989); J. Edward Chamberlain and Sander L. Gilman (eds.), *Degeneration: The Dark Side of Progress* (New York, NY: Columbia University Press, 1985); Ian Dowbiggin, "Degeneration and Hereditarianism in French Mental Medicine 1840–90," in William F. Bynum, Roy Porter and Michael Shepherd (eds.), *The Anatomy of Madness: Essays in the History of Psychiatry*, vol. 1 (London/New York, NY: Routledge, 1987), 188–232.

was established in Berlin and began laying the foundations for genealogical research and for popularizing its results among the German public. One year later, a parallel society was founded in Vienna. Both societies were nonprofit organizations whose main task was to promote research in historical genealogy, heraldry and related fields such as sigillography (the study of seals) and vexillology (the study of flags).[14] During the 1870s, in registered societies all around Germany, distant relatives gathered and established archives devoted to the documentation of their particular familial histories.[15] The beginning of the twentieth century saw the foundation of the Central Office for German Personal and Family History in Leipzig and the parallel "Roland" in Saxony; local branches of the national societies soon followed.

Despite this institutional expansion and professionalization of the genealogical vocation, genealogists were frustrated that their scholarship never gained the esteem of a respected scientific field, and worked hard to carve an independent niche for their academic pursuit. They invented their own methods for numerating pedigree charts, insisted on certain styles for documenting familial relations, and sought to establish connections with the more respectable scientific disciplines. And like anthropologists and psychiatrists, they also insisted that their work had immense importance for the life of the nation, not by characterizing racial types or stemming the tide of mental degeneration but by fortifying the ties between different social strata and cementing the links between one's family, community and nation.[16]

The rediscovery of Mendel's laws at the turn of the century had diverse repercussions for each of these fields of scholarship: physical anthropology, psychiatry and genealogy. This chapter will examine the first attempts to apply Mendelian theory in all three fields. As we will see, between 1908 and 1913, the overall impression was that among humans, too, the study of hereditary phenomena was on the verge of a new era. No less important, for all three fields, Mendelian thinking seemed to be opening new paths of scientific inquiry, paths that ambitious scholars

---

[14] Both societies still exist today. More information on their history and on their past (and present) activities can be found on their websites: www.herold-verein.de/herold/kurzportraet.html; http://www.adler-wien.at/index.php/de/internal/history/1-foundation.

[15] David Warren Sabean and Simon Teuscher, "Kinship in Europe: A New Approach to Long-Term Development," in Sabean, Teuscher and Jon Mathieu (eds.), *Kinship in Europe: Approaches to Long-Term Development (1300–1900)* (New York, NY: Oxford University Press, 2007), 1–32, 16.

[16] Amir Teicher, "'Ahnenforschung macht frei': On the Correlation between Research Strategies and Socio-Political Bias in German Genealogy, 1898–1935," *Historische Anthropologie* 22, no. 1 (2014): 67–90.

were eager to exploit. Before these ventures are analyzed, however, it is worthwhile to look briefly at some of the debates on the nature and essence of Mendelian theory itself.

Defining Mendel's Laws

The so-called rediscovery of Mendel's laws was, in reality, an act of political compromise. In the late nineteenth century, three different botanists – the Dutch Hugo de Vries, the German Carl Correns and the Austrian Erich von Tschermak – noted certain regularities in their own studies on hybridized plants and recognized their potential significance. To avoid emerging priority disputes, they eventually accredited Mendel with primacy.[17] But the question of what constituted Mendelian theory remained unsettled, and the precise definition of its principal components could only be decided upon through ongoing negotiations – and experimentation. Furthermore, as several historians have already pointed out, various parts of Mendelian theory were initially treated with reservations by some of the most influential figures in the international scientific community. Protagonists of the British biometric school, who labored on developing novel statistical tools for evaluating the strength of heredity, saw Mendelism as incommensurable with their research program and as incompatible with the Darwinian idea of gradual evolution.[18] Others saw it as, at best, representing one hereditary mechanism among many; Mendel's rules, they argued, did not pertain to heredity in general, but only to the unique case of hybridization.[19] In 1906, German zoologist Heinrich Ernst Ziegler declared that Mendel's laws were obviously not applicable to humans: for if they were, he argued, given the extent of human crossing and racial mixture throughout the ages, scholars would have noted them long ago.[20]

---

[17] Augustine Brannigan, "The Reification of Mendel," *Social Studies of Science* 9, no. 4 (1979): 423–454, esp. 428–9.

[18] William B. Provine, *The Origins of Theoretical Population Genetics* (Chicago, IL: University of Chicago Press, 1971), supplies the classical account of this alleged incompatibility of Mendelian heredity and Darwinian evolution.

[19] See, for example, Ludwig Plate's comment on Heinrich Ernst Ziegler's book *Die Vererbungslehre in der Biologie* in Ludwig Plate, "Kritische Besprechungen und Referate," *Archiv für Rassen- und Gesellschaftsbiologie* [hereafter *ARGB*] 2, no. 5–6 (1905): 852.

[20] Heinrich Ernst Ziegler, "Die Chromosomen-Theorie der Vererbung in ihrer Anwendung auf den Menschen," *ARGB* 3, no. 6 (1906): 804.

Yet as time went by, evidence from the gardens of botanists and laboratories of experimental zoologists pointed more and more clearly to a general applicability of Mendel's principles, and these became a new basis for studying heredity in plants, animals and, later, also humans. At the same time, no clear consensus existed on what precisely Mendel's laws were, since Mendel did not summarize his findings in that form. Throughout his paper Mendel mentioned various regularities, rules, principles, conclusions, theories and hypotheses, some of which he high-lighted more than others; but the formulation of his scientific notions in the format of a condensed, concrete list of laws was left to his followers.[21]

The discussions on the content of Mendel's theorem gradually coalesced around two different formulations, which continue to exist side by side to this very day. In one of them, there are only **two** Mendel-ian laws. The first law is the **Law of Segregation**, which states that in sexually reproducing organisms, the hereditary factors received from both parents for any given trait do not blend into each other but remain separate. When the organism reproduces, it passes on to each of its descendants, with equal probability, either the factor it received from its father or the one from its mother. As a result, when conducting controlled hybridization experiments, it is possible (in principle) to com-pute the chances that a certain trait will appear among the progeny. The second law, the **Law of Independent Assortment**, states that factors for different traits are inherited independently of each other. For example, the color of the pea and its height are different traits; therefore, their respective hereditary factors are mutually independent.[22]

In the alternative version, there exist **three** Mendelian laws. The first is the **Law of Uniformity** (***Uniformitätsregel***), which states that all the

---

[21] This has allowed for considerable scope of interpretation regarding Mendel's real objectives in constructing his experiments as well as in compiling his paper, and consequently also regarding the essence of Mendel's work in Mendel's own eyes. The scholarly debate on these matters has yielded a range of thought-provoking studies and unorthodox historical interpretations. For example, see Robert Olby, "Mendel no Mendelian?" *History of Science* 17 (1979): 53–72; Alain F. Corcos and Floyd V. Monaghan, "The Real Objective of Mendel's Paper," *Biology and Philosophy* 5 (1990): 267–292; Raphael Falk and Sahotra Sarkar, "The Real Objective of Mendel's Paper: A Response to Monaghan and Corcos," *Biology and Philosophy* 6 (1991): 447–451; Floyd V. Monaghan and Alain F. Corcos, "The Real Objective of Mendel's Paper: A Response to Falk and Sarkar's Criticism," *Biology and Philosophy* 8 (1993): 95–98; Robert Olby, "Mendel, Mendelism and Genetics," 1997, www.mendelweb.org/MWolby.intro.html.

[22] The formation of these two Mendelian laws in the English-speaking world is surveyed in Jonathan Marks, "The Construction of Mendel's Laws," *Evolutionary Anthropology: Issues, News, and Reviews* 17, no. 6 (2008): 250–253. Unfortunately, Marks' survey completely ignores essential developments in the non-English-speaking world.

direct descendants (F1) of the crossing of two pure types would look the same. All of them would exhibit the same variety of any particular trait, whether it is the paternal one, the maternal one, an intermediate one or even a novel variant. For example, the crossing of yellow peas and green peas would yield only yellow peas. The second law is the **Law of Segregation** (*Spaltungsregel*): when these first-generation hybrids are crossed, their own descendants (F2) display the traits of each of the original parents, and possibly also the variant displayed by the hybrids, according to certain ratios (1:3 or 1:2:1). Finally, the third law is the **Law of Independence (*Unabhängigkeitsregel*)** or the **Law of the Autonomy of the Features (*Regel der Selbstständigkeit der Merkmale*)**, which, just like the Law of Independent Assortment, refers to the independent heredity of different traits.[23]

There is a delicate yet fundamental difference between these two formulations. In the first version, the laws, as they are defined, pertain to the behavior of gametes (sex cells), or genes. It is the hereditary factors that break away from one another during reproduction (First Law), and they do so irrespective of other hereditary factors (Second Law). Conversely, in the second version, the uniformity, segregation and autonomy describe occurrences on the visible, phenotypic, external level: the hybrids of the first generation all **look alike** (First Law), but the parental **traits reappear** in the following generation (Second Law) irrespective of other **traits** (Third Law). The first set of laws describes the hereditary mechanism, or underlying reasons for observed phenomena; the second set deals with regularities in visible outcomes of certain experimental procedures. These differences have further implications. For instance, in the first version, "segregation" describes general gametic behavior that characterizes sexual reproduction; in the second version, in contrast, "segregation" is the name of a much more limited phenomenon, one which specifically describes the second generation of hybridization (F2) – but not, for instance, the first generation (F1).

Both of the formulations described in the foregoing are themselves generalizations, and temporal end points, of a long and gradual process of shaping and reshaping that took place in different scientific communities throughout the first half of the twentieth century. Of the three "rediscoverers" of Mendel, Hugo de Vries was the first to speak of a law of segregation (*"la loi de disjonction des hybrides"*), which he initially did not attribute to Mendel at all until pressured to do so by another

---

[23] There are also many intermediate versions, which mix elements of both formulations. As an illuminating exercise, readers are encouraged to look at the entry on Mendelian inheritance in Wikipedia and switch between languages.

"rediscoverer," Carl Correns.[24] Correns also published a paper titled "G. Mendel's Law Concerning the Behavior of Progeny of Varietal Hybrids." In this paper, Correns outlined certain regularities that were observed in the second filial generation (F2) when hybrids were re-crossed; he named the totality of these phenomena "Mendel's law." This single law encompassed both de Vries' *loi de disjonction* (the future Law of Segregation) and the observation on the independent assortment of different traits (the future Law of Independence). According to Correns, all other attributes of the Mendelian theory could be derived from it.[25] However, this combined "Law" was too inclusive and thus quickly denied general applicability by Correns himself.[26] In a later article from the same year (1900), Correns, somewhat hesitatingly, also coined the "Prevalence Rule" (*Prävalenz-Regel*).[27] The Prevalence Rule was a precursor of the later Law of Uniformity: it stated that with respect to any given trait, in the first hybrid generation all the progeny would be similar to one of the parents, whose trait would therefore "dominate." And so, by the end of 1900, two rules seemed to have stemmed from Mendel's work: the Prevalence Rule and the Segregation Rule, the latter encompassing both the phenomenon of the 1:3 segregation ratio in the F2 generation and the Independence Rule.[28]

---

[24] See Hugo de Vries, "Sur la loi de disjonction des hybrides," *Comptes Rendus de l'Académie des Sciences, Paris* 130 (1900): 845–847. Mendel's name did not appear in this original French version of the paper, but did pop up in a comment in the German translation: Hugo de Vries, "Das Spaltungsgesetz der Bastarde," *Berichte der deutschen botanischen Gesellschaft* 18 (1900): 83.

[25] Carl Correns, "G. Mendels Regel über das Verhalten der Nachkommenschaft der Rassenbastarde," *Berichte der deutschen botanischen Gesellschaft* 18 (1900): 158–168, reprinted in Carl Correns, *Carl Correns gesammelte Abhandlungen zur Vererbungswissenschaft aus periodischen Schriften 1899–1924*, ed. Fritz von Wettstein (Berlin: Julius Springer, 1924), 9–18, and electronically available at www.esp.org.

[26] See the postscript to Correns' article, where he emphasizes "that Mendel's law of segregation cannot be applied universally," in ibid., 18.

[27] Carl Correns, "Gregor Mendel's 'Versuche über Pflanzen-Hybriden' und die Bestätigung ihrer Ergebnisse durch die neuesten Untersuchungen," *Botanische Zeitung* 58, no. 15 (1900): 229–235. Reprinted in Correns, *Carl Correns gesammelte Abhandlungen*, 19–24. The two laws are defined only after the relevant phenomena are described, and Correns distances himself from their naming: "Aus dieser 'Prävalenz-Regel,' wie man sie nennen kann ..." and later "... die man die Mendel'sche 'Spaltungsregel' nennen kann." See ibid., 20–21.

[28] See Carl Correns, "Ueber Levkojenbastarde," *Botanisches Centralblatt* 84 (1900): 97–113. The article begins with the section: "Zur Kenntnis der Grenzen der Mendel'schen Regeln" (Note on the limitations of the Mendelian rules), and proceeds to define these rules, before moving to examine Correns' own work. See also Valentin Haecker, "Über die neueren Ergebnisse der Bastardlehre, ihre zellengeschichtliche Begründung und ihre Bedeutung für die praktische Tierzucht (Vortrag, gehalten im Verein für vaterländische Naturkunde am 10. März 1904)," *ARGB* 1, no. 3 (1904):

The content of the laws and their titles continued to alter in the years that followed. In 1905, Correns separated the Independence Law from the Segregation Law.[29] Experimental evidence on the limited validity of the Prevalence Rule also led to its substitution by the more general Uniformity Law. "For a while," explained the zoologist Valentin Haecker in a 1911 textbook on heredity, "following the example of Correns, the first rule was designated the 'Prevalence Rule.'... However, it was soon realized that such an exclusive dominance or prevalence of one feature in the first generation is only a special case, and that as a rule for the first generation [of hybrids] only its uniformity can apply."[30] For any given trait, first-generation hybrids all looked alike; but they did not necessarily look like either of their parents.

Various formulations for the numbering, phrasing and content of the laws continued to circulate in biologically related disciplines throughout the following decades. In the eugenicist Hermann Muckermann's 1909 *Outline of Biology*, the *Spaltungsregel* was itself divided into three subrules: (1) the purity and independence of the gametes, (2) the resplitting of parental traits and (3) the possible dominance of one feature over another.[31] In a 1922 book on the applicability of hereditary research to the sphere of mental traits, psychiatrist Hermann Hoffmann offered another articulation of Mendel's laws: following the zoologist Heinrich Prell, the first and second laws he listed were the *Spaltungsregel* and the *Unabhängigkeitsregel* (usually considered to be the second and third laws, respectively); as the third law, instead of the *Uniformitätsregel*, he defined the "Equiproportionality rule" (*Äquiproportionalitätsregel*). According to this rule, the relative proportions of gametes containing the factors *AB*, *Ab*, *aB* and *ab* should all be the same. One could speak of Mendelian inheritance, claimed Hoffmann, only when all three rules applied.[32]

321–338, esp. 324, where Haecker mentions Mendel's rules as including "the so-called Prevalence Rule and the Segregation Rule."

[29] Carl Correns, *Über Vererbungsgesetze* (Berlin: Gebrüder Borntraeger, 1905).

[30] Valentin Haecker, *Allgemeine Vererbungslehre* (Brunswick: Friedrich Vieweg & Sohn, 1911), 219.

[31] Hermann Muckermann, *Grundriss der Biologie. Erster Teil: Allgemeine Biologie* (Freiburg i. B.: Herdersche Verlagshandlung, 1909), 103.

[32] Hermann Hoffmann, *Vererbung und Seelenleben* (Berlin: Julius Springer, 1922), 7–8. Prell's own formulation was unique. He listed the *Unabhängigkeitsregel* first and the *Spaltungsregel* second in order to ease the comparison between Mendelian heredity and two other kinds of "heredities": (1) "*Kroßvererbung*," compiled of "Batesonian" and "Morganian" heredities, equal to Mendelian heredity in the first two rules but differing from it in the third, which he called the *Disproportionalitätsregel*, and (2) "*Wechselvererbung*," characterized by a slightly modified *Spaltungsregel* and a third *Superdisproportionalitätsregel*. See Heinrich Prell, "Die Grundtypen der gesetzmäßigen Vererbung," *Naturwissenschaftliche Wochenschrift, Neue Folge* 20 (1921): 289–297.

In another textbook from 1932 regarding human heredity, the identical third law was named "The Law of Free Combination of the Genes."[33] One can still find various versions of Mendel's three, four and even five laws in textbooks from the mid-1930s.[34] Practically speaking, each scholar had his own articulation and emphases, which often also varied over time.[35] Some clearly drew on an interpretation of Mendel's own ideas as reported in his original paper; others were inspired by further developments in the emerging field of genetics, such as the study of chromosomal behavior and the mapping of hereditary factors onto specific loci on certain chromosomes. The majority of the resulting formulations, however, revolved, with minor modifications, around the three laws described here.

The lack of clarity regarding the phrasing of the laws also generated a controversy with regard to the meaning of the novel verb "to mendelize" (*mendeln*) or to "mendel out" (*herausmendeln*). Coined by Hugo de Vries in 1903, these verbs usually denoted the fact that traits of one of the two parental varieties (e.g., green seeds) reemerged in the second generation of crosses after being concealed in the first hybrid generation. Some biologists, however, thought the verb "to mendelize" should only be applied when all the (three) Mendelian rules came into effect, including the independence rule (which, in many cases, did not apply, since some traits were correlated with each other). The matter was never settled.[36] Another disputed issue was the status of Mendelian regularities as

---

[33] Karl Saller, *Einführung in die menschliche Erblichkeitslehre und Eugenik* (Berlin: Julius Springer, 1932), 3: "Das Gesetz der freien Genkombination."

[34] See Hermann Otto and Werner Stachowitz, *Abriß der Vererbungslehre und Rassenkunde, einschließlich der Familienkunde, Rassenhygiene und Bevölkerungspolitik* (Frankfurt a. M.: Moritz Diesterweg, 1934), 19–25; Otto Steche, *Leitfaden der Rassenkunde und Vererbungslehre, der Erbgesundheitspflege und Familienkunde für die Mittelstufe* (Leipzig: Quelle & Meyer, 1934), 19. See also Bundesarchiv Berlin-Lichterfelde [hereafter BArch], R1501/5513, Bl. 40.

[35] Cf. the differences between Erwin Baur, *Einführung in die experimentelle Vererbungslehre* (Berlin: Gebrüder Borntraeger, 1911) throughout the consecutive editions from 1911, 1914, 1919, 1924 and 1930 (pp. 45ff, 65ff, 62ff, 65ff and 64ff, resp.), and compare these with Ludwig Plate's 1932 textbook on heredity and Mendelism, which lists six basic formulas and statements that constitute the core of Mendelism: Ludwig Plate, *Vererbungslehre, mit besonderer Berücksichtigung der Abstammungslehre und des Menschen* (Jena: Gustav Fischer, 1932), 199–200.

[36] A contemporary's analysis of the differences in the phrasing of the laws and the usage of the verb "*mendeln*" can be found in Ernst Lehmann, "Zur Terminologie und Begriffsbildung in der Vererbungslehre," *Induktive Abstammungs- und Vererbungslehre* 22 (1919): 237–260. Lehmann surveys the views of Correns, Baur, Lang, Haecker, Plate, Bateson, Ziegler, de Vries and other prominent biologists of the time. His own view is that "*mendeln*" should be used only when all three Mendelian rules apply.

"laws." In 1907, psychiatrist Robert Sommer insisted on clarifying to his readers that

One can only speak of Heredity **Laws** in the stricter sense when hereditary facts, under certain conditions, appear time and again in the same way, and [when] a similar cause can be designated for a series of phenomena. Judging by this criterion, much of what is called "Heredity **Laws**" appears to be merely a **fact**, at the most a **rule**, but not a **law** in the sense of natural sciences.[37]

In the same year, Muckermann stressed that "[t]he Mendelian results are not laws, but rules, which, according to the available observations may not yet claim general validity."[38] But unlike Sommer, for whom the distinction between laws and rules had to do with the ability to identify underlying biological causes, for Muckermann the difference was only a matter of scope: if Mendelian regularities were found to be valid in a large enough number of cases, they could be considered laws. Ten years later, Muckermann became convinced that this condition was fulfilled, and he awarded the Mendelian phenomena the status of laws.[39]

The fact that such a wide variety of terms, formulations and phrasings existed for a rather narrow set of principles and regularities does not necessarily point to a disciplinary confusion with regard to the content of Mendelian teaching itself. In fact, it could just as well testify to the

---

[37] Sommer further argued that "[i]t is expedient to adhere also in this area to these three grades of cognition, in accordance with the methodology of natural-scientific research: the perception of particular facts, the formation of rules, [and] the acknowledgement of laws." Robert Sommer, *Familienforschung und Vererbungslehre* (Leipzig: Johann Ambrosius Barth, 1907), 66. Significantly, Sommer could not have made the same remark, had he been referring to the formulation of these laws, which addressed underlying causes from the outset. See a similar view expressed by Valentin Haecker and repeated by Wittermann in Ernst Wittermann, "Psychiatrische Familienforschungen," *Zeitschrift für die gesamte Neurologie und Psychiatrie* [hereafter *ZgNP*] 20, no. 1 (1913): 153–278, 160. See also Kurt Dresel, "Inwiefern gelten die Mendelschen Vererbungsgesetze in der menschlichen Pathologie?" *Virchows Archiv für pathologische Anatomie und Physiologie und für klinische Medizin* 224, no. 3 (1917): 258–264; Dresel, however, concludes that Mendel's laws are entitled to the status of "laws." The discussion on the difference between laws and rules has a long history in German natural philosophy and goes at least as far back as Alexander Baumgarten and Immanuel Kant. See Alexander Gottlieb Baumgarten, *Metaphysics* (Halle a. S./Magdeburg: C. H. Hemmerde, 1779), §432 and Kant's *Third Critique*, particularly, Ak. 5: 184, in Immanuel Kant, *Critique of Judgement*, trans. James Creed Meredith (Oxford/New York, NY: Oxford University Press, 2007), 19–20. I wish to thank James Messina and Anat Schechtman for referring me to the relevant material.

[38] Muckermann, *Grundriss der Biologie*, 104.

[39] See his allusion to *Mendelschen Vererbungsgesetz* in Hermann Muckermann, *Die Erblichkeitsforschung und die Wiedergeburt von Familie und Volk* (Freiburg i. B.: Herdersche Verlagshandlung, 1919), 5.

deep internalization of Mendelian theory by practicing scholars. The constant reshaping of the laws by scientists, researchers, university professors and high school teachers could not have taken place unless each of these scholars had been confident that he knew what these laws/rules were all about. Indeed, while the formulations suggested by different scholars vary considerably, many (though not all) of them reduce to the same basic ideas and are practically equivalent. And so, paradoxically, the freedom to manipulate and dispute the articulation of the laws went hand in hand with a growing agreement on most of their fundamental characteristics and on the overall validity of the phenomena these laws described.

The elasticity in formulating Mendel's laws was made possible also thanks to the relative constancy and stability of other, nonverbal media through which these laws were perceived, communicated and explained. Mendel's laws were understood, discussed and propagated with the help of a relatively well-defined set of algebraic denotation methods, visual demonstration techniques and standardized examples. Where words differed, drawings and calculations agreed. Whether as rules or as laws, Mendelian notions proliferated in both scientific and popular media as a relatively fixed set of images and equations, some of which we will encounter throughout this book.

First Steps in Mendelizing Racial-Anthropology: Fischer's Rehoboth Study

Facing growing numbers of experimental results showing the validity of Mendel's principles for various traits of animals and plants, anthropologists were also tempted to test these principles' applicability to their own object of study: human races. In 1907, an article in the *Political-Anthropological Review* declared that "there is no doubt that these [i.e., Mendel's] breeding experiments shed a bright light on the mixture and segregation of human races and individual attributes."[40] At the time, however, there were but few concrete examples of Mendelian traits among humans to support such a claim. As late as 1910, one could still maintain that Mendel's laws were actually not valid for human diseases or human traits in general, and to support this assertion by describing hemophilia as an explicitly non-Mendelian disease – a rather ironic assertion, in light of hemophilia's later status as a "classic" Mendelian

---

[40] Wilhelm Haacke, "Das Gesetz der Rassenmischung (Bericht und Notizen)," *Politisch-Anthropologische Revue* 6, no. 4 (1907): 269.

disease.[41] But in 1910 such a statement was already becoming a minority view; the prevailing notion in Germany was now that "Mendel's laws are valid for humans just as they are for other organisms," and that furthermore "the validity of the 'Segregation rule' is expressed in the plurality of European types."[42]

The principal reason for this latter shift was the gradual accumulation of proofs for Mendelian mechanisms not only in plants and animals but also in humans: positive evidence simply seemed to be piling up. Notably, the vast majority of these proofs were related to the hereditary patterns of pathological traits and not to those of normal human attributes. The champion of Mendelian research in England, William Bateson, stated in 1909, "Of Mendelian inheritance of **normal** characteristics in man there is as yet but little evidence."[43] Within only a few years, this changed as well. In the introduction to his 1913 textbook on heredity, the zoologist Ludwig Plate declared that "already around 60 hereditary features of man, both neutral variations and pathological changes, can be traced back to Mendel's law and in this sense lend themselves to be understood from the hereditary perspective." Plate also acknowledged that "this fact will be new and surprising for many physicians."[44] This accumulation of medical findings alone was nevertheless not enough. To truly capture the attention of racial anthropologists, Mendelism had to be capable of supplying answers to the questions with which anthropologists were preoccupied. These had to do with the delineation of human races, their characteristics and their biological essence, their historical and evolutionary development as well as the results of their cross-breeding.

During the 1910s Mendelism began to yield results to these questions. This was largely due to the work of an anatomist from Freiburg by the name of Eugen Fischer. According to the later account in his unpublished autobiography, written in 1946, Fischer's moment of Mendelian

[41] Reinhard Frhr. von den Velden, "Gelten die Mendelschen Regeln für die Vererbung menschlicher Krankheiten?" *Politisch-Anthropologische Revue* 9, no. 2 (1910): 91–97. See, however, Fehlinger's reponse: Hans Fehlinger, "Die Giltigkeit [sic] der Mendelschen Vererbungsregeln für den Menschen," *Politisch-Anthropologische Revue* 9, no. 7 (1910): 374–379.

[42] Hans Fehlinger, "Menschenarten und Menschenrassen," *Politisch-Anthropologische Revue* 9 (1910): 198–204 (quote from p. 201).

[43] William Bateson, *Mendel's Principles of Heredity* (Cambridge: Cambridge University Press, 1909), 205.

[44] Ludwig Plate, *Vererbungslehre: Mit besonderer Berücksichtigung des Menschen, für Studierende, Ärzte und Züchter* (Leipzig: Wilhelm Engelmann, 1913), VI. See also the table in ibid., 397–398.

epiphany occurred around 1907–8. Those years saw an intense debate among the German public over German rule in African colonies, a debate that became especially bitter in 1907, when the death toll that accompanied colonization started to soar. This prompted Fischer to prepare a lecture, "The Anatomy of Human Races with Special Consideration of the Residents of Our Colonies." When searching for relevant literature, a small booklet titled *The Bastards of German South-West Africa* drew his attention. The Bastards in question were the offspring of Boer settlers and local Hottentot women, who, following the Dutch spelling, referred to themselves as *Basters*.[45] If the booklet Fischer encountered could be trusted, the bodily and mental indices of those *Basters* stood between those of the Europeans and the locals. "All at once it appeared to me perceptibly, that this must be an ideal example for Mendelian crossing-rules among humans." The next day Fischer stumbled upon his old school friend Max Bartenstein, who was a captain in the military forces in Africa. Fischer asked Bartenstein if he had seen the *Basters*, and inquired about their physical appearance. Bartenstein replied that "there were dark and bright, cute European-looking and beastly (*garstige*) Hottentot-looking young women." At that moment, Fischer recalled, "it was undoubtedly clear to me that these crosses had to mendelize and that I must determine that [by direct observation]."[46] Fischer made up his mind to go to the town of Rehoboth (in today's Namibia), where the *Basters* resided.

Fischer joined an expedition to Rehoboth in 1908; the results of his study of the native population would take five more years to be published. In the meantime, he was trying to determine whether his initial intuition was correct. In 1910 he observed that Mendel's laws had "immense significance and wide-ranging – one would almost like to say universal validity in the plant kingdom, but also in the animal kingdom." Fischer therefore attempted to evaluate their applicability for a case of a family whose members suffered from hair loss (*Haararmut*). Following Mendelian conventions, he designated different familial generations using the P1, F1, F2 denotation. In Mendelian experiments, these terms referred to the parental generation of "pure lines" and the subsequent

---

[45] While the word "Bastard" sounds offensive to an English speaker, the members of this community take pride in their heritage and readily identify themselves as *Basters*. See their website, www.rehobothbasters.org.

[46] Max-Planck-Gesellschaft Archiv [hereafter MPG Archive] III Abt., Rep. 94, Nr. 45. "Fünfzig Jahre im Dienste der menschlichen Erbforschung und Anthropologie. Lebenserinnerungen und Einblicke in die Entwicklung dieser Wissenschaften. Von Eugen Fischer," Bl. 36–37.

self-fertilized filial generations. Importantly, the F2 did not simply stand for the grandchildren of P1, but represented specifically the results of self- or mutual-fertilization of the hybrid members of the F1 generation, who were themselves children of the same parents (P1). This was hardly the case with human families, where children of the same parents did not normally copulate. Fischer nevertheless explained that two of the five F1 family members married distant relatives, so, in this sense, they approximated the Mendelian re-crossing of F1 hybrids. Fischer then examined whether Mendelian suppositions could account for the frequency of hair loss in different siblings' generations. Many of his observations seemed to fit the Mendelian scheme, but he could not yet reach any clear-cut conclusions on the matter.[47]

Fischer's work on "The Rehoboth Bastards," published in 1913, was much more conclusive. The book was rich in empirical and theoretical observations and methodologically complex, with more than 300 pages of text, 10 tables summarizing the physical measurements of 310 people, 23 extensive family lineages and 19 pages of facial photographs with 4 images on each page. The study was meant to address the problem of racial crossing, on which "our true knowledge … [is] close to zero!"[48] The future of racial-anthropology, claimed Fischer, did not lie in the methods of his predecessors – who aggregated data on populations and computed average types – but in the scrutiny of separate family lines and their fates when crossed. Unique historical circumstances made the *Baster* nation a perfect object for racial inquiry, since all of its members were descendants of male Dutch settlers and local Hottentot women and of further intermixtures of their offspring.[49] Here, the *Basters'* history seemed to neatly mirror the fundamental contours of Mendelian experimentation: "since he [the young Bastard male] could not obtain pure-white girls, he preferred marrying [girls] of his own kind, that is Bastard girls, to pure Hottentots; the son of a Boer who lived with a Hottentot woman took the respective daughter of the neighbor, [and so] a Bastard married a Bastard (two Bastards 1st degree) and created a new Bastard generation (Bastards 2nd degree)."[50]

Fischer took extensive physical measurements of the *Basters'* bodily and facial features and computed standardized ratios (indices) between some of these measurements. He also collected physiological data on

---

[47] Eugen Fischer, "Ein Fall von erblicher Haararmut und die Art ihrer Vererbung," *ARGB* 7, no. 1 (1910): 50–56 (quote from p. 50).

[48] Eugen Fischer, *Die Rehobother Bastards und das Bastardierungsproblem beim Menschen* (Jena: Gustav Fischer, 1913), 1.

[49] Ibid., 2, 16–23.   [50] Ibid., 20.

growth and development, such as the age of first menstruation (for women) or of the first appearance of facial hair (for men). He simultaneously classified his data in two different ways: according to family lineages and according to racial groups. In the first case, families were put into the P1-F1-F2 scheme and the hereditary nature of various traits was examined. In the second case, Fischer divided the *Basters* into three different clusters in accordance with the presumed proportions of European/Hottentot blood in their ancestry and analyzed the differences between the minimal, maximal and mean values of these clusters with respect to various features. His familial examinations convinced Fischer that "quite a number of features in the crossing of Europeans and Hottentots are inherited according to the Mendelian rules. ... Thus it has been determined – probably for the first time on a somewhat broader basis – that human races intersect according to Mendel's rules, precisely like countless plant and animal races."[51]

To emphasize the last point, Fischer attached a table reviewing the traits he examined and detailing their dominant and recessive manifestations. All of the traits, from eye color through facial index to finally "many mental traits," "probably," "very probably" or "certainly" followed the Mendelian segregation law (*Aufspaltung*) (Figure 1.1).[52]

In his Rehoboth study, Fischer also coined a new term, which designated the only proper route through which, in his opinion, anthropological research could make progress: it was called Anthropo-biology. Fischer wished to strengthen the ties between anthropological work and the biological sciences, which for him meant, above all, Mendelian teaching. The term Anthropo-biology stressed the futility of earlier racial studies that, according to Fischer's judgment, had either been full of data but poor in biological analysis or simply failed to distinguish hereditary from environmental influences. At the same time, the term reasserted the ostensible biological innovation in Fischer's own work. The Bastards study was initially received with some reservation but gradually came to be seen in Germany as the first large-scale proof for the general applicability of Mendel's laws for human racial crosses.[53]

[51] Ibid., 171. The entire statement in emphasized in the original.   [52] Ibid., 172.

[53] On the initial responses to Fischer's Rehoboth study, see Niels Lösch, *Rasse als Konstrukt: Leben und Wirken Eugen Fischers* (Frankfurt a. M.: Peter Lang, 1997), 76–81. Given Fischer's pivotal scientific role during the Nazi period, the Rehoboth study has received much attention from historians in another context: the debate over continuities between the genocidal policies characterizing German colonialism and the Holocaust. See, for example, Fatima El-Tayeb, *Schwarze Deutsche. "Rasse" und nationale Identität 1890–1933* (Frankfurt a. M./New York, NY: Campus, 2001); Kundrus, "Von Windhoek nach Nürnberg? Koloniale 'Mischehenverbote ' und die nationalsozialistische Rassengesetzgebung," in Kundrus (ed.), *Phantasiereiche. Der deutsche Kolonialismus aus*

Merkmalgruppe	Dominant	Rezessiv	Ergebnis
Augenfarbe · · · · · · ·	dunkel	hell	ziffernmäßig an Familien erwiesen.
Haarfarbe · · · · · · · ·	„	„	„ „ „ „
Hautfarbe · · · · · · · ·	(hell?)	—	nicht geklärt, Spaltung sicher, aber Dominanz ungewiß.
Gelber Hautton · · · · ·	—	gelb	vielleicht! —
Körpergröße · · · · · · ·	große?	?	Aufspaltung sicher erwiesen.
Gliederproportionen · · · ·	?	?	„ wahrscheinlich, nach den Gesamtziffern.
Kopfform (L.-Br.-Ind.) · ·	?	?	Aufspaltung sicher erwiesen.
Gesichtsindex · · · · · ·	?	?	„ „ „
Index fronto-jugalis · · · ·	größerer	kleinerer	„ „ „
Morphol. Gesichtshöhe · ·	?	?	„ „ „.
Physiogn. Obergesichtshöhe ·	?	?	„ „ „
Jochbogenbreite · · · · ·	?	?	„ „ „
Haarform · · · · · · · {	gebogen dicht kraus	gerade } locker }	„ „ „
Augenspaltenform · · · ·	gerade	schief	„ „ „
Augenspaltenweite · · · ·	?	?	wahrscheinlich spaltend.
Lippendicke · · · · · · ·	?	?	„ „
Form der Nasenlöcher · ·	?	?	„ „
Form des Nasenrückens · ·	konkav??	konkav?	„ „
Nasenindex · · · · · · ·	nied. Index (schmale hohe Nase)	hoher Index	sicher spaltend (gewisse Unklarheiten).
Viele Einzelheilen der Physiognomie · · · · · ·	?	?	sehr wahrscheinlich spaltend.
Viele geistige Eigenschaften	?	?	„ „ „

Figure 1.1 Fischer's table summarizing various anthropological features and their "probable/very probable/certain" Mendelian character, 1913.
Source: Eugen Fischer, *Die Rehobother Bastards und das Bastardierungsproblem beim Menschen* (Jena: Gustav Fischer, 1913), 172

kulturgeschichtlicher Perspektive (Frankfurt a. M.: Campus, 2003), 110–131; Jürgen Zimmerer, "The Birth of the *Ostland* out of the Spirit of Colonialism: A Postcolonial Perspective on the Nazi Policy of Conquest and Extermination," *Patterns of Prejudice* 39, no. 2 (2005): 197–219; Benjamin Madley, "From Africa to Auschwitz: How German South West Africa Incubated Ideas and Methods Adopted and Developed by the Nazis in Eastern Europe," *European History Quaterly* 35, no. 3 (2005): 429–464. For an overview of the wider context of the debate, see Roberta Pergher, Mark Roseman, Jürgen Zimmerer, Shelley Baranowski, Doris L. Bergen and Zygmunt Bauman, "The Holocaust: A Colonial Genocide? A Scholars' Forum," *Dapim: Studies on the Holocaust* 27, no. 1 (2013): 40–73.

Fischer did not work in a vacuum, nor was he the sole promoter of the idea that Mendelism provided the key for solving the mysteries of hereditary transmission among human races. The Mendelian direction of inquiry also captured the attention of the eugenicist Fritz Lenz at least as early as 1912. Reviewing a work by another scholar, Lenz insisted, "There is only one heredity and it is based on the continuity of the units of germ plasma, which necessarily follow the Mendelian laws." He further added that "it seems more and more that one needs to perceive all heredity as a Mendelian one, and that therefore mendelizing and heredity in sexual reproduction are identical terms."[54] What precisely this meant for the future of anthropology was still to be seen.

The Mendelization of Mental Disorders: Rüdin's "Routes and Aims"

How could a psychiatrist move beyond the tedious treatment of individual cases and seek a more comprehensive solution to the rising tide of mental disturbances in the population? One way to do that was by studying the hereditary elements of mental abnormalities. By noting the occurrence of diseases among patients' relatives, some psychiatrists hoped they could better assess the different elements causing a disease and devise the treatment accordingly. During the second half of the nineteenth century, mental asylums became centers for the collection of statistical data on patients and for documenting the various influences that may have contributed to the outbreak of mental anomalies; and between 1880 and 1910, a few psychiatrists took the initiative to use these databases for conducting in-depth genealogical studies of the familial histories of select patients. In the same period, several others attempted to evaluate the hereditary burden of mental anomalies by independently gathering dozens of pedigrees of their patients' families and by calculating the percentage of patients with ill relatives.[55] Yet throughout the first decade of the twentieth century, Mendelian reasoning was by and large absent from such attempts at hereditary analyses.

---

[54] Fritz Lenz, "Über die idioplasmatischen Ursachen der physiologischen und pathologischen Sexualcharaktere des Menschen," *ARGB* 9, no. 5 (1912): 545–603 (quotes from p. 555, 564).

[55] Bernd Gausemeier, "Pedigrees of Madness: The Study of Heredity in Nineteenth and Early Twentieth Century Psychiatry," *History and Philosophy of the Life Sciences* 36, no. 4 (2015): 467–483; Theodore M. Porter, *Genetics in the Madhouse: The Unknown History of Human Heredity* (Princeton, NJ: Princeton University Press, 2018).

In 1908, this absence was noted by an ambitious disciple of Kraepelin by the name of Ernst Rüdin. No scholar, Rüdin remarked, has yet tackled seriously and systematically the issue of Mendelism in psychiatry.[56] The following year Rüdin habilitated at the medical faculty of the Munich Ludwig-Maximilian-University, and in 1911 he met his own challenge and published a long, programmatic treatise, bearing the title "Several Routes and Aims of Family Research, with Regard to Psychiatry."

Rüdin's treatise epitomized the attraction that the Mendelian model had for psychiatric research, and more generally for eugenic socio-biological thought. Rüdin explained that earlier attempts to evaluate the force of heredity provided merely average ratios, and were therefore general, imprecise and non-applicable for individual cases. In contrast, Mendel's laws enabled exact quantifications and in many cases allowed for reliable conclusions on the exact hereditary constitution of one's parents or offspring.[57] Rüdin went to great lengths to explicate the Mendelian laws and their relevance for the psychiatric and prophylactic sphere. Using a large number of illustrations, he demonstrated different facets of Mendel's theory and compiled a long list of animal and human characteristics that were already recognized as inherited in either a dominant or a recessive Mendelian fashion. Following his extensive exposition, Rüdin made it clear that "the tricky question of whether, in the field of psychiatry, disorders are already known of which one could safely say that they follow Mendelian regularity, I must answer negatively." The reason for that, however, was probably that "as far as I know, [this question] has not yet been examined seriously with the proper means."[58] His own data indicated that certain forms of dementia praecox (schizophrenia) were inherited as Mendelian recessive, while manic-depressive disorder and many forms of "psychopathic, degenerate and defective states" had the characteristics of dominant traits.

Somewhat like Fischer in his Rehoboth study, Rüdin's treatise set forth a new vision and work plan for the entire profession. In Rüdin's view, the only way for psychiatry to make progress was by conducting extensive

---

[56] Rüdin made this note while commenting on David Heron's *A First Study of the Statistics of Insanity and the Inheritance of the Insane Diathesis*, published in 1907. Heron denied the applicability of Mendel's laws to mental disorders; Rüdin emphasized that Mendel's laws could neither be denied nor confirmed, for they hadn't yet been properly, thoroughly and genealogically investigated in psychiatry. See Ernst Rüdin, "Kritische Besprechungen und Referate," *ARGB* 5, no. 1 (1908): 133–135.

[57] Ernst Rüdin, "Einige Wege und Ziele der Familienforschung, mit Rücksicht auf die Psychiatrie," *ZgNP* 7, no. 1 (1911): 487–585.

[58] Ibid., 519.

genealogical studies and by collecting data from every imaginable source on the diseases of patients and their families. Court and police records, schools, institutions for the poor or drunkards, hospitals, asylums, military recruitment offices and governmental bureaus, as well as historical documentation from biographies, family chronicles, certificates of birth, marriage or death, letters, diaries, drawings – all of these should be mined for the gathering of relevant anthropological and psychiatric data. Rüdin envisaged a governmental health office (*Reichsgesundheitsamt*) with a special department for psychiatric-genealogical research that would be responsible for coordinating the assembly of these data in local and regional registries.[59]

Once gathered, Rüdin explained, the accumulated data could help answer questions that had long been troubling psychiatrists: What effect did alcohol have on the hereditary substance? What significance did the order of birth, the age of the parents, the month of delivery or the interspaces between births bear on the health of the progeny? Was there a difference between male and female hereditary influence? What were the relations between hereditary and environmental effects? Answering these questions was essential for coping properly with the dangers of degeneration and for supporting a eugenic countercurrent of regeneration. "He who prevents, need not cure!" exclaimed Rüdin.[60] With this ambitious vision of the power of science, Rüdin ended his treatise.

Rüdin's ideas on the potential of Mendelism to advance the causes of both psychiatry and eugenics were appealing. Nevertheless, it was not entirely clear how individual psychiatrists could translate his ideas into actual research policies. The massive collection of genealogical data that Rüdin described required either an institutional framework or at least a scholar personally devoted to such a project. There was one scholar who had already been actively locating, collecting and analyzing large amounts of genealogical-medical data from state and church records – the physician and statistician Wilhelm Weinberg – but he was not a

---

[59] Rüdin was not the first to advocate for the foundation of public bureaus that would gather comprehensive data on the population for the purpose of psychiatric research. Kraepelin himself called for extensive studies of whole regions by trained psychiatrists; but Kraepelin saw clinical observation, not genealogical investigation, as the focal point of such population surveys. At the International Congress of Mental Health Care in Berlin in 1910, he proposed – together with Rüdin – to build an international network of psychiatrists who would collect statistical data. Robert Sommer made a similar proposal at the same congress. See Alois Alzheimer, "Ist die Einrichtung einer psychiatrischen Abteilung im Reichsgesundheitsamt erstrebenswert?" *ZgNP* 6, no. 1 (1911): 242–246. See also Weber, *Ernst Rüdin. Eine kritische Biographie* (Berlin: Springer, 1993), 107.

[60] Rüdin, "Einige Wege," 571.

psychiatrist and did not (yet) have direct access to psychiatric information.[61] An institutional framework was to be established and headed by
Rüdin himself only six years later.

Nevertheless, Rüdin's treatise had a clear and immediate impact:
following its publication, psychiatrists writing on issues of heredity
could no longer ignore potential Mendelian interpretations of their
results. For example, Wilhelm Strohmayer, one of the founders of child
and youth psychiatry in Germany, did not mention Mendel's name in a
paper published in 1910, although he noted that "chromosomal studies" could potentially be relevant to his work. After Rüdin's treatise had
been printed, Strohmayer began to include explicit references to possible Mendelian readings of his results – in one case, by assuming that
the Habsburg protruding lip was a dominant Mendelian trait, in
another by postulating that various mental diseases were caused by a
single dominant psychopathological factor. Strohmayer was reticent
with such Mendelian interpretations. The small number of children in
each family made the computation of Mendelian ratios impractical, he
wrote; furthermore, one could never know if a certain individual was a
homozygous "DD" or a heterozygous "DR." All in all, he declared,
such postulations took him astray from the "area of proved facts" to the
"land of hypothesis." He therefore insisted on using ancestral charts for
studying heredity and, as he put it, was reluctant to force every finding
into "the procrustean bed" of Mendelism. However, in another article
published in 1912, Strohmayer seems to have partially surrendered:
he adopted the Mendelian interpretive framework with no apparent
reservation.[62]

Similarly in 1912, Dr. Schuppius, head of the Clinic for Psychiatric
Disorders and Neuroses at the University of Breslau, used the Mendelian

---

[61]   See Wilhelm Weinberg, "Die württembergischen Familienregister und ihre Bedeutung
als Quelle wissenschaftlicher Untersuchungen," *Württembergische Jahrbücher für Statistik
und Landeskunde* 1 (1907): 174–198; Weinberg, "Aufgabe und Methode der
Familienstatistik bei medizinisch-biologischen Problemen (Vortrag, gehalten in der 8.
Sektion des XIV. Kongresses für Hygiene und Demographie in Berlin)," *Zeitschrift für
soziale Medizin, Medizinalstatistik, Arbeitsversicherung, soziale Hygiene und die Grenzfragen
der Medizin und Volkswirtschaft* 3 (1908): 4–26. For his earlier works, and a reappraisal of
a slightly later work of his, see Alfredo Morabia and Regina Guthold, "Wilhelm
Weinberg's 1913 Large Retrospective Cohort Study: A Rediscovery," *American Journal
of Epidemiology* 165, no. 7 (2007): 727–733.
[62]   See Wilhelm Strohmayer, "Die Ahnentafel der Könige Ludwig II. und Otto I. von
Bayern," *ARGB* 7, no. 1 (1910): 65–92; Strohmayer, "Die Vererbung des Habsburger
Familientypus," *ARGB* 8, no. 6 (1911): 775–785; Strohmayer, *Psychiatrisch-genealogische
Untersuchung der Abstammung Königs Ludwig II. und Ottos I. von Bayern* (Wiesbaden: J. F.
Bergmann, 1912); Strohmayer, "Die Vererbung des Habsburger Familientypus (Zweite
Mitteilung)," *ARGB* 9, no. 2 (1912): 150–164.

framework to support his thesis regarding the polymorphic nature of mental disorders. Schuppius analyzed hereditary relations in a family, where the father was healthy and the mother fell ill during her fourth pregnancy. The couple's first three children were healthy, while all their other children suffered from mental disorders. Schuppius thus postulated that both parents were carriers of a recessive disposition for mental illness, which had somehow been transformed into a dominant trait following the mother's illness. He calculated the expected proportions of sick, healthy and "odd" offspring and claimed that they conformed to his Mendelian hypothesis (Figure 1.2).[63] The "traits" or "illnesses" Schuppius discussed were various sorts of mental disorders clumped into a vague "unit-disorder" factor. Schuppius' aim was to show that different mental disorders had a common single origin, a uniform hereditary disposition with alternating manifestations; with several twists and turns, he managed to make use of Mendel to support this view.

Yet Mendelism could just as well be mobilized to support the exactly opposing view. In 1913, an assistant in the psychiatric clinic at the University of Halle, Philip Jolly, published a 285-page study on the heredity of psychoses. Based on a systematic collection of psychiatric-genealogical data and on exhaustive interviews with patients' family members, house physicians and officials of the local authorities, Jolly's study probably came the closest one could get to fulfilling Rüdin's earlier mentioned vision.[64] Jolly wished to investigate the heritability of various mental illnesses and to determine whether they alternated, degenerated or remained unchanged when inherited.[65] Mendel's laws backed up Jolly's conclusions: different mental illnesses did not have a

---

[63] The mother (left, on the upper row) was, according to Schuppius' description, "aa" whereas the father was an "Aa" ("a" being the pathological factor). After she gave birth to the third child, her disposition became dominant. That is why "all the children born after the sickening of the mother are necessarily sick" (all of them receiving the mother's now dominant pathological factor), "but to more or less equal parts homozygous and heterozygous abnormal" ("Aa" and "aa" in equal proportions). Notably, Schuppius seems to have been completely at ease with transforming a recessive trait into a dominant one after the mother's third delivery, something that would be entirely incompatible with our modern understanding of these terms as expressing a fixed attribute of the alleles in question. Schuppius, "Über Erblichkeitsbeziehungen in der Psychiatrie," ZgNP 13, no. 1 (1912): 218–284, esp. 243.

[64] For Matthias Weber, Jolly's study constitutes "a transformation within psychiatric human genetics, drawing its principal epistemological boundaries and making the first steps towards developing a methodological procedure which was later extensively used by Rüdin." See Weber, Ernst Rüdin. Eine kritische Biographie, 100.

[65] Judging by the interim summaries Jolly provides throughout his work, it appears that the Mendelian chapter was a last-minute addition, and that the original principles guiding him while collecting and analyzing the data were not part of a Mendelian approach. But after 1911, a neglect of Mendelian considerations when dealing with heredity was no

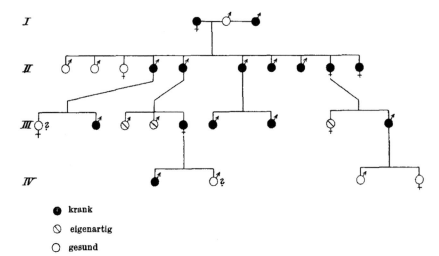

● krank

⊘ eigenartig

○ gesund

Figure 1.2 A pedigree of the mentally ill, analyzed in a Mendelian fashion (see text), 1912.
Reprinted by permission from Springer Nature: Schuppius, "Über Erblichkeitsbeziehungen in der Psychiatrie," Zeitschrift für die gesamte Neurologie und Psychiatrie 13, no. 1 (1912): 230. (license no.: 4507040477813). https://doi-org.rproxy.tau.ac.il/10.1007/BF02865610

single origin; each mental illness stemmed from an independent hereditary disposition.[66]

Rüdin's treatise therefore achieved its primary task. In the years immediately following its publication, Mendelism became for psychiatrists an

longer considered proper; they became sine qua non, and – so it appears – Jolly added a Mendelian interpretation to his findings. Probably because the part on Mendel appeared at the conclusion of Jolly's paper, one contemporary considered Mendelism to be the goal of the entire work; see W. Medow, "Zur Erblichkeitsfrage in der Psychiatrie," *ZgNP* 26, no. 1 (1914): 494. As explained, I believe that the opposite is true.

[66] In an article published in 2009, Anne Cottebrune has shown that in U.S. research (mostly – Davenport and Rosanoff and Orr), Mendelian investigations were performed with an underlying assumption on a single polymorphic neuropathic constitution, whereas in Germany Mendelian studies were dominated by Kraepelinian ideas on distinct disease entities. This description is generally valid, as the next chapter will show; but at least in the first half of the 1910s, contrasting voices were still heard in Germany, and, as the case of Schuppius shows, were supported on the basis of Mendelian teaching as well. Mendelism was therefore not intrinsically bound up with Kraepelin's perception of distinct and clearly defined disease entities; it could just as well serve to substantiate the idea of a unitary and polymorphic psychosis. See Anne Cottebrune, "Zwischen Theorie und Deutung der Vererbung psychischer Störungen," *NTM Zeitschrift für Geschichte der Wissenschaften, Technik und Medizin* 17, no. 1 (2009): 35–54.

obligatory framework that had to be addressed, and one that was also flexible enough to legitimize competing claims on the nature of mental disorders. As one psychiatrist stated in 1912, dominance and recessiveness could provide "the key for the language that our family trees and ancestors' charts speak."[67] Scholars who did not refer to Mendelian interpretations of their findings added a kind of apology to their work, commenting that their investigations were too premature for proper Mendelian analysis.[68] Methodologically speaking, psychiatrists continued to lean on earlier methods of analysis – the examination of ancestral charts of famous historical figures (like Strohmayer), pedigrees of their patients (like Schuppius), or even on more extensive statistical data (like Jolly). Mendelian notions were not therefore used to reshape the methods of investigations, but primarily to define a legitimate interpretative framework for the findings on the patterns of inheritance of mental diseases.

Searching for the Ultimate Pedigree Format: Genealogy and Mendelian Thinking

A year before the rediscovery of Mendel's work, the genealogical community in Germany was already in a state of excitement. This was due to the publication, in 1898, of the historian Ottokar Lorenz' *Textbook of the Entire Scientific Genealogy*. With its first chapter bearing the title "Genealogy as Science," the prevailing feeling in the genealogical community was that the book had finally set genealogy on its own firm ground. No longer could familial research be dismissed as merely an amateurish hobby; it was a proper discipline with its own research methods and valuable contributions to other disciplines, from history and sociology through law and political science to the natural sciences, zoology, psychology and psychiatry.[69] In the decade that followed, the academic relevance of

[67] Hans Roemer, "Über psychiatrische Erblichkeitsforschung," *ARGB* 9, no. 3 (1912): 292–329 (quote from p. 302).
[68] See Paul Albrecht, "Gleichartige und ungleichartige Vererbung der Geisteskrankheiten," *ZgNP* 11, no. 1 (1912): 541– 580; A. Luther, "Erblichkeitsbeziehungen der Psychosen," *ZgNP* 24, no. 1 (1914): 12–81; Medow, "Zur Erblichkeitsfrage in der Psychiatrie"; Hermann Krueger, "Zur Frage nach einer vererbbaren Disposition zu Geisteskrankheiten und ihren Gesetzen," *ZgNP* 25, no. 1 (1914): 113–182, esp. 116, 179–181.
[69] Ottokar Lorenz, *Lehrbuch der gesammten wissenschaftlichen Genealogie* (Berlin: Wilhelm Hertz, 1898). See also Bernd Gausemeier, "Auf der 'Brücke zwischen Natur- und Geschichtswissenschaft': Ottokar Lorenz und die Neuerfindung der Genealogie um 1900," in Florence Vienne and Christina Brandt (eds.), *Wissensobjekt Mensch: Humanwissenschaftliche Praktiken im 20. Jahrhundert* (Berlin: Kulturverlag Kadmos, 2008), 137–164; Gausemeier, "In Search of the Ideal Population: The Study of Human Heredity before and after the Mendelian Break," in Staffan Müller-Wille and

genealogy seemed to have been confirmed when several German universities listed genealogy as a course in the curriculum of historical studies. At the same time, popular guides on how to conduct genealogical investigations were published and found large enough audiences to justify further editions and publications.[70]

In his *Textbook*, aside from systematically analyzing the relations of genealogy to other scientific disciplines, Lorenz thoroughly scrutinized the visual formats available for the representation of familial relations. Two basic pedigree formats had evolved in Europe over centuries. The first was the "family lineage," or "family tree" (*Stammtafel*), which depicted the male descendants of a reputable progenitor. Female descendants who got married were, as a rule, excluded from such family trees, whose focus lay on those carrying a common family name. A competing genealogical format was the "ancestral chart" (*Ahnentafel*). This type of chart, which became popular during the fifteenth century, was used primarily to verify the noble origins of aristocrats or of those who wished to marry into aristocratic families. Ancestral charts depicted the two parents, four grandparents, eight great-grandparents, sixteen great-great-grandparents and so forth of a chosen individual. They therefore expanded "backward" in time, unlike the family lineages that progressed "forward," that is, from the ancient forefather to his more recent descendants. Another notable difference between the two formats was that while the family lineages varied greatly, according to the number of descendants that each and every couple (or father) along the family chain had borne, ancestral charts had a fixed format, since the number of biological parents at every generation followed a strict geometric order.

While Lorenz stated at the outset that both the *Stammtafel* and the *Ahnentafel* were of equal importance for genealogical work, it was clear that he preferred the latter. If the genealogical discipline aspired to tackle biological questions, if it aimed at becoming truly scientific, it had to take into account the latest biological knowledge, Lorenz argued. Cytological research had shown that hereditary substances came from both paternal and maternal progenitors; both were therefore essential also for

Christina Brandt (eds.), *Heredity Explored: Between Public Domain and Experimental Science, 1850–1930* (Cambridge, MA: MIT Press, 2016), 337–364.

[70] Friedrich Wecken, *Taschenbuch für Familiengeschichtsforschung* (Leipzig: Degener & Co., 1919), 23–28; Friedrich von Klocke, *Die Entwicklung der Genealogie vom Ende des 19. bis zur Mitte des 20. Jahrhunderts, Prolegomena zu einem Lehrbuch der Genealogie* (Schellenberg bei Berchtesgaden: Degener & Co., 1950). In English, see Eric Ehrenreich, *The Nazi Ancestral Proof* (Bloomington: Indiana University Press, 2007), 16–19 (Ehrenreich's account relies heavily on von Klocke).

genealogical inquiries. As a result, the male-biased family lineage was inadequate for examining hereditary phenomena; only ancestral charts could provide biologically worthwhile results. "Indeed," Lorenz claimed, "all [genealogical] presentations, which did not result from the fixed format of the *Ahnentafel*, are totally worthless."[71]

Lorenz' book, which genealogists embraced with overt enthusiasm, gave a major impetus to turning genealogists' attention to the compilation of ancestral charts. In a speech given before the attendees of the first meeting of the Leipzig Central Office in November 1904, genealogist Adolf von den Velden continued to campaign for the same cause: "... I would like to attempt to draw the attention of wider circles, and especially of family historians and biographers, and direct them more than has been done so far to the genealogical value that the ancestral chart has next to the family lineage, and in many respects even the advantage [that the ancestral chart] has over it."[72] In the following years, genealogists invested efforts in standardizing methods for presenting and numerating ancestral charts, methods that later facilitated their ability to share genealogical data and publish their findings. In genealogical publications, the balance between family lineages and ancestral charts palpably began to shift in favor of the latter.[73]

One of the explicit motivations for focusing on ancestral charts was that they corresponded to contemporary biological notions. In particular, ancestral charts gave visual expression to the idea that maternal and paternal hereditary influences were equally important, a premise shared by Mendelian theory. But for all other purposes of Mendelian analysis, ancestral charts were quite useless. What lay at the core of Mendelian inquiries were crossing experiments and the examination of the distribution of traits among progenies. If any equivalent to such an experimental method existed in the human domain, it lay in descendants' lineages, not in ancestral charts. The *Ahnentafel* was valueless for Mendelian analysis since it presented only one child of any given couple; the ability to

---

[71] Lorenz, *Lehrbuch*, 85–86, 347–348, 385–386, 414 (quote from p. 414). On additional ideological motivations that induced Lorenz to prefer the *Ahnentafel* over the *Stammtafel*, see Bernd Gausemeier, "From Pedigree to Database: Genealogy and Human Heredity in Germany, 1890–1914," in *Conference: A Cultural History of Heredity III: 19th and Early 20th Centuries*, Max-Planck Institute Preprint Series 294 (2006): 179–192, as well as his "Pedigree vs. Mendelism: Concepts of Heredity in Psychiatry before and after 1900," in *Conference: A Cultural History of Heredity IV: Heredity in the Century of the Gene*, Max-Planck Institute Preprint Series 343 (2008): 149–162.

[72] Adolf von den Velden, "Wert und Pflege der Ahnentafel (Vortrag, gehalten in der ersten Hauptversammlung am 21. November 1904)," *Mitteilungen der Zentralstelle für deutsche Personen- und Familiengeschichte* 1 (1905): 17–22 (quote from p. 17).

[73] Teicher, "'Ahnenforschung macht frei.'"

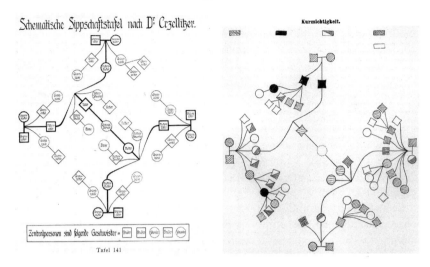

Figure 1.3 Left: Crzellitzer's 1909 kinship chart (*Sippschaftstafel*) template. Right: the same chart colored to highlight the purebred nature of parents or grandparents (here with respect to shortsightedness), 1919.
Source: Friedrich Wecken, *Taschenbuch für Familiengeschichtsforschung* (Leipzig: Degener & Co., 1919), 41; Arthur Crzellitzer, "Methoden der Familienforschung," *Zeitschrift für Ethnologie* 41 (1909): 193

discern Mendelian ratios among progeny was denied from the outset. Realizing this difficulty, the psychiatrist and genealogist Robert Sommer made it clear that "the prerequisites for the recognition of the facts of heredity within the family are realized … only by combining both methods, namely through research of family trees (*Stammbäume*) … and ancestors' rows (*Ahnenreihen*)."[74]

But how precisely were ancestral charts and family lineages to be combined? In 1908, the Jewish ophthalmologist Arthur Crzellitzer offered an innovative amalgamation of the two formats in what he termed a kinship chart (*Sippschaftstafel*). The kinship chart depicted not only the proband's parents, grandparents and great-grandparents, but also the proband's siblings, uncles and aunts, cousins and second-degree cousins, all encircling the proband (Figure 1.3).[75] Crzellitzer considered

[74] Sommer, *Familienforschung und Vererbungslehre*, 26.
[75] Arthur Crzellitzer, "Methoden der Familienforschung," *Zeitschrift für Ethnologie* 41 (1909): 181–198.

this format especially suitable for hereditary research. First, he noted, it did not depict ancestors beyond the great-grandparental generation, on whom reliable data were anyhow difficult to obtain. Moreover, Crzellitzer claimed that the hereditary influence of distant relatives was negligible, as indicated already in the Old Testament, where the iniquity of the fathers was visited upon "the children unto the third and fourth generation" – but not further than that. It was probably no coincidence, Crzellitzer added, that natural terms for designating relatives more distant than great-grandparents or cousins did not exist. In this respect, he thought, popular notions mirrored biological realities, and the kinship chart depicted only those relatives who were generally perceived to be "part of the family circle."[76]

But linguistic and biblical argumentation served Crzellitzer mainly as preliminary rhetoric, paving the way for his more substantial claim. According to Crzellitzer, the biological objectives of family research were to examine which human traits followed Mendel's Dominance Rule (*Prävalenzregel*) and which followed the Segregation Rule (*Spaltungsregel*), characterizing recessive traits in the second generation of hybridization. Such an inquiry was possible only if the examined traits had been constant in each of the parents originally crossed. If they were, those parents could be justly regarded as purebred ("*rein/konstant gezüchtet*"), and correspond to Mendel's pure lines of peas. The kinship chart made it possible to inspect the families of each of the proband's parents, and conclude, accordingly, whether or not certain traits bred true in each of these families.[77]

To facilitate the ability to note racial purity, Crzellitzer offered to color the circles in the pedigree according to the following rule: For any examined trait, from musical talent to shortsightedness, the circles and squares in the chart should be filled with colors beginning with white (symbolizing the lack of trait) and gradually darkening to black (prominent manifestation of the trait). After such coloring, "one glance at the chart shows the abundance and intensity of the hereditary burden."[78] Coloring pedigrees to note and demonstrate such burdens was already established procedure, but Crzellitzer's use of this method fashioned

---

[76] Ibid., 189–190.

[77] Ibid.; see also Arthur Crzellitzer, "Methodik der graphischen Darstellung der Verwandtschaft, mit besonderer Berücksichtigung von Familien-Karten und Familien-Stammbüchern," in Sommer, *Bericht über den II. Kurs mit Kongreß*, 25–37, as well as the report on Crzellitzer's presentation of his *Sippschaftstafel* in *Naturwissenschaftliche Wochenschrift*, Neue Folge 12, no. 14 (1913): 222–223.

[78] Crzellitzer, "Methoden der Familienforschung," 195.

this practice into a particular aid for Mendelian analysis.[79] As he explained,

Through appropriate coloring or shading, the distribution of certain properties among the members of the family becomes apparent. One sees at once whether the father or the mother of the central person are **purebred** (*reinrassig*) = homozygous; if so the entire half of the chart carries **similar** color. If only the grandparents are purebred, then only a quarter of the chart is uniformly shaded or colored.[80]

According to Crzellitzer, his chart's unique attribute was therefore not that it allowed discerning the distribution of traits among progeny, but that it enabled one to ascertain that the parents of the proband were homozygous. It is this particular part of the Mendelian teaching that Crzellitzer saw as essential for aligning genealogical work with the requirements of biological teaching.

★ ★ ★

Crzellizer's 1908 kinship chart, Rüdin's 1911 treatise and Fischer's 1913 Rehoboth study all pointed to the potential promise that Mendelian reasoning had for the study of heredity among humans, whether this was done in the framework of genealogy, psychiatry or anthropology. In the scholars' respective circles, these works were recognized as expanding new scientific horizons. Years later, Rüdin would reflect on his career in a public talk on "20 Years of Research into Human Heredity." In his opening words, Rüdin would identify his 1911 publication as giving him the first impetus to deal with psychiatric hereditary research. "The main idea of this essay was that German psychiatry absolutely needs to address and deal with Mendelian hereditary teaching." In his subsequent work, he would explain, he attempted to do so in a systematic fashion; and the goal of his Genealogical-Demographic Department in the German Research Institute for Psychiatry (*Deutsche Forschungsanstalt für Psychiatrie*), established in 1917, was to align psychiatric research with the study of splitting, or Mendelian heredity.[81] One year after the conclusion of World War II, Fischer would also enforce his quasi-metaphysical bond with Mendelian theory on the very first page of his

---

[79] Cf. Robert G. Resta, "The Crane's Foot: The Rise of the Pedigree in Human Genetics," *Journal of Genetic Counseling* 2, no. 4 (1993): 235–260.

[80] Arthur Crzellitzer, "Der gegenwärtige Stand der Familienforschung," *Sexual-Probleme* 8, no. 4 (1912): 221–243 (quote from p. 235).

[81] Historisches Archiv des Max-Planck-Instituts für Psychiatrie [hereafter MPIP], GDA 98, "20 Jahre menschlicher Erbforschung an der Deutschen Forschungsanstalt für Psychiatrie in München, Kaiser Wilhelm Institut."

autobiography, by pointing out that he "habilitated in precisely the year that Mendel's laws experienced their resurrection!"[82] Fischer would also recollect his memories from the 1910s and claim that "looking back today, I believe that my whole scientific and external course of life relied in a large part on this work [on the Rehoboth Bastards]. ... My whole essence was now oriented to questions of heredity."[83] Crzellitzer would not survive to reflect on his life course with such satisfaction. In 1938 he attempted to flee the Nazis, leaving Germany with his family, and in the years that followed he moved constantly until he was captured, interned, released, deported, transported and finally murdered in Sobibór extermination camp on July 13, 1943.[84] As we will see in the following chapters, these different personal fates, career paths and scientific visions were intimately intertwined.

---

[82] MPG Archive III Abt., Rep. 94, Nr. 45. Vorwort.   [83] Ibid., S. 45, 49.

[84] Horst A. Reschke, "Arthur Czellitzer," in *Encyclopedia of Jewish Genealogy* (Salt Lake City, UT: Center of Jewish History, 1987), available at http://digital.cjh.org:1801/ webclient/DeliveryManager?pid=523964.

2   Mendelism Maturing: From Experimental to Interpretative Framework, 1913–1933

During the 1910s, Mendelism gained momentum. Even an infectious disease as complex as tuberculosis was examined from a Mendelian perspective: instead of focusing on the bacillus causing the inflammation, the body's overall immunity was analyzed and found to be determined by two Mendelian hereditary factors, one controlling the structure of the thorax, the other the body's capacity to fight the disease.[1] Applications of Mendelian thinking spread also beyond the medical domain. In 1917, zoologist Valentin Haecker published a small booklet titled *The Heritability in the Male-line and the Patriarchal Family Concept*. Haecker wished to offer parents whose sons perished in the ongoing war, "a small consolation ... that the physical and spiritual virtues, which the family is proud of and through which the family members appear internally and externally related, don't cease abruptly [with the death of their sons], but continue to live on also among the children of the daughters." Haecker explained that Mendel's laws disproved the popular belief that the male types were dominant over the female types. Consequently, Haecker declared, even if the family name was extinguished, familial traits could continue to prosper through the daughters.[2]

The following year, zoologist Heinrich Ernst Ziegler, who twelve years earlier had argued that Mendelism could not be valid for humans, opened the chapter on human heredity in his book *The Study of Heredity*

---

[1] Jens Paulsen, "Über die Erblichkeit von Thoraxanomalien mit besonderer Berücksichtigung der Tuberkulose," *ARGB* 13, no. 1 (1921): 10–31. For a larger perspective on tuberculosis research at the boundary between heredity and infection, see Michael Worboys, "From Heredity to Infection? Tuberculosis, 1870–1890," in Jean-Paul Gaudillière and Ilana Löwy (eds.), *Heredity and Infection: The History of Disease Transmission* (London/New York, NY: Routledge, 2001), 81–100, and Bernd Gausemeier, "Borderlands of Heredity: The Debate about Hereditary Susceptibility to Tuberculosis, 1882–1945," in Bernd Gausemeier, Staffan Müller-Wille and Edmund Ramsden (eds.), *Human Heredity in the Twentieth Century* (London: Pickering and Chatto, 2013), 13–26.

[2] Valentin Haecker, *Die Erblichkeit im Mannesstamm und der vaterrechtliche Familienbegriff* (Jena: Gustav Fischer, 1917) (quote from p. 30).

in Biology and Sociology by asserting authoritatively that "all human dispositions, favorable or unfavorable, are inherited according to the laws of biological heredity. Talents and other intellectual capabilities are inherited in the same manner as physical traits. Pathological dispositions and deformities are also subject to the same laws of heredity."[3] Extrapolating from these principles, a booklet published in the mid-1920s stated that different social strata were also sharply distinct in the hereditary, Mendelian sense. Class division, explained the author, initially began as a division of labor, which in itself was the result of hereditary capabilities differentially distributed among the population. Since every man preferred to marry a woman of his own social class, the children born from such marriages inherited the same tendencies from both of their parents. This led to intensive inbreeding that consolidated the hereditary material in each class. Studying the results of crosses between distinct social classes – which were hereditarily distinct as well – therefore should be conducted via Mendelian reasoning, explained the author.[4]

If mental components were biologically inherited, it was only natural that specific professions would be grounded in biology as well. In 1929, the physician Karl von Behr-Pinnow looked into the hereditary mechanism underlying artistic violin making. Five specific traits were required for such a profession: (1) the ability to evaluate the quality of wood; (2) the ability to evaluate and use polish; (3) carving talent; (4) construction talent; and (5) a sense of form. In addition, (6) musical talent was required, but musicality in itself was dependent on many different components. On the basis of family lineages, von Behr-Pinnow felt that he could confirm his hypothesis, namely, that violinmaking was a polymeric hereditary trait, composed of six separate hereditary units, most of them dominant.[5]

By the second half of the 1920s, there was hardly a human trait left that was not shown to be Mendelian in one way or another – from the clockwise/counterclockwise direction of the hair whorl, through the bass/soprano/baritone pitch of the voice, to mental capacities and learning skills.[6] Psychologists, medical practitioners and animal breeders took

---

[3] Heinrich Ernst Ziegler, *Die Vererbungslehre in der Biologie und in der Soziologie* (Jena: Gustav Fischer, 1918), 239. On his earlier stance, see page 28 above.

[4] Sandor Kaestner, *Was muss der Familiengeschichtsforscher von der Vererbungswissenschaft wissen?* Praktikum für Familienforscher 5 (Leipzig: Degener & Co., 1924), 12.

[5] [Karl] von Behr-Pinnow, "Eine Sonderuntersuchung für Berufsvererbung," *Volksaufartung, Erbkunde, Eheberatung* 4 (1929): 59–63.

[6] Felix Bernstein, "Beiträge zur mendelistischen Anthropologie: Quantitative Rassenanalyse auf Grund von statistischen Beobachtungen über den Klangcharakter der Singstimme," *Sitzungsberichte der Preussischen Akademie der Wissenschaften, Physikalisch-mathematische Klasse* (1925): 61–70; Walter Schwarzburg, "Statistische Untersuchungen über den menschlichen

the liberty to experiment with Mendelian reasoning and to apply it to solve a diverse range of questions. But in the scholarly communities of genealogists, psychiatrists and anthropologists, the engagement with Mendelian principles led to different results. As we will now see, genealogists were reluctant to subject themselves to the dictates of Mendelian theory. Psychiatrists and anthropologists, on the other hand, embraced Mendelian reasoning, while not confining themselves to simple Mendelian schemes. As a result, during the 1920s Mendelism changed its status and its function within these scientific communities, a change of great significance for what was to follow.

Genealogy Rejects Mendelism

The wish to set genealogy on a solid scientific ground and to relate it to biological teaching was expressed clearly by Lorenz in his 1898 *Textbook* and was shared by many scholars both within and outside the genealogical profession. Two scholars in particular invested great efforts in furthering the interconnections between genealogy and biology: Hans Breymann, who was head of the Leipzig Central Office for German Personal and Family History, and the lawyer and heraldist Kekulé von Stradonitz. Von Stradonitz played a key role in the preparations for the 1912 Second Congress for Family Research, the Science of Heredity and Regeneration Study, where genealogists, psychiatrists, anthropologists and biologists met to discuss the coordination of their work; Breymann, the genealogist Adolf von den Velden and Crzellitzer were also active in helping the congress to take shape, as was the physician and statistician Wilhelm Weinberg.[7] The inner pressure to "biologize" research was accompanied by external pressure. As exemplified by Rüdin's 1911 treatise, leading scholars outside the genealogical community had marvelous visions of using genealogical data in their own studies and projects, visions that could materialize only if genealogists learned how to subordinate their work to the needs of genetic and eugenic research.[8]

Scheitelwirbel und seine Vererbung," *Zeitschrift für Morphologie und Anthropologie* [hereafter *ZMA*] 26, no. 2 (1927): 195–224; W[ilhelm] Peters, *Die Vererbung geistiger Eigenschaften und die psychische Konstitution* (Jena: Gustav Fischer, 1925).

7   Robert Sommer, *Bericht über den II. Kurs mit Kongress für Familienforschung, Vererbungs- und Regenerationslehre in Giessen von 9. bis 13. April 1912* (Halle a. S.: Carl Marhold, 1912), 1.

8   Ernst Rüdin, "Einige Wege und Ziele der Familienforschung, mit Rücksicht auf die Psychiatrie," *ZgNP* 7, no. 1 (1911): 487–585; Walter Scheidt, "Erbbiologische und bevölkerungsbiologische Aufgaben der Familienforschung," *Archiv für Sippenforschung und alle verwandten Gebiete* 9 (1928): 289–315.

Though vehemently promoted, such a reorientation of genealogy had its fierce opponents. In a long footnote in an article published in 1919, genealogist Friedrich von Klocke insisted that,

Genealogy, as an independent science, does not belong to the natural sciences! This must be stated in the most definite manner. ... Lorenz' textbook, under the influence of the then almighty Darwinism ... was regrettably shaped according to this [biological] orientation; the end of the previous century was indeed the time when the all-encompassing attempt ... was made to raise history to the "rank of science," through the introduction of natural-scientific methods. Genealogy may accept "teachings" from natural science – that goes without saying; yet from this one cannot develop a natural-scientific "side" of genealogy to correspond to the humanistic [side of it].[9]

Von Klocke further admitted that genealogical data may provide helpful assistance to hereditary research, but also quoted Weinberg as saying that "attempts to tackle heredity solely from genealogical perspectives have always led to a fiasco."[10] To end with a softer tone, he clarified that he did not principally oppose genealogical interest in hereditary questions; he wished only to determine properly the fundamental standpoint which genealogy should have on these matters.

Von Klocke was not alone: his colleague, the archivist Friedrich Wecken, similarly insisted that genealogy was mainly oriented toward personal, familial, historical and sociological research, and as such, clearly part of social, not natural, sciences.[11] Judging by the content of the articles published in genealogical journals, this was the majority view. The intellectual motivation underlying the work of most genealogists stemmed mainly from the thrill of diving into archives, putting together historical puzzles and discovering ancient family and communal curiosities.[12] Biological or eugenic trajectories were a nice supplement with a

---

[9] Friedrich von Klocke, "Vom Begriff Genealogie und den Verdeutschungen des Wortes," *Familiengeschichtliche Blätter* [hereafter *FB*] 17, no. 12 (1919): 217–228 (quote from p. 224).

[10] Weinberg's quote slightly alters the original meaning of his claim. Weinberg did not oppose the use of genealogy, but emphasized that the research of heredity cannot be carried out solely on the basis of genealogical charts, that recognizing the fundamental laws of biology was a preliminary precondition if meaning was to be extracted from these charts, and that mass statistics were also required. See his comments in Sommer, *Bericht ueber den II. Kurs mit Kongress*, 59. Obviously, quoting Weinberg (albeit out of context) – who by that time was already an indisputably authoritative figure in the field of genetic research – as opposing the use of genealogy, was meant to strengthen von Klocke's theoretical standpoint against the biological trend.

[11] Friedrich Wecken, *Taschenbuch für Familiengeschichtsforschung*, 3rd ed. (Leipzig: Degener & Co., 1924).

[12] A fine example of this puzzle-solving urge is found in an article titled "Ein genealogisches Rätsel" (Genealogical puzzle), which contained a riddle on the nature of the genealogical

clear advantage in the field of scientific prestige, but they were never meant to take over the entire scholarly field.

The biologization of genealogy was frustrated however, not because of professional proclivities, but due to the fundamental inability of genealogists to "mendelize" their field. When, following Lorenz, genealogists shifted their focus from the construction of family lineages to the compilation of ancestral charts, this did not entail major changes in their methods of data retrieval from archival sources, nor did it challenge their reliance on the historical study of written documents. It mainly entailed a certain format for representing their findings and suggested an additional potential layer of analysis to their data once retrieved. Seen from that perspective, the *Ahnentafel* was in reality not that different from the *Stammtafel* and left intact many traditional genealogical research practices. The Mendelian model was a different story. Not only did Mendelian theory suggest that ancestral charts were less important for the advancement of biological knowledge than originally expected; worse still, Mendelian considerations showed that the examination of many ancestral generations – both in the form of an *Ahnentafel* and of a *Stammtafel* – was of very limited value. In this respect, Mendelian thought broke sharply with earlier assumptions on the significance of genealogical inquiry for understanding heredity. Unlike previous perceptions of heredity that assumed that ancestral influences gradually accumulated (e.g., in Galton's work), the Mendelian model claimed that the traits of an individual were dependent solely upon the factors actually inherited from one's parents. The identification of those factors did not necessitate the inspection of the parents' own parents; an examination of their other children could be just as illuminating. Therefore, in medical and biological circles, a pedigree containing three thoroughly researched generations was considered more valuable for Mendelian analysis than any lineage, whatever its format may be, which documented many ancestral generations but contained little verifiable medical data. In the words of eugenicist Lenz, which must have been discomforting for genealogists:

To put great time and effort into the research of distant forefathers, as is customary in historical genealogy, is usually not worth it for the heredity researcher. Knowledge of the characteristics of offspring is no less valuable than

ties between certain persons. The solution came in the form of a complex descendants' chart that explained and shed light on the historical documentation presented along with the riddle. See Franz Carl Frhr. von Guttenberg, "Ein genealogisches Rätsel," *FB* 11, no. 3 (1913): 24. See also Ludwig Flügge, "Die Bedeutung der Genealogie für die allgemeinere Wissenchaft und für das praktische Leben," *FB* 20, no. 3 (1922): 3–8.

that of ancestors, and the same applies for relatives on the side-lines. A man shares on average the same amount of common hereditary mass with his child as with one of his parents, with his cousin as with his great-grandfather; and since it is naturally much easier to determine something certain on the living than on the dead, the examination of collateral-relatives is even more important than the complete exploration of ancestors. For most purposes it is sufficient if the relatives are examined up to the grandparents and their descendants.[13]

Mendelian logic therefore undermined the most fundamental premise of the genealogical discipline – the search for distant ancestors – and challenged the relevance of the genealogical expertise to biology *in toto*. Mendelian research demanded not only the abandonment of ancestral charts, but the transformation of the entire orientation, method and character of the genealogical pursuit. Such a transformation could hardly have found favor in the eyes of most genealogists.

Consequently, the majority of practicing genealogists never paid more than lip service to the importance of Mendel's teaching.[14] And they showed only limited interest in Crzellitzer's kinship chart, which through its collateral character followed (or, anticipated) precisely Lenz' dictum. The demise of the kinship chart was gradual, and at first sight may seem purely contingent. The beginning was promising: initially, after its introduction in several forums by Crzellitzer, the kinship chart was incorporated into genealogical textbooks and presented in articles on genealogical methodology as a basic form of genealogical presentation, side by side with the *Stammtafel* and the *Ahnentafel*.[15] One minor issue needed to be resolved: Crzellitzer completely disregarded the denotational practices and visual conventions customary among genealogists. For example, he used straight connecting lines instead of curly ones – an issue of great symbolic value for the practitioners in the genealogical community.[16] In 1911 Breymann suggested this deficiency could be corrected, commenting that "a greater compatibility in its graphical treatment with the existing genealogical presentation methods would have been possible, and

[13] *BFL* (1927), 419.
[14] For instance, Wecken, *Taschenbuch für Familiengeschichtsforschung* (1924), 220.
[15] See, for example, Wecken, *Taschenbuch für Familiengeschichtsforschung* (1919); Eduard Heydenreich, *Handbuch der praktischen Genealogie* (Leipzig: Degener, 1913), 49; Gottfried Roesler, "Genealogie als Grundlage der Familienpolitik," *Volksaufartung, Erbkunde, Eheberatung* 5, no. 5 (1930): 101–107.
[16] Furthermore, Crzellitzer also did not use Kekulé's ancestral numeration method, which by that time had become standard among genealogists. See Stephan Kekulé von Stradonitz, "Die Genealogie auf der Internationalen Hygiene-Ausstellung zu Dresden," *FB* 10 (1912): 3–4, 19–20, 39–40; Wecken, *Taschenbuch für Familiengeschichtsforschung* (1919), 40; (1924), 61. Whether anti-Jewish sentiments played a role in the overall disregard that the (mostly nationalistic) genealogical community displayed toward Crzellitzer's chart will remain a matter of speculation.

if such treatment could still be undertaken, it would be highly welcome."[17] Von Klocke took this upon himself and put forward a "relatives chart" (*Verwandtschaftstafel*), which was almost similar to Crzellitzer's, but formatted "properly," according to the genealogical standards of the time. But the presentation of this new chart also brought a change in the explicit motivations for using it: according to von Klocke, its principal utility was that it depicted the legal consequences of family-relatedness as defined in article 1589 of the German Civil Code (BGB). Hereditary applications were mentioned only as a supplement; the argument regarding the ability to note parental homozygosity was omitted entirely.[18]

Revealingly, even von Klocke's properly formulated chart never came to be used by genealogists. As the years passed, the original motivations underlying Crzellitzer's construction of his chart were forgotten, as was its specific format; only the name remained. In 1930, in a presentation of the available formats of genealogical charts, Gottfried Roesler listed the "relatives' chart" as the fourth and final basic format of genealogical presentation, after the family lineage (*Stammtafel*), the descendants' chart (*Nachfahrentafel*) and the ancestors' chart (*Ahnentafel*). The relatives' chart, he explained, was "principally a combination of descendants' and ancestors' charts, so there is nothing additional to say about it"; to this succinct description he attached a drawing of a pedigree far removed from both Crzellitzer's and von Klocke's formats.[19] Outside the confines of the genealogical community, a kinship chart that may be regarded as a derivative of Crzellitzer's did prosper: it was used during the 1920s as an aid for physicians counseling couples before marriage. According to a new law enacted in Prussia in the mid-1920s, prospective marriage partners were advised to fill out a health certificate with data on their parents, grandparents, uncles and aunts as well as their brothers and sisters; state officials were to inquire whether such certificates had been exchanged before proceeding to handle marriage requests.[20]

[17] Hans Breymann, "Über die Notwendigkeit eines Zusammengehens von Genealogen und Medizinern in der Familienforschung (Vortrag, gehalten anläßlich der Hauptversammlung der internationalen und der Deutschen Gesellschaften für Rassenhygiene zu Dresden am 6. August 1911.)," *ARGB* 9, no. 1 (1912): 18–29 (quote from p. 25).

[18] See pp. 39–41 in the 1919 edition, pp. 80–83 in the 1924 edition and pp. 58–60 in the 1930 edition of Wecken, *Taschenbuch für Familiengeschichtsforschung*.

[19] Roesler, "Genealogie als Grundlage der Familienpolitik," 103. See also Helmut Benecke, "Erläuterungen zu dem schematischen Grundriß einer Verwandtschaftstafel," *Zeitschrift der Zentralstelle für Niedersächsische Familiengeschichte* 8, no. 10 (1926): 191–195. Benecke criticized the available formats of kinship charts and offered his own version of such a chart while completely ignoring the original heredity-oriented motivations underlying Crzellitzer's format.

[20] See the wording of the Prussian decree and example of a health certificate in "Eheberatungsstellen und Gesundheitszeugnisse in Preussen," *Kultur und Leben.*

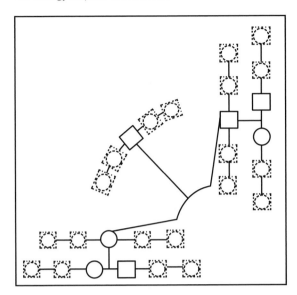

Figure 2.1 Kinship chart intended to assist doctors performing marriage counseling, 1927.
Source: Th[eobald] Fürst, " Wie kann die Tätigkeit des Schularztes der Erblichkeitsforschung und Rassenhygiene dienen?" *ARGB* 19, no. 3 (1927): 313

Such kinship charts were intended to accompany written evaluations, according to predetermined criteria, of one's health and of his/her suitability for marriage. The spread of these charts among physicians did not change the fact of their absolute disregard among genealogists.

As argued above, the essential conflict between Mendelism and genealogical research was unresolvable and, perhaps consequently, largely unaddressed by those in the genealogical community who favored the biological orientation. For them, the continuing failure to bring both fields together was a source of constant frustration. In the early 1920s, Breymann lamented the fact that all efforts to combine genealogy with biology were only partially successful, and affected only a few researchers. Biologists and physicians, he claimed, did not realize the potential benefits of genealogical studies, whereas genealogists lacked sufficient knowledge in medicine and biology to properly exhaust their data and turn their findings into valuable scientific products. Day by day, Breymann exclaimed, genealogists overlooked important facts that could have

Monatsschrift für Kulturgeschichte und biologische Familienkunde 8 (1926): 230–236; [ed.], *ARGB* (1927), 313; BArch R/86 5625.

become a treasure for hereditary science had they been documented correctly. Breymann hoped that a change was still possible and hypothesized that the postwar economic and material shortage would compel biologists and genealogists to join forces. As he saw it, the Leipzig Central Office, which he himself headed, was to be pivotal in organizing and managing such an integrated effort.[21]

Breymann's hopes may have been raised when, in 1923, the Prussian Ministry of Interior convened a meeting to discuss the establishment of a research institute that would advance the biological-hereditary study of the German population. Four scholars were invited to express their professional opinions on the matter. From Munich came the psychiatrist Rüdin; from Berlin, the hereditary researcher Heinrich Poll; from Leipzig, Breymann and von Stradonitz.[22] It was the latter two who opened the discussion and who outlined the intimate links between genealogical research and eugenic studies. Characteristically, the biologist Poll objected to their views and insisted that the advancement of hereditary knowledge could be achieved only through experimental studies, not by studying human pedigrees.[23] The discussion became one of many that eventually culminated in the establishment, four years later, of Fischer's Institute for Anthropology, Human Heredity and Eugenics.

Such efforts notwithstanding, for all the reasons detailed above, in the genealogical community itself the constant pressure toward biologization did not bear much fruit, and the attempts to offer an alternative family chart, more suited to the needs of hereditary research, failed miserably. "It is disastrous," lamented Rüdin in 1911, "that in certain medical-genealogical circles the ancestral chart is one-sidedly recommended and cultivated."[24] Twenty years later, the eugenicist Hermann Werner Siemens still complained about the same thing: "there is little support for the common assumption that it [the *Ahnentafel*] is

---

[21] Hans Breymann, "Genealogie und Vererbungslehre," *FB* 20, no. 9/10 (1922): 193–196; Arthur Crzellitzer, "Anleitung zu biologischen Untersuchungen für Genealogen," *FB* 21, no. 4 (1923): 33–40.

[22] See the notes of the discussion in BArch R1501/109421 Bl. 30–32.

[23] Poll, a physical anthropologist, eugenicist and pioneer of twin research, was of Jewish origin. On his life and career, see James Braund and Douglas G. Sutton, "The Case of Heinrich Wilhelm Poll (1877–1939): A German-Jewish Geneticist, Eugenicist, Twin Researcher, and Victim of the Nazis," *Journal of the History of Biology* 41, no. 1 (2008): 1–35.

[24] Rüdin, "Einige Wege," 531. See also earlier critical remarks of Rüdin on the exaggerated emphasis on the *Ahnentafeln* in Ernst Rüdin, "Kritische Besprechungen und Referate," *ARGB* 5, no. 2 (1908): 272–275; Ernst Rüdin, "Kritische Besprechungen und Referate," *ARGB* 3, no. 5 (1906): 750.

hereditarily-biologically so much more valuable than the descendants' chart."[25] Yet unfortunately for eugenicists, genealogists' outright rejection of alternative charts, their frustrating inability to absorb Mendelian methods and their insistence on cultivating *Ahnentafeln* in spite of their shortcomings were neither capricious nor coincidental. They were deeply rooted in the genealogical professional identity.

Psychiatry Embraces Mendelism

Five years after the publication of his treatise calling for the Mendelization of psychiatric work, in the midst of World War I, Rüdin published another monograph, this time on the inheritance and genesis of dementia praecox. This publication played a pivotal role in subsequent scientific developments and was instrumental in Rüdin's later career and academic reputation.

In contrast to his earlier treatise, which used pedigrees extensively, Rüdin began his monograph by denying the utility of individual pedigrees for studying heredity. As he saw it, it was possible to back up any theory on the basis of a selection of one pedigree or another. Worse still, many researchers became "hypnotized" by pedigrees heavily loaded with mental defects, thus selecting in advance families with especially salient hereditary influences. He therefore intentionally refrained from providing pedigrees. Instead, he used statistics. By accumulating and processing large amounts of familial data, statistics could compensate for the small number of children that characterized human copulations and assist in uncovering regularities that otherwise could not be detected. Moreover, statistical methods could correct for sampling bias such as the preselection of families with conspicuous hereditary overload; they could overcome difficulties in assessing heritability that stemmed from the fact that mental disorders made themselves apparent only at specific age intervals; and they could help evaluate the susceptibility to mental illnesses of first-, second-, third- or last-born children by correcting for family size and for the ages of parents at the time of conception.[26]

---

[25] Hermann Werner Siemens, "Bedeutung und Methodik der Ahnentafelforschung," *ARGB* 24 (1930): 185–197 (quote from p. 188).

[26] Ernst Rüdin, *Zur Vererbung und Neuentstehung der Dementia praecox (Studien über Vererbung und Entstehung geistiger Störungen I.)* (Berlin: Julius Springer, 1916), III, 2–3, 19–20. On the rejection of pedigrees in human genetics see also Bernd Gausemeier, "In Search of the Ideal Population: The Study of Human Heredity before and after the Mendelian Break," in Staffan Müller-Wille and Christina Brandt (eds.), *Heredity Explored: Between Public Domain and Experimental Science, 1850–1930* (Cambridge, MA: MIT Press, 2016), 337–363.

Rüdin did not devise the needed statistical procedures by himself. His advisor on these matters was Wilhelm Weinberg. Rüdin was not the first to exploit Weinberg's methods, many of which had been published from 1907 onward in the *Archive for Racial and Social Biology*, which Rüdin co-edited.[27] In 1912, the Swedish racial anthropologist Herman Lundborg applied Weinberg's methods in his studies of mental deficiencies, and a year later they were used by psychiatrist Ernst Wittermann, who stated that "only through the works of Weinberg, the application of biological hereditary rules for psychiatric family study was made possible."[28] The French William Boven used them as well in his 1915 dissertation on Mendelian heredity of dementia praecox and manic-depressive insanity.[29] But it was only after the publication of Rüdin's monograph, which had been delayed due to the circumstances of the war, that they became widely known.[30]

With the help of Weinberg's statistical techniques and on the basis of data gathered on 701 families of schizophrenic patients (including 755 probands and their 2,732 siblings), Rüdin embarked on his hereditary analysis. It was obvious, he noted, that dementia praecox was not a dominant Mendelian trait. Was it recessive? Rüdin examined families in which both parents were healthy and at least one child was ill. If the disease was recessive, the ill child should have been a homozygous "RR" type (or otherwise he would not manifest the disease), and should have received his ill factors from both his parents (or otherwise he could not

---

[27] Wilhelm Weinberg, "Weitere Beiträge zur Theorie der Vererbung. 4. Ueber Methode und Fehlerquellen der Untersuchung auf Mendelsche Zahlen beim Menschen," *ARGB* 9 (1912): 165–174; Weinberg, "Über neuere psychiatrische Vererbungsstatistik," *ARGB* 10, no. 3 (1913): 303–312; Weinberg, "Auslesewirkungen bei biologisch-statistischen Problemen," *ARGB* 10, no. 4 (1913): 557–581. For a reappraisal of Weinberg's methods and their significance, see James F. Crow, "Hardy, Weinberg and Language Impediments," *Genetics* 152 (1999): 821–825. A general review on the age-of-onset challenge in statistical analysis, including reference to Weinberg's solution to the problem, can be found in N[eil] Risch, "Estimating Morbidity Risks with Variable Age of Onset: Review of Methods and a Maximum Likelihood Approach," *Biometrics* 39, no. 4 (1983): 929–939.

[28] Herman Lundborg, *Medizinisch-biologische Familienforschungen innerhalb eines 2232-köpfigen Bauerngeschlechtes in Schweden* (Jena: Gustav Fischer, 1913); Ernst Wittermann, "Psychiatrische Familienforschungen," *ZgNP* 20, no. 1 (1913): 153–278 (quote from p. 158).

[29] William Boven, *Similarité et mendélisme dans l'hérédité de la démence précoce et de la folie maniaque-dépressive (Thése-Univ. de Lausanne)* (Vevey: Impr. Säuberlin & Pfeifer S.A., 1915).

[30] On the reception of Rüdin's work, see Matthias M. Weber, *Ernst Rüdin: Eine kritische Biographie* (Berlin: Springer, 1993), 113–114; Volker Roelcke, "Die Etablierung der psychiatrischen Genetik in Deutschland, Großbritannien und den USA, ca. 1910–1960. Zur untrennbaren Geschichte von Eugenik und Humangenetik," *Acta Historica Leopoldina* 48 (2007): 173–190.

become homozygous); but since the parents did not manifest the disease, both had to be heterozygous "DR." In such families with two "DR" parents, a certain proportion of ill children was expected, namely, 25%. However, the real proportion of ill children born to such healthy parents was far removed from this and stood at 4.48%. The simple monohybrid assumption therefore had to be ruled out.[31] However, the low proportion of ill children could be explained by an adapted assumption that schizophrenia was determined by two recessive factors ("xxyy"). If all the healthy parents were dihybrid heterozygous ("XxYy") then not one-quarter but one-sixteenth of the parents' progeny were expected to become ill. In fact, that ratio was not far from the actual one (1/16=6.25%). Rüdin therefore posited that a dihybrid recessive assumption could plausibly account for the inheritance of schizophrenia.

The dihybrid assumption had an additional merit. Mental disturbances frequently observed among relatives of schizophrenics could now be interpreted as stemming from some of the children inheriting just one of the two deleterious alleles (e.g., "Xxyy"). Additional mental disturbances could thereby also be explained on the basis of Mendelian reasoning. Rüdin went on to examine different aspects of schizophrenia, such as its frequency among half-siblings and cousins of those affected, its relation to birth order (were first-borns more affected than third-borns?) and family size, and its reported tendency to manifest itself at earlier ages in successive generations, a phenomenon called "anticipation" or "anteposition." Rüdin repeatedly referred to the Mendelian framework, although many of his calculations relied on a statistical toolkit that did not depend on Mendelian suppositions.[32]

For psychiatrist Eugen Bleuler, Rüdin's former teacher, this novel statistical treatment seemed like a clear aberration from Rüdin's earlier Mendelian aspirations. Bleuler rejoiced: "Finally, Rüdin has put an end to the naive attempts to study the heredity of psychoses according to Mendelian viewpoints." Mental disturbances, in Bleuler's view, lay on a continuum, the delimitation of which was necessarily arbitrary. "If other pairs of traits of plants and animals had not forced upon us the Mendelian theory, we would have never considered, with regard to dementia praecox (and many other psychoses) anything else than what we now call intermediate heredity, that is, a heredity with endless gradations concerning the intensity of the disease." Bleuler listed ten preconditions necessary for making heredity research applicable to the psychiatric field;

---

[31] Rüdin, *Zur Vererbung und Neuentstehung der Dementia praecox*, 27–52.   [32] Ibid.

only a few of them, according to his view, could be said to have been realized.[33]

In 1917, one year after the publication of his treatise on schizophrenia, Rüdin became the director of the Genealogical-Demographic Department of the newly established German Research Institute for Psychiatry in Munich. Throughout the 1920s, Rüdin and his disciples – most notably, Hermann Hoffmann, Adela Juda, Bruno Schulz and Hans Luxenburger – continued to develop and refine an array of statistical techniques, which were mobilized to answer two principal questions: First, given that a person had a certain mental disease, such as schizophrenia or manic-depressive insanity, what were the chances that his/her relatives would manifest the same – or another – disease? Second, how high were these chances compared with the general morbidity risk in the overall population, that is, to the chances of non-relatives of the mentally ill acquiring the same disease? Answering these questions properly demanded an intimate familiarity with sampling techniques, the correction of sampling bias, methods for compiling control groups and fine-grained analysis of statistical data.

Rüdin called his new methodological approach "empirical hereditary prognosis" (*empirische Erbprognose*). The adjective "empirical" signified not only that the results were based on actual observations, but also that they did not rely on any specific assumptions regarding the mechanism of heredity – not even on Mendelian suppositions.[34] In this respect, Bleuler was correct: in the new statistical direction propagated by Rüdin, the acquisition of knowledge on mental illnesses was no longer based on their categorization as Mendelian factors. When collecting and analyzing data, Rüdin and his followers made no principal distinction between environmental and hereditary effects, and did not confine themselves to Mendelian suppositions of whatever kind; their focus was on gaining insights into the heritability of mental illness, whatever the mechanism underlying such heritability may be. When measured against the visions outlined in his 1911 treatise, it certainly seemed like Mendelism was formally abandoned as a research strategy.

Yet Mendelism continued to thrive in psychiatric scholarship also during the heydays of "empirical hereditary prognosis," to the extent that in 1923, the Viennese psychiatrist Josef Berze would declare, "Under the influence of Rüdin, psychiatric heredity research stands

---

[33] Eugen Bleuler, "Mendelismus bei Psychosen, speziell bei der Schizophrenie," *Schweizer Archiv für Neurologie und Psychiatrie* 1 (1916): 19–40 (quote from pp. 19, 25–26).
[34] See Ernst Rüdin, "Empirische Erbprognose," *ARGB* 27, no. 3 (1933): 271–283.

totally under the flag of 'Mendelism.'"[35] Although Mendelism ceased to
provide research guidelines and did not serve as the sole measure for the
success of a study, it was nevertheless widely used and often referred to.
These references came via two principal channels: research on schizophrenia and psychiatric nosology.

In the study of schizophrenia, Rüdin's 1916 hypothesis that schizophrenia was a dihybrid Mendelian trait became a cornerstone for later
research. It surfaced repeatedly in psychiatric literature throughout the
1920s, was slightly altered, and prompted the phrasing of more
nuanced hypotheses which were all Mendelian *par excellence*. This
was of significant symbolic value, because schizophrenia had a special
status in psychiatry, considered to be paradigmatic for understanding
and treating nervous and mental disorders. Practically all the prominent psychiatrists of the time, from Kraepelin and Bleuler through
Freud and Jung, Jaspers, Husserl, Kretchmer and Birnbaum, saw
schizophrenia as a focal point of psychiatric theory, while schizophrenia also had prominence in the cultural and popular sphere as "the
most horrifying and most frequent hereditary disorder."[36] The
founder of the Heidelberg school of psychopathology, Karl Wilmanns,
stated in a 1922 talk that "[t]he history of dementia praecox is the
history of psychiatry during the last 30 years."[37] It was therefore of
great importance that schizophrenia continued to be defined and
analyzed using simple Mendelian terms.

Even those who did not adhere to the letter of Rüdin's dihybrid
hypothesis agreed that schizophrenia was recessive, a fact reaffirmed time
and again by the inspection of various pedigrees.[38] In 1921, Tübingen
psychiatrist Hermann Hoffmann examined relatives of schizophrenics
and found that some of them had abnormal mental attributes, which he
defined as constituting "schizoid personalities."[39] Following Rüdin,
Hoffmann suggested that schizoids were simply those with only one of

---

[35] Josef Berze, "Beiträge zur psychiatrischen Erblichkeits- und Konstitutionsforschung,"
ZgNP 87, no. 1 (1923): 94–166 (quote from p. 94).

[36] Ernst Rüdin, "Ueber die Vorhersage von Geistesstörung in der Nachkommenschaft
(Vortrag im Februar 1928 in der Gesellschaft für Rassen-Hygiene in München),"
ARGB 20, no. 4 (1928): 394–407 (quote from p. 399).

[37] Karl Wilmanns, "Die Schizophrenie," *ZgNP* 78, no. 1 (1922): 325–372 (quote from
p. 325).

[38] For instance, Wittermann, "Psychiatrische Familienforschungen"; Erwin Zoller, "Zur
Erblichkeitsforschung bei Dementia praecox," *ZgNP* 55, no. 1 (1920): 275–293; Hans
Heise, "Der Erbgang der Schizophrenie in der Familie D. und ihren Seitenlinien," *ZgNP*
64, no. 1 (1921): 229–259.

[39] Hermann Hoffmann, *Studien über Vererbung und Entstehung geistiger Störungen* (Berlin:
Julius Springer, 1921), 28.

the two recessive factors for schizophrenia. Hoffmann therefore declared with satisfaction that Mendelian phenotypes and psychiatric categories could fit elegantly together. He took for granted the idea that the causes of the disorders he examined lay in Mendelian heredity: "[W]e naturally have to do our best to analyze also the manifestation of character anomalies in a Mendelian manner, i.e., to make phenotypes and genotypes coincide [...] Just like any other attribute, any other property, the schizoid temperament must also follow a certain mode of inheritance." Hoffmann also examined the possibility that schizophrenia was based on a hereditary mechanism more complex than the two-factor hypothesis – namely, a trihybrid hereditary modus – and showed it was implausible and incompatible with empirical data.[40]

These ideas were further developed by the Munich psychiatrist Eugen Kahn. Schizophrenia, Kahn noted, combined two essential elements. One was a certain kind of mentality or personal character that existed before and regardless of the onset of the disease – a complex but steady state of psychiatric disorders, called "schizoid personality." The other was a process of mental deterioration (*Verblödung*) that appeared at a certain age and had a lasting impact on an individual's personality. After studying a collection of pedigrees, Kahn noted that (1) schizophrenia was usually transmitted indirectly (schizophrenic children having non-schizophrenic parents); (2) "schizoid personality" was directly inherited, i.e., characterized only individuals who had at least one schizoid parent; and (3) not all the descendants of two schizophrenic parents became schizophrenics themselves. This led Kahn to refine Rüdin's supposition and suggest that schizophrenia consisted of a dominant factor for schizoid personality accompanied by a recessive factor for a degenerative process (*dominante Schizoidanlage und recessive Prozeßanlage*). Only the presence of both factors resulted in complete schizophrenia.[41]

These ideas did not go unchallenged. In an article from 1925, Bruno Schulz, who was a disciple of Rüdin, asserted that "even though the dihybrid recessive hereditary mechanism of dementia praecox has been rendered fairly probable by Rüdin, it would nevertheless be going too far if one proposed to see it as irrefutably certain." The intensity of mental disorders among patients' offspring, for instance, was positively correlated with the number of ill ancestors among the families of their parents,

[40] Ibid., 91–96.
[41] Eugen Kahn, *Schizoid und Schizophrenie im Erbgang*, Studien über Vererbung und Entstehung geistiger Störungen IV (Berlin: Julius Springer, 1923).

a fact that did not fit simple Mendelian assumptions.[42] For Schulz, the results of his study were of interest precisely because they were unlike what would have been expected from crude Mendelian suppositions. Yet also in the case of Schulz, the dihybrid hypothesis still functioned as the default hypothetical framework.[43]

Similar Mendelian assumptions informed studies on other disorders as well. In 1925, Hamburg psychiatrist Friedrich Meggendorfer described his studies on dementia senilis, which were based on a combination of pedigree analysis and anatomical examinations of affected patients. Because in some families the disease had been transmitted directly from parents to children, simple dominance seemed to Meggendorfer a reasonable initial supposition. Nevertheless, not all of his pedigrees fit this simple-dominance assumption. Recessiveness was similarly ruled out. The next logical step, Meggendorfer explained, was to assume a mutual action of two or more "probably dominant" factors. But this did not match his data either. Further investigations into the clinical symptoms prevalent among the family members of senile individuals convinced Meggendorfer that "one of the two hypothetical factors for dementia senilis is a nervous, irritable, unrestrained, maybe even schizoid disposition, and in general a weak and fragile constitution of the central nervous system, while the other one is a process-disposition, maybe a disposition for the ordinary aging process."[44] He believed that this supposition, very similar to Kahn's, found reasonable support in his examined pedigrees.

As some of the works cited above attest, Mendelian thought had an impact on psychiatric theory that went beyond the confines of heredity research. This was primarily because the categorization and classification of different psychiatric phenomena into distinct disease entities was bound up with, and dependent upon, hereditary and genealogical insights. That was true well before the rediscovery of Mendel's laws, in the works of nineteenth-century scholars such as Heinrich Schüle

---

[42] Bruno Schulz, "Zum Problem der Erbprognose-Bestimmung. Die Erkrankungsaussichten der Neffen und Nichten von Schizophrenen," *ZgNP* 102 (1925): 1–37 (quote from p. 4).

[43] Bruno Schulz, "Zur Frage einer Belastungsstatistik der Durchschnittsbevölkerung. Geschwisterschaften und Elternschaften von 100 Hirnarteriosкleritiker-Ehegatten," *ZgNP* 109, no. 1 (1927): 15–48. See the footnote on p. 39: "For the sake of simplicity, I have depicted dementia praecox everywhere as a **simple**-recessive inherited trait, which is most probably not true." The emphasis on "simple" (which is negated at the end of the sentence) is Schulz', the implication being that dementia praecox was recessive, but not monogenic. See also ibid., p. 47.

[44] Friedrich Meggendorfer, "Ueber die hereditaere Disposition zur Dementia senilis," *ZgNP* 101 (1925): 387–405 (quote from p. 401).

and Richard von Krafft-Ebing; it was true also for Kraepelin himself.[45]
Even when a certain disease entity was considered to be firmly established, the results of hereditary investigations often encouraged
the refinement, change or even neglect of previously assumed categories.[46] Such changes, in their turn, had implications also for clinical
work.[47] Epilepsy, for instance, which had been initially perceived as a
single disease entity with diverse manifestations, was later divided
into several distinct disorders; the independent heritability of each of
these newly defined disorders confirmed the credibility of the new
classification.[48] On the basis of similar hereditary considerations,
"Pfropfschizophrenia," wherein people with mild or moderate mental
retardation experienced schizophrenia, was shown to be "a purely
random concurrence of imbecility and dementia praecox in an individual. A [statistical, hereditary] correlation between imbecility and
dementia praecox cannot be proven. Clinically and biologically it is
therefore not to be regarded as an entity, but merely as a result of
summation of a purely external nature."[49]

Heredity research, and Mendelian considerations in particular, also
provided an elegant solution to the debate on the existence of the
so-called intermediate/mixed psychoses (Mischpsychosen). These disorders, which combined symptoms belonging to different disorder

---

[45] Eric J. Engstrom and Matthias M. Weber, "Making Kraepelin History: A Great
Instauration?" History of Psychiatry 18, no. 3 (2007): 270; Weber and Engstrom,
"Kraepelin's 'Diagnostic Cards': The Confluence of Clinical Research and
Preconceived Categories," History of Psychiatry 8, no. 31 (1997): 375–385.

[46] In addition to the upcoming examples, especially illuminating for this process, are Hans
Roemer, "Eine Einteilung der Psychosen und Psychopathien, für die Zwecke der
Statistik vereinbart zwischen der psychiatrischen Klinik Heidelberg und den Heil- und
Pflegeanstalten Illenau und Wiesloch," ZgNP 11, no. 1 (1912): 69–90; Roemer, "Über
psychiatrische Erblichkeitsforschung," ARGB 9, no. 3 (1912): 292–329; Bleuler,
"Mendelismus bei Psychosen, speziell bei der Schizophrenie"; Eugen Kahn,
"Konstitution, Erbbiologie und Psychiatrie," ZgNP 57, no. 1 (1920): 280–311.

[47] "We depart from cases of schizophrenia with equal or similar course [of disease]," Eugen
Kahn explained in 1923, "to acquire first an overview on their mode of heredity and then
search … by considering equal or similar schizophrenics, to find heredity-determined
common relations. Initially we do not count questionable cases, but do that only once we
have gained a firm footing. As a proof for overall heredity we then consider the biological
relations of cases and, taking into account the high pathogenetic estimation of heredity,
expect to also gain insight into the clinical-systematic viewpoint." Eugen Kahn, Schizoid
und Schizophrenie im Erbgang, 21. See also the extensive study by Bruno Schulz, "Zur
Erbpathologie der Schizophrenie," ZgNP 143 (1932): 175–298.

[48] K. Gerum, "Beitrag zur Frage der Erbbiologie der genuinen Epilepsie, der epileptoiden
Erkrankungen und der epileptoiden Psychopathie," ZgNP 115 (1928): 320–422
(esp. 321).

[49] Carl Brugger, "Die erbbiologische Stellung der Pfropfschizophrenie," ZgNP 113 (1927):
348–378 (quote from p. 377).

complexes (usually manic-depressive insanity and schizophrenia), severely challenged the standard Kraepelinian classification of mental illnesses. Consequently, there was hardly a scholar of reputation who did not try to tackle the issue of mixed psychoses at one time or another.[50] Mendelism gave a simple and persuasive solution to the problem: diagnostic difficulties were simply the result of random combinations of Mendelian factors.[51] Many studies of Rüdin's school addressed this issue, searching for the roots of mixed psychoses in the latent hereditary dispositions scattered among patients' family members. Looking back in 1939, the esteemed psychiatrist and neurologist Robert Gaupp succinctly summarized this development: "Opposition to the [notion of] mixed psychoses has become gradually smaller the more hereditary biology progressed. The **mixture of symptoms** from **different spectrum disorders** as a result of the combination of various hereditary factors seems to us today a clinical triviality, and many clinical manifestations of acute psychosis, in which Kraepelin's classification seemed to fail, find their proper interpretation by precise description of the observations and of the hereditary findings."[52] Mendelian theory therefore supplied a convenient way out of cases that did not allow for classification to be carried out according to the accepted categories.

There were naturally also vocal opponents to the heredity-oriented research, those who considered this entire branch of science as "*Erbmythologie*" – mythology of heredity – with an explicit reference to the former disciplinary overemphasis on brain research, the *Gehirnmythologie* of the Griesinger school.[53] As we have seen, Bleuler disputed the basic nosological assumptions on the character of schizophrenia; others similarly challenged the presumption to clearly delineate mental disorders to

---

[50] See Ferdinand Adalbert Kehrer and Ernst Kretschmer, *Die Veranlagung zu seelischen Störungen* (Berlin: Julius Springer, 1924); Robert Gaupp and Friedrich Mauz, "Krankheitseinheit und Mischpsychosen," *ZgNP* 101, no. 1 (1926): 1–44; and the short review offered by Sven Stenberg, "Zur Frage der kombinierten Psychosen," *ZgNP* 129 (1930): 724–738.

[51] "When we encounter great differential-diagnostic difficulties in a case, we will nevertheless also have to consider the possibility of a combination of two dispositions," wrote Josef Berze as early as 1910. Berze's statement, taken from his *Die hereditären Beziehungen der Dementia praecox: Beitrag zur Hereditätslehre* (Leipzig & Wien: F. Deuticke, 1910), appears in Hoffmann, *Studien über Vererbung und Entstehung geistiger Störungen*, 40. See also Oswald Bumke's statements on the same page.

[52] Robert Gaupp, "Die Lehren Kraepelins in ihrer Bedeutung für die heutige Psychiatrie," *ZgNP* 165, no. 1 (1939): 47–75 (quote from p. 53).

[53] Wilmanns, "Die Schizophrenie," 359. For the possibility of dealing with combinations of symptoms without regressing to hereditary influence, see Maurycy Bornstein, "Zur Frage der kombinierten Psychosen und der pathologischen Anatomie der Landryschen Paralyse," *ZgNP* 13, no. 1 (1912): 1–16.

begin with.[54] Still others, while accepting Kraepelin's delineation of mental illnesses, had doubts about the ability of Mendelian hypotheses to capture biological reality correctly. Occasional criticism existed also on the scientific quality of hereditary analysis as it had been performed in psychiatry. In 1923, a sharp and well-reasoned attack on the Mendelian analysis of mental disorders came from the pen of the Viennese schizophrenia expert Berze, who drew attention to the fact that the clinical demarcation of mental disturbances did not necessarily need to agree with that of hereditary factors. Researchers, wrote Berze, tended to look at both issues as part of the same complex, but they were not. Mendelian research as it was implemented in psychiatry, he argued, was overly simplistic; hereditary-minded psychiatrists clung to the simplest forms of Mendelian thinking, absorbing only belatedly and partially recent advances in genetic research. As one example, Berze referred to the problematic lack of distinction between dominant-recessive and epistatic-hypostatic relations. Whereas dominance and recessivity referred to different variants of the same gene, epi/hypostasis had to do with the shared impacts of different genes on a particular phenotype. Berze claimed, "A more attentive reading of the works that appeared in the Mendelian era reveals that this important point did not always receive its appropriate consideration; this is also true for psychiatric heredity research." He was especially critical of a work by Hermann Hoffmann, where, according to Berze, "this neglect has already evolved into a system."[55]

Such criticism notwithstanding, the use of hereditary considerations to delineate disease entities was (and still is) pervasive – in psychiatry as well as in other fields of medicine. In 1926, Gaupp observed that "hereditary-biological research became one of the most important aids for recognizing the inner regularities in the appearance and course of psychiatric diseases." Furthermore, he argued, this development was only made possible through the connection of Kraepelinian nosology and modern,

---

[54] See Alfred E. Hoche, "Die Bedeutung der Symptomenkomplexe in der Psychiatrie," *History of Psychiatry* 2, no. 7 (1991): 334–343, and the concise analysis offered in Jules Angst, "Historical Aspects of the Dichotomy between Manic-Depressive Disorders and Schizophrenia," *Schizophrenia Research* 57, no. 1 (2002): 5–13.

[55] Berze, "Beiträge zur psychiatrischen Erblichkeits- und Konstitutionsforschung," 118. Berze's criticism was directed at Hermann Hoffmann's *Die individuelle Entwicklungskurve des Menschen* (Berlin: Springer, 1922), where Hoffmann attempted to apply Richard Goldschmitt's theory to the human psychic sphere. See another critical review of Hoffmann's work in "'Referate,'" *Psychologische Forschung* 6, no. 1 (1925): 208, where Wolfgang Köhler referred to Hoffmann's work as "rather cheap pseudo-theories" (*etwas billige Pseudotheorien*).

Mendelian-oriented heredity research.[56] Mendelian thought provided a convenient conceptual framework of imagined hereditary factors that could be equated with phenotypic disease entities. By doing so, it became an important framework of reference also for psychiatrists working outside of the Rüdin school, and it remained that way long after Rüdin abandoned his earlier attempts to "mendelize" mental disorders directly.

Physical Anthropology Redefines Itself

As we have seen, the Mendelian method relied on dividing the inspected object of inquiry into discrete, independently transmitted traits, each studied on its own. This division was one of the breakthroughs that enabled Mendel himself to devise his theory: he looked at every pea trait and examined it in isolation, irrespective of other traits. It was not the plant as a whole, but the plant's different characteristics that came under scrutiny. Such an approach fit neatly with the efforts of psychiatrists to divide mental diseases into distinct disease entities. In physical anthropology, such division seemed to suit the research style that was already well developed by the beginning of the twentieth century, wherein anthropologists measured different parts of the human body with ever-finer tools and tabulated, distributed and analyzed the results of such detailed examinations. In this sense, it was long before the rise of Mendelian genetics that human populations were transformed by anthropologists into collections of (presumably distinct) measurable traits with characteristic mean, maximal and minimal values. These latter computations of average and extreme values, however, also reveal a crucial point: beyond the seemingly separate measurements, anthropologists still clung to their assumption that there was order beneath the disorder, and that there was such a thing as race, or racial type – a holistic entity that existed, in the past if not now, and that should be studied and described as a totality. This latter assumption was fundamental for racial-anthropology and functioned as one of its defining paradigms.

However, it was not entirely clear how such a holistic paradigm could be accommodated within Mendel's theory, and specifically how it could coexist with Mendel's third law, which stressed the autonomy or independency of traits. The need to rethink the idea of race in the Mendelian age found clear expression in Fischer's Rehoboth study. Addressing the popular assumption that primitive races were prepotent when intermixed with civilized races, Fischer argued that none of the original races in his

---

[56] Gaupp and Mauz, "Krankheitseinheit und Mischpsychosen," 8.

study could be said to be prepotent. Several exceptions notwithstanding, the *Basters* tended to display intermediate values compared with those of their original races. But when they didn't, the reason, according to Fischer, was Mendel's laws, where "there exists no racial prepotency, but a prepotency of features!" Indeed, Fischer thought, it was always "the predominance of a feature, not of a race as such with all its features!"[57] Furthermore, although the average features of the *Basters* were unalike those of the original races, a new Bastard race was not created: the different qualities of the parental races simply did not amalgamate, but resegregated according to the Mendelian segregation rule. What was created was therefore a mixing of racial features (*Rassenmerkmalsgemisch*) but not a mixed race (*Mischrasse*).[58] Accordingly, races needed to be reconceptualized as combinations of traits, held together by some kind of biological mechanism, such as selection pressure, inner-genetic compatibility, or inbreeding. Whatever the force that held certain traits together may be, it was individual traits that needed to be studied; indeed, understanding the mechanisms governing the heritability of different traits was a prerequisite for any sensible discussion on the larger issue of racial typology.

One human feature that was particularly important for racial anthropologists was the human skull. Three principal factors contributed to the elevated disciplinary status of skulls among anthropologists, apart from their obvious Shakespearean quality. First, as long as racial anthropologists aspired to correlate existing populations with earlier phases of human evolution, they had to rely on craniology. On almost no other human body part could data be extracted both from living populations and from ancient ones.[59] Second, as often happens in science, measuring practice had its own inertia: the skull's three-dimensionality allowed for measuring of multiple kinds. Each measuring technique, once applied, became increasingly more precise; debates over which was the most meaningful index led to the spread of quantification from one angle to another, from length to volumes, from distances to ratios. Disputes over accuracy triggered the invention of new tools that, in turn, generated more data that demanded new processing tools, and so on.[60] Third,

---

[57] Eugen Fischer, *Die Rehobother Bastards und das Bastardierungsproblem beim Menschen* (Jena: Gustav Fischer, 1913), 205.

[58] Ibid., 223–226.

[59] Gustav Michelsson, "Kritische Betrachtungen über Methoden und Aufgaben der Anthropologie," *ZMA* 23, no. 2 (1923): 263–294, esp. 264.

[60] From the 1870s to the 1910s, the practice of skull measuring reached frenzied proportions. As one example, an article from 1917 documented the morphological study of 618 human skulls; of each of these skulls, 37 different magnitudes, 26 indices

building on the phrenological tradition and on the fact that the skull was the repository of the human brain, different skull shapes or volumes seemed to mean something regarding the mental capacities of the examined racial types. And so, anthropologists in Germany (and elsewhere) performed studies on the head diameter, skull volume or skull shape and examined the relations of craniometric data to school achievements, professional occupations, social strata and mental deficiencies. Often, they found that the skull types recognized as belonging to their own nation or race were associated with abilities that ranked on top of foreign races. For example, several German studies showed that dolichocephaly, or long-headedness, which characterized the Nordic race, was correlated with higher intelligence, and that other skull shapes could indicate mental inferiority. And in more than one study, an alleged affinity of the African skull and the skull of various primate apes substantiated European evolutionary superiority over non-European peoples.[61]

Being such a favorable object of anthropological examination, it was only a matter of time until the human skull would become a target also for Mendelian analysis. Already in 1908, the racial theorist Leo Sofer, while discussing the Nordic and Alpine races in Swabia, argued that in certain areas, the majority of the population belonged to the Alpine race, and a minority was characterized by a combination of Alpine brachycephalic (short-head) skulls, Nordic blond hair and blue eyes. "Ever since Mendel, this independent heredity of individual racial features is no longer a mystery to us," Sofer explained. Sofer also claimed that brachycephaly was dominant with respect to dolichocephaly, but he did not support this assertion with a detailed Mendelian analysis.[62] Four years later, the German-Jewish emigrant Elias Auerbach sent to the *Archive for Racial and Social Biology* a short paper on the plasticity of the skull and the cephalic index. Auerbach proclaimed that "[i]t is high

---

and 12 angles were measured – and this was far from exceptional. See Eugen Adams, "Über postembryonale Wachstumsveränderungen und Rassenmerkmale im Bereiche des menschlichen Gesichtsschädels," *ZMA* 20, no. 3 (1917): 551–628. See also Rudolf Martin, *Lehrbuch der Anthropologie* (Jena: Gustav Fischer, 1928); Martin listed more than 80 different measurements and numerous indices one could calculate from any given skull.

[61] Carl Röse, "Beiträge zur europäischen Rassenkunde," *ARGB* 2, no. 5–6 (1905): 689–798; Fritz Bachmaier, "Kopfform und geistige Leistung. Eine Betrachtung an Münchner Volksschülern," *ZMA* 27, no. 1 (1928): 30–36; Sei Hara, "Vergleichende Untersuchungen über einen planimetrischen Cranio-Facialindex," *ZMA* 30, no. 3 (1932): 571–585. Stephen Jay Gould, *The Mismeasure of Man* (London/New York, NY: W. W. Norton, 1981), offers invaluable insights into this research program.

[62] Leo Sofer, "Über die Plastizität der menschlichen Rassen," *ARGB* 5, no. 5/6 (1908): 660–668 (quote from p. 662); Sofer, "Auf den Spuren der mendelschen Gesetze," *Politisch-Anthropologische Revue* 7, no. 7 (1908): 349.

time that anthropometry takes hereditary thought as the overruling principle in its methodology, since only in such a way can it become an exact science." The only reliable way for studying cephalic indices was through Mendelian analysis, Auerbach argued. The cephalic index in itself was not a Mendelian factor, he explained, because it was composed of two independent factors – the skull length and the skull width. As complicated as it might have been to analyze each of these through a Mendelian perspective, such an inquiry was essential, Auerbach believed, as it served the higher purpose of combating theories of environmental predominance, which threatened to undermine much of the anthropological project in its totality.[63]

In 1916, one of Fischer's students, Max W. Hauschild, finally presented a concrete model for mendelizing skulls.[64] After examining approximately 700 skulls, Hauschild argued that, in accordance with the studies of Fischer, the shape of the skull must conform to Mendel's laws. Which of the two forms was dominant, dolichocephaly or brachycephaly? Phrasing the question this way was misleading, Hauschild explained, because, "traits mendelize, [but] combinations of such [traits] split again in racial crossing. It is not 'the' hair, for instance, which mendelizes, but the hair color and hair form."[65] As Hauschild saw it, in a long skull, it was the length that was dominant, and in a wide skull, the width. Denoting dominance with a capital letter, he defined that

Long skull = Lb
Wide skull = lB

But since skulls also had a third dimension – their height – one should in fact consider four types: LbH, Lbh, lBH and lBh.[66] "These four basic types of skull forms correspond almost precisely to the skulls of the four races which resided in western Germany before the migration of peoples," Hauschild observed.[67] He further explained that, as a consequence of Mendel's laws, (1) dominant features were inherited according

---

[63] Elias Auerbach, "Zur Plastizität des Schädels, mit Bemerkungen über den Schädelindex," *ARGB* 9, no. 5 (1912): 604–611 (quote from p. 609).

[64] Max Wolfgang Hauschild, "Das Mendeln des Schädels," *Zeitschrift für Ethnologie* 48, no. 1 (1916): 35–40. The paper was presented at a meeting held in February 1916 by the Berlin Society for Anthropology, Ethnology and Prehistory.

[65] Ibid., 36.

[66] There is a typing error in the denotation in Hauschild's text, the two last forms denoted both with LB (instead of lB). I corrected it above according to Hauschild's intentions as apparent in his literal description. Readers who already have a hard time following Hauschild's style of reasoning should not be discouraged; as will be shown below, the problem is not theirs, but Hauschild's.

[67] Ibid., 36.

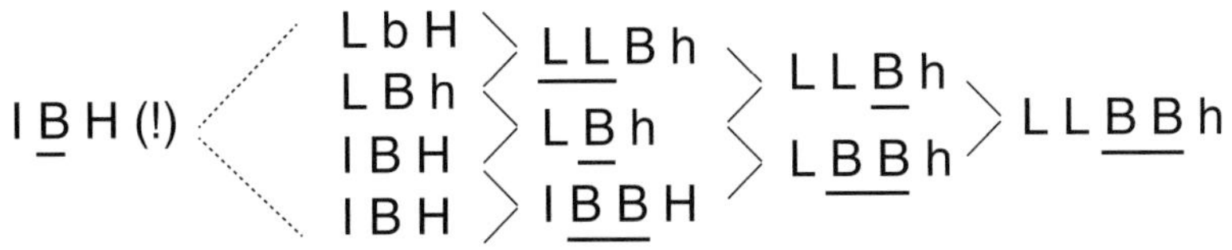

$$I \underline{B} H (!) \begin{array}{c} L b H \\ L B h \\ I B H \\ I B H \end{array} \begin{array}{c} \underline{L L B h} \\ L \underline{B} h \\ I \underline{B} \underline{B} H \end{array} \begin{array}{c} L L B h \\ L B \underline{B} h \end{array} L L \underline{B} \underline{B} h$$

Figure 2.2 Reproduction of Hauschild's scheme of mendelized skulls and their crossing. Max Wolfgang Hauschild, "Das Mendeln des Schädels," *Zeitschrift für Ethnologie* 48, no. 1 (1916): 37.

to the formula $L \times B = L + 2LB + B$; (2) when the same component was inherited in its dominant form from both parents, the crossing created an even larger skull; (3) "if the height is inherited in a latent dominant manner from both parents, then in the offspring generation, this characteristic cannot be smaller (recessive) [of either parent], and so the offspring will inherit the height of the taller parent at the cost of the length or width"; and (4) finally, "in a state of equal hereditary valence of the width, length and height, B was dominant over [both] L and H, and L [dominant] over H."[68] Hauschild summarized his results with the help of the following scheme, which was meant to facilitate the understanding of his model (Figure 2.2).

Five years later, armed with a more complex statistical toolbox, Hauschild refined his model and tested it on a collection of skulls from graves in the vicinity of Göttingen.[69] Everything seemed to fit smoothly. But then Hauschild found an academic rival: a Dutch scholar by the name of Gerrit Pieter Frets, who was similarly interested in the hereditary nature of skulls. Between 1921 and 1926, Frets offered two competing Mendelian models to account for his own findings on the heritability of skull shape. According to his analysis, it was either 19 independent Mendelian factors or else 3 allelomorphic pairs that were influencing the headform.[70] In 1927, Frets wrote to the *Journal of Morphology and Anthropology* that published Hauschild's latest treatise on the topic. In an unmistakable tone, he criticized Hauschild's utterly confused understanding of

---

[68] Ibid., 36–37. In the original, two more rules appeared, but their omission here is immaterial for the present discussion. The phrasing of the third rule (in the original text, rule number 4; here, the first quote) in the German origin is at least as unintelligible as it is in its translated form.

[69] Max Wolfgang Hauschild, "Die Göttinger Gräberschädel. Ein Beitrag zur Anthropologie Niedersachsens," *ZMA* 21, no. 3 (1921): 365–438. See also Max Wolfgang Hauschild, "Die kleinasiatischen Völker und ihre Beziehungen zu den Juden," *Zeitschrift für Ethnologie* 52/53, no. 6 (1920): 521.

[70] G[errit] P[ieter] Frets, *Heredity of Headform in Man* (The Hague: Martinus Nijhoff, 1921); Frets, *The Cephalic Index and Its Heredity* (The Hague: Martinus Nijhoff, 1926).

Mendelism and specified the myriad of deficiencies in Hauschild's analysis. He pointed out that Hauschild's denotation Lb was meaningless, since in Mendelian thinking it was pairs of factors which were analyzed. Did Hauschild mean LLbb? This was unclear. In addition, the L × B formula used by Hauschild seemed to be an unfortunate conglomeration of two unrelated expressions: namely, Mendel's A + a binomial expansion, on the one hand, and the results expected in dihybrid crossing (LLbb × llBB), on the other. In general, Hauschild's work entirely lacked the distinction between the inheritance of one pair of alternative factors (with an intermediate hybrid form) and the inheritance of two independent pairs of factors. Hauschild's suggestion that extreme hybrid vigor (*Luxurieren*) came as a result of conjoining two dominant factors similarly betrayed a total misapprehension of two unrelated and incommensurable biological phenomena.[71]

Hauschild had already passed away by the time Frets' criticism was published. Aside from the studies of Hauschild and those of Frets, it is difficult to find studies on the Mendelian nature of skulls or skull indices in German anthropological literature. Even in the international literature these were extremely rare, a fact that may be attributed to the practical impossibility of molding skulls into simple Mendelian schemes when these are rigorously defined.[72] The only scholar who continued to pursue Mendelian research on skulls was Frets, who published on the topic, reiterating the claim already expressed by Sofer in 1908, that brachycephalic skulls were dominant over dolichocephalic skulls.[73] This dominance was considered to be firmly established and found its way also into the popular writings of Günther, thus reaching a larger audience interested in racial matters.[74]

Nevertheless, by 1924 Fischer could declare that the studies of Frets and Hauschild had proved that the human skull form, as expressed in the cephalic index, was inherited according to Mendel's laws. To an extent, this finding was trivial for Fischer, because "Mendelian heredity is *the* heredity. For all the traits that interest us anthropologically, there exists no other [form of heredity]. And Mendelian heredity applies to all traits under all circumstances. All human racial traits are inherited,

---

[71] Frets, "Die Auffassungen M. W. Hauschild's über die Erblichkeit der Kopfform," *ZMA* 26, no. 2 (1927): 257–260.

[72] An exception is Halfdan Bryn, "Research into Anthropological Heredity. II. The Genetic Relation of Index Cephalicus," *Hereditas* 1 (1920): 198–212.

[73] G[errit] P[ieter] Frets, "Über die Dominanz des brachycephalen Kopfindex," *ZMA* 29, no. 2/3 (1931): 512–517.

[74] Hans F. K. Günther, *Der nordische Gedanke unter den Deutschen* (Munich: J. F. Lehmanns, 1925), 34.

and, as said, without exception according to Mendel's laws."[75] Fischer further claimed that reexaminations of older studies of skulls by scholars such as Otto Ammon, Johannes Ranke, Romedius Wacker and Franz Boas revealed that they were all compatible with Mendelian teaching; occasional deviations from Mendelian expectations were probably the result of environmental interferences. In Fischer's view, the fact that among the offspring of brachycephalic and dolichocephalic parents, the distribution of skull shapes diverged from Mendelian expectations, did not disprove the Mendelian hypothesis; it simply proved that additional factors were influencing the skull shape as well, and Fischer encouraged further investigations into this area.[76] Glossing over deviations, incongruences and methodological flaws, the Mendelization of the skull, as tendentious as it might have been, did for anthropology what schizophrenia provided for psychiatry: it showed that in matters of heredity, the discipline's most essential building block matched contemporary biological requirements.

Another human feature whose Mendelization seemed promising for racial-anthropology was blood. In 1900, in the same year that Mendel's laws were rediscovered, Austrian physician Karl Landsteiner reported on the agglutination effect of blood serum, and distinguished three main blood types: A, B and C (later to be labeled O). For this discovery, which made blood transfusion a life-saving, and not life-threatening, practice, Landsteiner would win the 1930 Nobel Prize.[77] As early as 1910, the Mendelian nature of the A and B blood types was firmly established by German internist Emil von Dungern and his Polish colleague, Ludwik Hirszfeld. Following extensive familial studies, these two scholars were able to demonstrate that A and B were both dominant with respect to O, and postulated that these serological attributes were induced by two independent factors with two alternative alleles each (A and a, B and b; only those with the genotype "aabb" would belong to the O group).[78] This view prevailed until 1924, when German mathematician Felix Bernstein showed on the basis of statistical considerations that a competing hypothesis of a single hereditary factor with three alternative forms

---

[75] Eugen Fischer, "Betrachtungen über die Schädelform des Menschen," *ZMA* 24, no. 1 (1924): 37–45 (quote from p. 38).

[76] See in this respect also the study of one of Fischer's students on mechanical and chemical influences on the shape of the skull: Gabriele Neubauer, "Experimentelle Untersuchungen über die Beeinflussung der Schädelform," *ZMA* 23, no. 3 (1925): 411–442.

[77] For a general introduction to the issue, cf. W[inifred] M. Watkins, "The ABO Blood Group System: Historical Background," *Transfusion Medicine* 11, no. 4 (2001): 243–265.

[78] Emil von Dungern and Ludwik Hirszfeld, "Über Vererbung gruppenspezifischer Strukturen des Blutes," *Zeitschrift für Immunitätsforschung und experimentelle Therapie* 6 (1910): 284–292.

(which he called A, B and R) could clarify the available familial findings. The genotypes were therefore not of the form AABB, AaBB, and so forth, but simply AA, AB, BB, AR, BR and finally RR (only the last inducing the O blood type).[79]

The mere fact that blood groups were inherited in a simple Mendelian fashion would not have aroused the interest of anthropologists if it did not also have racial bearing. Throughout the first two decades of the twentieth century, observations gradually accumulated on the frequencies of blood types in different parts of the world. During World War I, Hanka and Ludwik Hirszfeld undertook a general survey of the geographical distribution of blood types, studying hundreds of individuals from each of the sixteen different nations that took part in the battles on the Macedonian front. They found that Europeans had a relatively high proportion of agglutinogen A; Asians and Africans, B; and Russians, Turks, Arabs and Jews had more or less equal proportions of both. In order to convey these differences effectively, they devised a "biochemical index," which was in essence the ratio between the percentage of A (including AB) in a given population to the percentage of B (including AB) in the same population. The European types had indices in the range of 2.5–4.5; Asian and African types, 0.5–1.09; and the rest between 1.3 and 1.8. The geographical distribution of the A and B genes, the Hirszfelds concluded, suggested that humans originated from two different "biochemical races," one (A) came from Central or Northern Europe, the other (B) from India.[80]

This discovery inspired great hopes among biologically minded racial anthropologists. A single Mendelian factor – the blood type – seemed to

[79] Felix Bernstein, "Ergebnisse einer biostatischen zusammenfassenden Betrachtung über die erblichen Blutstrukturen des Menschen," *Klinische Wochenschrift* 3 (1924): 1495–1497. Interestingly, a fair amount of evidence had been incompatible with this triple-allele hypothesis; but none survived critical examination after Bernstein had published his thesis. Following their reevaluation, the inconsistent cases were reinterpreted as stemming from faulty observation methods or sloppy record keeping; and when none of these applied, the source of mismatch was attributed to false parenthood claims. See Laurence H. Snyder, "Human Blood Groups: Their Inheritance and Racial Significance," *American Journal of Physical Anthropology* 9, no. 2 (1926): 233–263; Pauline Mazumdar, "Two Models for Human Genetics: Blood Grouping and Psychiatry in Germany between the World Wars," *Bulletin of the History of Medicine* 70 (1996): 609–657. For a fuller description of the story, see also Tanemoto Furuhata, "A Summarized Review on the Gen-Hypothesis of Blood Groups," *American Journal of Physical Anthropology* 13, no. 1 (1929): 109–130, esp. 125: "Summing up all the reports on investigation of families given to this day, we learn that hardly any reports inconsistent with the new hypothesis have been given since 1926, in spite of the fact that many were given before 1925, the year when the theory was advanced."
[80] Ludwik Hirschfeld and Hanka Hirschfeld, "Serological Differences between the Blood of Different Races," *The Lancet* 194, no. 5016 (October 18, 1919): 675–679.

provide the key for studying racial classification, population history, human evolution and heredity. Articles suggesting divisions of human groups according to their blood genotypes, blood phenotypes, the frequencies of those or the ratios between them began to proliferate in German (and also American) literature.[81] In 1926, anthropologist Otto Reche, together with the naval physician Paul Steffan, founded the German Society for Blood Group Research in Vienna; two years later, the society began publishing the *Journal for Racial Physiology*, whose main task was to correlate blood group studies with racial-anthropology. Throughout the pages of this journal, the various distribution of blood types in the European population were illustrated with the help of novel visual techniques, substantiating the difference between the Nordic race (high frequency of A) and the Eastern/Slavic elements (high frequencies of B). Studies were also conducted to show that Blood type B characterized criminals and inferior social elements of various kinds.[82]

In 1929, one scholar addressed the great initial enthusiasm surrounding blood types research. He explained, "[I]f we raise the question why specifically the research of blood group affiliation arouses such wide interest ... we could, aside from the technical simplicity, clarity and speed of reaction [of blood type tests], specify the following reasons: with agglutination, we deal with a process that, whether by merit or by fault of L. and H. Hirszfeld, has the reputation of shedding some light on the mystery of the origin of human races."[83] The following year, another scholar explained that "ever since it has been determined that with relation to blood group there are four types of humans, and that the affiliation of each individual to a specific blood group remains unchanged throughout his entire life despite external influences, the extraordinary

[81] See Snyder, "Human Blood Groups"; Reuben Ottenberg, "A Classification of Human Races Based on Geographic Distribution of the Blood Groups," *Journal of the American Medical Association* 84, no. 19 (May 9, 1925): 1393; and the analysis offered in Jonathan Marks, "The Legacy of Serological Studies in American Physical Anthropology," *History and Philosophy of the Life Sciences* 18 (1996): 345–362. For examples of articles in German, see Paul Steffan, "Die Bedeutung der Blutuntersuchung für die Bluttransfusion und die Rassenforschung," *ARGB* 15, no. 2 (1924): 137–150; W. Klein and H. Osthoff, "Haemagglutinine, Rasse- und anthropologische Merkmale," *ARGB* 17, no. 4 (1926): 371–378; W[alther] Kruse, "Ueber Blutzusammensetzung und Rasse," *ARGB* 19, no. 1 (1927): 20–33; Herbert Leveringhaus, "Die Bedeutung der menschlichen Isohämagglutination für Rassenbiologie und Klinik," *ARGB* 19, no. 2 (1927): 1–17.
[82] Pauline Mazumdar, "Blood and Soil: The Serology of the Aryan Racial State," *Bulletin of the History of Medicine* 64, no. 2 (1990): 187–219.
[83] Max Berliner, "Blutgruppenzugehörigkeit und Rassenfragen," *ZMA* 27, no. 2 (1929): 161–170 (quote from p. 161). Incidentally, Berliner's own study, performed on cattle, showed that blood types were actually not a reliable criterion of racial diagnosis.

importance of blood groups for anthropology has become clear."[84] In 1930, when anthropologist Karl Saller publicized a new "measuring case for anthropological journeys," it included a kit for determining blood type – a new addition to the customary anthropological measuring instruments, the tables of eye colors and skin colors and the photography apparatus.[85] By 1932, world literature contained nearly 3,000 bibliographic items dealing with blood types; of those that appeared after 1920, roughly 12 percent were devoted specifically to the relationship between blood types and race.[86]

Celebratory declarations aside, the actual willingness of most anthropologists to put aside available resources and existing research frameworks for the sake of analyzing blood groups was rather limited. At the end of 1927, Bernstein applied to the German *Notgemeinschaft*, the main supporter of scientific projects in Germany after World War I, to receive funding for extensive blood group surveys in Germany, stating that these might also become helpful in shedding light on ancient racial migrations, as the studies of Hirszfeld and others had suggested.[87] In a meeting held to discuss Bernstein's application, the general atmosphere was conspicuously unsupportive. Fischer, who set the tone, thought that indeed, when it came to hereditary factors, "Germany should be covered with a network of observations that should become ever tighter." But studying blood groups alone, Fischer thought, was akin to advocating a survey that would document solely the distribution of nose shapes in the overall population. The anticipated costs of Bernstein's surveys would be better invested in the mapping of both normal and pathological anthropological attributes in German localities.[88] Almost unanimously the other discussants,

---

[84] Lasas, "Ueber die Blutgruppen der Litauer, Letten und Ostpreußen," *ARGB* 22, no. 3 (1930): 270–274 (quote from p. 270). For more on the initial enthusiasm that the issue had raised and its possible relations to the romantic perceptions of blood and its connotations, see a contemporary's point of view in Walter Scheidt, *Rassenunterschiede des Blutes* (Leipzig: Georg Thieme, 1927), 16–17.

[85] Karl Saller, "Ein Meßkoffer für anthropologische Reisen," *ZMA* 27, no. 3 (1930): 492–496. "Saller's measuring suitcase" cost 50 marks.

[86] According to the analysis offered by Mazumdar, of the items that were published between 1901 and 1920, 70 percent concerned blood transfusion. Of those published after 1920, 50 percent dealt with transfusion and transfusion-related problems, 15 percent were studies of blood groups and diseases, 12 percent were race studies and 6 percent were studies of the genetics of blood groups (the remaining 17 percent were distributed among various topics). See Mazumdar, "Blood and Soil," 187–188.

[87] BArch R1501/126242, Bl. 24–40,"Vorschläge zur Organisation der Bestimmung der menschlichen Blutgruppen in Deutschland."

[88] BArch R1501/126242, Bl. 48–55, "Bericht über eine Besprechung am 17. Dezember 1927 in den Räumen der Notgemeinschaft über Rassenforschung, Blutgruppenforschung und anthropologische Untersuchungen" (quote from Bl. 49, p. 2). Walter Scheidt expressed

including prominent scholars such as Walter Scheidt, Karl Saller, Otto Reche, Carl Correns and Erwin Baur, concurred. Geneticist Richard Goldschmidt agreed: "In the study of human heredity one does not get any further with statistics." Reche, who had special interest in the topic, also emphasized that general statistics had led to nothing, and, accordingly, what was now needed was not extensive surveys, but the serological method combined with studies on other traits.[89]

The research of blood types remained rather isolated and ineffectual in German anthropology, even in the Nazi period, a fact worthy of note given the intimate connections its protagonists (most notably, Otto Reche) had to nationalistic and Nazi circles.[90] There were several reasons for this. First and foremost, unlike skulls, skeletons or bones, serological diagnosis was a research tool bound to living populations; it could not supply information on, or draw data from, formerly existing human groups. Hypotheses on evolutionary or racial history could be raised and tested only on the basis of statistical-cum-geographical considerations – a clear divergence from the prevalent anthropological practices. In addition, the racial theories that blood type research could sustain – or refute – had to do with much earlier evolutionary phases in human development than those that interested German scholars. The bipolar A vs. B elements enabled the phrasing of theories about ancient Western (Nordic?) vs. Eastern (Slavic? Indian?) tribes, but did not shed light on the later wanderings and mixtures of the more ingrained racial classifications, such as the Alpine, Dinaric, Eastern or Baltic races from the teachings of Chamberlain, Deniker or Günther.[91]

Second, unlike all other anthropological markers, blood indices were population-based quotients, calculated according to the frequencies of blood types among groups of persons; they could not be computed for individuals. They could therefore be used for raising hypotheses on racial evolution, migration and mixture, but could never help in the racial diagnosis of any particular individual. To the anthropologists' dismay,

similar opinions, which appeared in print before the discussion took place. See Scheidt, *Rassenunterschiede des Blutes*, 60.
[89] BArch R1501/126242, Bl. 51–53 (p. 4–6).   [90] Mazumdar, "Blood and Soil," 210.
[91] This was noted as early as 1927 by Walter Scheidt. After reviewing the many studies already gathered by that time on blood group distributions, he concluded that the available data frustrated the initial hopes of using blood types as useful racial markers with respect to the existing racial divisions. See Scheidt, *Rassenunterschiede des Blutes*, 42. On the fact that blood types gave information on periods in human evolution that extended much further backward than anthropologists had wished for (or could handle), see ibid., 61. On the fundamental problems of using present-day correlations to deduce information on past racial mixtures, see his "Zur Theorie der Auslese," *Zeitschrift für induktive Abstammungs- und Vererbungslehre* 46, no. 1 (1928): 318–332.

despite insistent efforts, a correlation between blood types and other anthropological markers such as skull shape, eye color, hair color and the like could not be detected. With this "lack of pronounced correlations between blood groups and other examined anthropological characteristics" (as a Dutch study from 1934, published in the *Journal for Morphology and Anthropology*, declared), the aspirations to distinguish Alpine and Nordic racial components or facial features through blood groups could not be fulfilled.[92]

And there were additional obstacles. Unlike visible physical traits, it was difficult to see how blood types could have undergone processes of selection, whether environmental or sexual. In the view of scholars such as Walter Scheidt, for whom races were principally the result of continuous selection, this significantly reduced the ability to consider blood types as genuine racial markers.[93] The explicitly statistical nature of blood type research also deterred many anthropologists. Finally, the remarkable fact that many of the leading scientists studying blood types in Germany and abroad were of Jewish origin did not play in its favor in the nationalistic circles of German anthropology.[94] Hence the study of blood types in racial-anthropology did not live up to early expectations. Despite initial enthusiasm, it fell into disrepute. In a sense, blood types stood at the opposite pole to skull types: their status as Mendelian traits was undeniable, but their relation to the larger domain of racial studies was found to be rather tenuous.

The year in which Bernstein's request for funding was refused – 1927 – also saw the establishment of Fischer's Kaiser Wilhelm Institute for Anthropology, Human Heredity, and Eugenics. The research methods and the objects of inquiry of the studies performed under the institute's umbrella were extremely diverse: from direct observation of the everyday activities of twin siblings in order to assess the affinities between their "personal tempo," to the evaluation of similarities in

---

[92] Floris Hers, M[arie] A[nna] van Herwerden and Th. J. Boele-Mijland, "Blutgruppen-Untersuchungen in der 'Hoeksche Waard,'" *ZMA* 33, no. 1 (1934): 84–95.

[93] Scheidt, *Rassenunterschiede des Blutes*, 17–18. In November 1933, an attempt was made under the auspices of Gerhard Wagner, leader of the Nazi physicians, to identify special characteristics of the Jewish blood. It failed. See Karl A. Schleunes, *The Twisted Road to Auschwitz* (Urbana: University of Illinois Press, 1970), 119.

[94] Mazumdar, "Two Models for Human Genetics," 635–639. Mazumdar also suggested that Hans F. K. Günther's dismissal of blood types contributed to their practical neglect in Nazi racial policy. That may indeed be so, but this dismissal in itself is not only a cause but also an effect of the unfavorable status of blood types in German anthropology. Mazumdar also stressed the facial-recognition orientation of racial-anthropology, which did not sit well with blood type research; see Mazumdar, "Blood and Soil," 211–216. For a more comprehensive account of the entire affair, see Rachel E. Boaz, *In Search of "Aryan Blood": Serology in Interwar and National Socialist Germany* (Budapest: CEU Press, 2012).

patterns of fingerprints, to analyses of the causes of tuberculosis, to examination of the effects of radiation on genetic material. The research agenda was broadly defined to allow for practically any study of human physical, physiological, morphological, pathological and psychological dispositions.[95] The common bond uniting these diverse projects was the attempt to analyze the relative share of genetic and environmental influences (broadly understood) on the resulting phenotype. Unlike the study of the skull or those of blood groups mentioned above, the majority of these studies were not "Mendelian" in any direct sense.

Thus, somewhat like in psychiatry, Mendelism, understood as a methodology, receded into the background of the institute's research program. This was mainly because the possibilities of gaining new knowledge on the basis of pure Mendelian analysis had exhausted themselves rather early, clearing the stage for other avenues of investigation into the heritability of traits.[96] Nevertheless, Fischer's rhetoric, both in his public statements and in his scientific papers, continued to lean, almost exclusively, on the authority of Mendelian theory. In 1921, Fischer joined forces with Lenz and with botanist Erwin Baur to publish what later became the most important textbook on human heredity in Germany for at least two decades: the two-volume *Outline of Human Heredity and Racial Hygiene* (the so-called *BFL,* after the names of its three authors).[97] Due to its success, it was republished and substantially expanded in further editions in 1923, 1927, 1931, 1936 and 1941, and an English translation appeared in 1931. Fischer wrote the second chapter of the book, which was devoted to "racial differences among humans." Discussing the shape of skulls, Fischer acknowledged the

---

[95] The institute's scientific work was meticulously studied by the German historian Hans-Walter Schmuhl; see his *The Kaiser Wilhelm Institute for Anthropology, Human Heredity, and Eugenics, 1927–1945: Crossing Boundaries,* Boston Studies in the Philosophy of Science, vol. 259 (Dordrecht: Springer, 2008).

[96] Even before the institute was established, Fischer did not confine himself solely to Mendelian analyses. For example, throughout the 1920s he developed his Domestication Theory, first postulated in 1914, which claimed that humans had undergone processes analogous to those of domesticated animals and that these had been the main source of hereditary mutations, and, eventually, of racial differentiation. Eugen Fischer, "Die Rassenmerkmale des Menschen als Domesticationserscheinungen," *ZMA* 18 (1914): 479–524. Fischer was not the first to raise this supposition, which goes at least as far back as Darwin. See Fischer's own references to other scholars who dealt with this theme in ibid., 482–483, and his later work *Rasse und Rassenentstehung beim Menschen* (Berlin: Ullstein, 1927).

[97] On the reception of the book in Germany and abroad, see the detailed study of Heiner Fangerau, *Etablierung eines rassenhygienischen Standardwerkes 1921–1941. Der Baur-Fischer-Lenz im Spiegel der zeitgenössischen Rezensionsliteratur* (Frankfurt a. M.: Lang, 2001).

many studies that pointed to the effects of environmental influences, and did not hesitate to list a range of anatomical factors that could directly and indirectly lead to changes in the skull form. Nevertheless, he insisted that different human populations had different characteristic skulls that were hereditary and that remained fairly constant throughout the ages. The fact that skull forms did not converge into an intermediate form, despite the many mixtures and wanderings of races, he attributed to Mendel's hereditary rules.[98] More generally, Fischer stressed that the most important traits for racial diagnosis and racial typology – namely, eye color, hair color, skin color, eye form, hair form, nose form and other facial markers, as well as mental qualities, body size and, again, the skull shape – were all proved to be hereditary; this meant that racial differences existed and that they were grounded in hereditary, Mendelian traits.[99] Similarly, in a talk given in 1924, Fischer explained that his Rehoboth study was in fact analogous to Mendel's crossing of white, red and pink flowers. The reporter of his lecture summarized that "there exists no other heredity than Mendelian heredity. And despite several disputed issues, it is strongly affirmed that, 'Humans mendelize as well.'"[100] Hauschild reiterated his mentor's ideas: when it came to human racial features, "man is no exception, the same features that mendelize in animals do so also in humans: the color of the skin, the hair, the eyes, hair form, body size etc."[101] In 1927, Lenz could authoritatively state that among humans, heredity was always Mendelian heredity.[102]

★ ★ ★

By the end of the 1920s, Mendel's theory was widely seen as providing the theoretical basis for understanding human heredity. On its basis, older notions such as accumulative or blending inheritance were rejected, new advancements in studying chromosomes and mutations

---

[98] Eugen Fischer, "Die Rassenunterschiede des Menschen," in *BFL* (1923), 83–94.
[99] Ibid., 117–118.
[100] Adolf Stoll, "Über Familienforschung und Vererbung," *Zeitschrift für Kulturgeschichte und biologische Familienkunde* 1, no. 2 (1924): 62–67 (quote from p. 62). Incidentally, the attribution of the red-white-pink experiment to Mendel himself was an anachronism, since Mendel did not distinguish pink from red flowers. This was performed only in later studies by Bateson (with *Primula*) and Carl Correns and Erwin Baur (with *Mirabilis jalapa*, the four o'clock flower) in which the possibility of an intermediate color was noted, oftentimes using a colorimetrical scale.
[101] Max Wolfgang Hauschild, *Grundriss der Anthropologie* (Berlin: Gebrüder Borntraeger, 1926), 38.
[102] Fritz Lenz, "Die Methoden menschlicher Erblichkeitsforschung," in *BFL* (1927), *Band I: Menschliche Erblichkeitslehre*, 411–468, esp. 411.

(in plants and animals) where obtained, and the association between specific traits or diseases and specific – even if still largely intangible – genes were substantiated. As we have seen, despite its prestigious status, Mendelian thinking found little support among genealogists: not only did it fail to answer the kind of questions family-historians were struggling with, it also made genealogical work of little value for biologists and for the medical profession. Thus, while acknowledging its importance, genealogists continued to follow the path they had outlined just before Mendel's theory was rediscovered, compiling ancestral and familial charts and even occasionally extracting quasi-hereditarian conclusions from them, regardless of Mendelian considerations.[103]

For psychiatrists, on the other hand, Mendelism fit exceedingly well with the successful new divisions of mental illnesses; in this sense, Rüdin's work was perfectly in line with that of his teacher Emil Kraepelin. The redefinition of inherited mental pathologies according to Mendelian principles simultaneously reinforced the validity of these divisions and the veracity of Mendelian thinking. Historian Pauline Mazumdar once made a distinction between two styles of human genetics in Germany between the wars, positioning Rüdin's statistical studies at the non-Mendelian end of the spectrum, as opposed to the Mendelian style that characterized blood type research. As we have seen, however, the discrepancy between the two is only partially valid: although the methods of "empirical hereditary prognosis" were not themselves "Mendelian," its results often were.[104]

Finally, in anthropology, applying Mendelian thinking to study the heritability of human traits was almost unavoidable if one wanted to keep pace with the advancement of biological knowledge. From that perspective, it may seem natural that Mendelism became part of mainstream physical anthropology. Nevertheless, Mendelian thinking also posed a

---

[103] Building on their ancestral charts, genealogists occasionally turned to computing the degree of "ancestral loss" (*Ahnenverlust*) by noting the number of ancestors who appeared more than once in one's lineage (being, for example, great-great-grandparents from two different sides). These computations, which were almost meaningless from the Mendelian perspective, provided interesting results, but, as one genealogist admitted, "Whether from these results any kind of conclusion can be drawn, [or] laws can be stated, I dare not say yet." See Wilhelm Karl Prinz von Isenburg, "Aus der Werkstatt eines Ahnentafelforschers," *FB* 23 (1925), 7–14 (quote from p. 11); Amir Teicher, "'Ahnenforschung macht frei': On the Correlation between Research Strategies and Socio-Political Bias in German Genealogy, 1898–1935," *Historische Anthropologie 22*, no.1 (2014): 67–90.

[104] Pauline Mazumdar, "Two Models for Human Genetics: Blood Grouping and Psychiatry in Germany between the World Wars," *Bulletin of the History of Medicine* 70 (1996): 609–657. For another criticism of Mazumdar's thesis, see Gausemeier, "In Search of the Ideal Population," 358 (fn. 75).

challenge to the anthropological profession: through its focus on particular traits, it moved away from the discussion on whole "races" or entire "types." Mendelism also required rethinking the meaning of principal notions with broader cultural implications, such as purity and hybridity. It is time to examine how these concepts were affected by Mendelian insights, and how Mendelian theory became entwined with the understanding of social problems in Germany, even before Hitler came to power.

3 Mendelism, Purity and National Renewal

One of the salient features of public discourse in late nineteenth-century Europe was its intense preoccupation with national purity. The search for methods to purify the nation, society, culture, religion, one's body or one's thoughts from malignant or alien elements was not an invention of modern times.[1] But it was particularly after the rise of nationalism in Europe that cultural changes that threatened hegemonic groups were described by those in power as alien, foreign to an imagined autochthonous character, subverting the longed-for national homogeneity. In the last third of the nineteenth century, the rise of modern medicine – the discovery of bacteria, the development of vaccination and disinfection methods, the study of epidemiology – rendered the discussion on impurity and pollution ever more concrete. The need to combat degeneration became part of a quest for "national resurrection," later to evolve into one of the defining features of fascist movements. This quest became simultaneously a medical one as well as a social, moral and political project.

Discussions on purity, disease and national character were therefore a meeting point for medical and moral concerns. Not only were terms, concepts and metaphors routinely exchanged between the medical and the moral and political domains; but also the very understanding of the essence of diseases and the danger lurking in heterogeneity were reshaped in response to scientific and cultural changes.[2] This was true both at the level of the individual as well as of the nation or collective.

---

[1] Cf., for example, Barrington Moore, Jr., *Moral Purity and Persecution in History* (Princeton, NJ: Princeton University Press, 2000).

[2] One of the most salient examples for this exchange of metaphors between the medical and the political domains is the militaristic language that accompanied the bacteriological revolution. See Christoph Gradmann, "Invisible Enemies: Bacteriology and the Language of Politics in Imperial Germany," *Science in Context* 13, no. 1 (2000): 9–30; Silvia Berger, *Bakterien in Krieg und Frieden. Eine Geschichte der medizinischen Bakteriologie in Deutschland 1890–1933* (Göttingen: Wallstein, 2009); Marianne Hänseler, *Metaphern unter dem Mikroskop* (Zurich: Chronos, 2009).

The Mendelization of anthropological and psychiatric work, analyzed in the previous two chapters, impinged directly on these changes, profoundly influencing the understanding of what racial purity and racial mixture harbored. In this chapter, we will review these transformations, and see how the Mendelian concept of recessivity became loaded with medical and social anxieties that drew upon earlier widespread fears. Furthermore, we will see how Mendelian thinking was applied to analyze the nature of the threat to the German nation posed by European society's ultimate "Other": the Jews. As this chapter will show, Mendelism reshaped the understanding of the threat posed by the Jews to their fellow citizens while simultaneously redefining the path that needed to be taken in order to build anew the nation, cleansed from damaging, destructive elements.

Racial Purity and Hybridization

The notion of purity was central to Mendelian thinking from very early on. Unlike previous conceptions of the nature of hereditary materials, Mendel's model, as it was understood after 1900, did not allow for hereditary elements to merge into each other or be influenced by external factors.[3] Many biologists therefore referred to Mendelian theory as "the theory of pure gametes." "The fundamental thought of Mendelism is not dominance," explained German zoologist Ludwig Plate in 1905, "[...] but the purity of the gametes of the bastard, wherein each egg and sperm cell receive only one disposition of an antagonistic pair." In Plate's mind, this gametic purity evolved into veritable sexual austerity: "to use a graphical image, it is as if two partners entered into an unlucky marriage and were quick to separate from each other as soon as possible. Each chooses a nucleus for itself and is pleased to be able to dwell there alone."[4]

---

[3] As Robert Olby pointed out four decades ago, Mendel himself believed that paternal and maternal factors did amalgamate, a fact expressed in his notation of the constant types stemming from hybridization with a single letter ("A," "a"). Mendelians of the early twentieth century were quick to transform this notation into the one still used today ("AA," "aa"), highlighting their perception of the factors' inability to blend. See Robert C. Olby, "Mendel no Mendelian?" *History of Science* 17 (1979): 53–72.

[4] Ludwig Plate, "Über Vererbung und die Notwendigkeit der Gründung einer Versuchsanstalt für Vererbungs- und Züchtungskunde (Vortrag, gehalten am 24. Oktober in der deutschen Gesellschaft für Züchtungskunde)," *ARGB* 2, no. 5/6 (1905): 777–796 (quote from p. 786). See similarly Erich Tschermak, "Die Mendelsche Lehre und die Galtonsche Theorie vom Ahnenerbe," *ARGB* 2, no. 5–6 (1905): 663–672, esp. 664.

With the progress of Mendelian thought, a novel definition for racial purity also began to surface. In 1909, Danish botanist Wilhelm Johannsen, who coined the term "gene," defined racial purity as the state of homozygosity, irrespective of the actual genealogical origins of the inspected individual.[5] In the same year, Bateson similarly stressed that "for the individual to be altogether purebred it must be homozygous in all respects."[6] It did not take long before the German term for homozygosity – literally, sameness of zygotes – became *reinerbig*, literally, purely heritable. Through the concept of homozygosity, purity therefore acquired a new meaning: it came to mean the uniformity of one's alleles.

What started as a theory of gametic purity was therefore applied to define anew the theoretical prerequisites for the purity of a developed organism. Pure organisms were homozygous ones, that is, they contained two identical copies of certain genes. But those prerequisites did not stop at the level of the individual organism, because, for anthropologists as well as for many biologists, it was neither gametes nor even individual organisms that were of primary interest, but entire populations. Different anthropologists soon started linking population purity to the individual level by extending the requirement of homozygosity to the entire people or race. According to this new perception, all of the members of a race had to be uniformly homozygous for the race to be considered genuinely pure. "The purer the race or the closer the proximity of the mixed races, the greater the number of traits inherited in a homozygous manner ... [whereas] the stronger and stranger the mixture, the more widespread the heterozygosity," explained one anthropologist in 1923.[7] Real purity demanded homozygosity from the entire population, asserted another.[8] "If we consider the anthropological structure of the ancient Caucasus countries, we do not find there self-contained, uniform and homozygous populations that could be in a position to impress their stamp on the descendants in racial crossing," wrote Felix von Luschan, contrasting Caucasians to the ancient Magyars who were probably "an extremely uniform and homozygous mass."[9] Ideally, a pure race was one whose genes were all identical in two respects: within each individual (i.e., each

---

[5] Wilhelm Johannsen, *Elemente der exakten Erblichkeitslehre* (Jena: Gustav Fischer, 1909), 128, 302.

[6] William Bateson, *Mendel's Principles of Heredity* (Cambridge: Cambridge University Press, 1909), 291.

[7] Ferdinand Wagenseil, "Beiträge zur physischen Anthropologie der spaniolischen Juden und zur jüdischen Rassenfrage," *ZMA* 23, no. 1 (1923): 33–150 (quote from p. 88).

[8] Walter Scheidt, *Allgemeine Rassenkunde als Einführung in das Studium der Menschenrassen* (Munich: J. F. Lehmanns, 1925), 352–357.

[9] Felix von Luschan, *Völker, Rassen, Sprachen* (Berlin: Welt-Verlag, 1922), 60, 165.

individual was homozygous) and across individuals (i.e., total population-genetic uniformity). Only such a race could remain truly unaltered throughout the ages.[10]

By definition, homozygosity meant similar factors (alleles) for the same trait. However, some scholars tended to confuse the segregation of different traits (the Independence Law) and the segregation of different factors for the same trait (the Law of Segregation), a confusion that left its mark also on the discussion of racial purity. For example, in 1920, Hauschild stated that "a type or trait-complex is inherited only in pure races, but not when these are crossed with other races, because here the trait-complex is segregated according to the Mendelian rules of crossing."[11] For Hauschild, when races were mixed, it was the law of independent assortment that destroyed their uniformity, not the segregation of the gametes.

Importantly, under the new conceptualization of racial purity, it was no longer sufficient for all individuals to look the same, or to look like their ancestors, for the race to be considered pure; the alleles of all those individuals also had to be equal, to ascertain that no deviant genes would be found in either the previous generations or the upcoming ones. Homozygosity was the only certain way to prevent the dreaded results of "mendeling out" (*herausmendeln*) – the reemergence of primeval racial types in a population, due to past intermixtures. In the biological sphere, the term "mendeling out" designated the reappearance of formerly existing traits due to the recoupling of recessive factors, a process that characterized the second (or later) generation of hybrid crosses. In the anthropological context, the term was also used to describe the reappearance of entire anthropological types in a mixed population; and

---

[10] There were alternatives to this position. Franz Boas, for instance, made a clear distinction between the variability of the entire population and that of certain lines of descent, or families, within it. Following this distinction, "homogeneity is not by any means identical with purity of race." Fischer's Rehoboth Bastards, according to Boas, were homogenous because every family in Rehoboth "represents practically the same line of descent." In such a stongly intermarried population, there is "a high degree of variability in the family, while all the families will be more or less alike." See Franz Boas, *Kultur und Rasse* (Leipzig: von Veit, 1914), 18–50; Boas, "On the Variety of Lines of Descent Represented in a Population," *American Anthropologist* 18, no. 1 (1916): 1–9; Boas, "Report of an Anthropometric Investigation of the Population of the United States (1922)," in Boas, *Race, Language and Culture* (New York, NY: Macmillan, 1940), 28–59 (quotes from pp. 32–33).

[11] Max Wolfgang Hauschild, "Die kleinasiatischen Völker und ihre Beziehungen zu den Juden," *Zeitschrift für Ethnologie* 52/53, no. 6 (1920): 518–528 (quote from p. 520).

anthropologists found in Mendel's theory a corroboration and explanation for this phenomenon.[12]

These theoretical considerations informed the writings of racial theorist Günther. In his *Racial Study of the German Volk*, he explained that when children exhibited traits different from both of their parents, this testified to the presence of foreign racial hereditary factors, temporarily concealed due to their recessive nature.[13] "Part of the definition of the term 'Race' is not only the phenotypic similarity of a group of persons, but above all their genotypic matching, not only the racial purity of appearance, but above all homozygosity. A race is therefore a group of people equal in kind and equal in heredity." Furthermore, it was only the racially pure person that was beautiful (*schön*): "his body and also his mental essence are uniform, and each part of his body and essence indicates equal physical and mental [hereditary] disposition."[14] Similar ideas also found their way into Hitler's thinking: in his *Mein Kampf* Hitler acknowledged that "[t]he result of this drive toward racial purity, which is universal in nature, is not only the sharp external demarcation of individual races, but also their own internally uniform nature."[15]

The notion of racial purity was inseparable from its counterconcept – racial mixture, hybridization or, as anthropologists usually called it, "bastardization." Like purity, hybridization also stood at the core of Mendel's work, as echoed by the title of his study: *Experiments in Plant Hybridization*. In 1904, still following nineteenth-century traditions, zoologist Valentin Haecker defined "Bastards" or "Hybrids" as the product of crossing two forms that differed substantially, that is, to a greater extent than that expected by individual variations. This definition, he explained, corresponded to the colloquial meaning of those terms, which

---

[12] See von Luschan, *Völker, Rassen, Sprachen*, 132; von Luschan, "Afrika," in Georg Buschan (ed.), *Illustrierte Völkerkunde* (Stuttgart: Strecker & Schröder, 1909), 378–379; von Luschan and Hermann Struck, *Kriegsgefangene: Ein Beitrag zur Völkerkunde im Weltkriege* (Berlin: D. Reimer, 1917), 93–94; Hans Fehlinger, "Koloniale Mischehen in biologischer Beziehung," *Sexual-Probleme* 8, no. 6 (1912): 377–378.

[13] Hans F. K. Günther, *Rassenkunde des deutschen Volkes* (Munich: J. F. Lehmanns, 1923), 211–212. In Günther's terminology, "concealed" and "recessive" are almost the same word, since he uses the German terms *überdeckend* and *überdeckbar* to describe dominant and recessive states. More on this below.

[14] Günther, *Rassenkunde des deutschen Volkes*, 211–2. See also Eugen Fischer and Hans F. K. Günther, *Deutsche Köpfe nordischer Rasse* (Munich: J. F. Lehmanns, 1927). For an analysis of Günther's ideas and ideals, see Christopher M. Hutton, *Race and the Third Reich: Linguistics, Racial Anthropology and Genetics in the Dialectic of Volk* (Cambridge: Polity Press, 2005); Amos Morris-Reich, "Race, Ideas, and Ideals: A Comparison of Franz Boas and Hans F. K. Günther," *History of European Ideas* 32, no. 3 (2006): 313–332.

[15] Adolf Hitler, *Mein Kampf* (Munich: Verlag Franz Eher Nachf., 1943), 312.

were reserved for persons born to parents of manifestly distinct races, not those with mild variations in skin color or nose shape.[16] This kind of racial hybridization, scholars agreed, was prevalent in Europe. In 1900, still ignorant of Mendel's laws, anthropologist Otto Ammon supplied a simple mathematical proof showing that human racial mixture was practically inevitable; a year later, Hungarian anthropologist Aurel von Török ridiculed the fact that "The 'pure-blooded' races live only in the imagination of unfortunately still too many anthropologists."[17] Most contemporary European peoples were racially mixed, Török argued, a view that was becoming consensual among professional anthropologists.

But under Mendelian suppositions, racial hybridization meant something different from what these scholars had in mind. Mendelian hybridization referred to any process through which heterozygous organisms were created.[18] In this latter sense, it quickly became agreed that in the human sphere every person was necessarily a "bastard" – that is, heterozygous – at least with respect to some of his features.[19] "In the sense of modern hereditary research, **every man is a bastard**; and every human reproduction is bastardization," clarified the 1926 *Handbook of Sexology* to its readers.[20] Such bastardization, understood as a heterozygous state, was not only historically, empirically and theoretically inescapable, but

---

[16] Valentin Haecker, "Über die neueren Ergebnisse der Bastardlehre, ihre zellengeschichtliche Begründung und ihre Bedeutung für die praktische Tierzucht (Vortrag, gehalten im Verein für vaterländische Naturkunde am 10. März 1904)," *ARGB* 1, no. 3 (1904): 321–338 (esp. 322).

[17] Aurel von Török, "Inwiefern kann das Gesichtsprofil als Ausdruck der Intelligenz gelten? Ein Beitrag zur Kritik der heutigen physischen Anthropologie," *ZMA* 3, no. 3 (1901): 351–484 (quote from p. 422); Otto Ammon, "Zur Theorie der reinen Rassetypen," *ZMA* 3, no. 3 (1900): 679–685. Ammon's line of reasoning was as follows: assuming random mating, if two-thirds of a certain founder population were Germanic and the remaining third belonged to a darker race, then the following generation would contain only 4/9 pure Germanics; in the next generation, the proportions of pure Germanics would decrease to 16/81. Generally speaking, with every generation the proportion of pure types would significantly decrease, the number of *Mischlinge* (hybrids) substantially increase. Therefore, Ammon explained, after seven generations almost no pure types would be left. On the overall consensus regarding the mixed character of the European population, see Benoît Massin, "From Virchow to Fischer: Physical Anthropology and 'Modern Race Theories' in Wilhelmine Germany (1890–1914)," in George W. Stocking (ed.), *Volksgeist as Method and Ethic: Essays on Boasian Ethnography and the German Anthropological Tradition* (Madison: University of Wisconsin Press, 1996), 79–154.

[18] Erwin Baur, "Bastardierung," in Eugen Korschelt et al. (eds.), *Handwörterbuch der Naturwissenschaften* (Jena: Gustav Fischer, 1912), 850–874.

[19] Fischer, *Die Rehobother Bastards und das Bastardierungsproblem beim Menschen* (Jena: Gustav Fischer, 1913), 139–141.

[20] Hermann Werner Siemens, "Bastard," in Max Marcuse (ed.), *Handwörterbuch der Sexualwissenschaft* (Bonn: A. Marcus & E. Webers Verlag, 1926) [reprint 2011], 45–46.

also biologically trivial, and occurred every time copulation took place. It may have harmed the uniformity of the population, but there was scant evidence that it had damaged the individual as such.

This Mendelian definition of hybridization encouraged a more morally neutral attitude toward it. However, the social fears associated with racial contamination that reverberated both in the public and in the scientific domains were not easily refuted. They found particular expression in two concrete medical anxieties. One, which was of special significance for anthropologists and racial-hygienists, was the hybrids' presumed reduced fertility. Different species were, by definition, mutually infertile. Was the same also true for the crossing of different races? Were hybrids doomed to extinction? There was empirical evidence that they were. It was statistically established, for instance, that Christian-Jewish marriages yielded fewer children than strictly Christian or strictly Jewish marriages; the same was true for the birth rates of "pure whites" and "pure blacks" compared with those of "Mulattos." The reasons underlying such reduced fertility, however, remained obscure and open to interpretation, including biological, social and economic explanations.[21]

In 1916, ethnologist Max Moszkowski, who was an admirer of Fischer's work, attempted to use Mendel's model to account for the infertility attributed to certain racial crossings.[22] Moszkowski noted that Mulattos from Jamaica who descended from Northern European and Negro couplings were infertile and gradually died out, whereas Mulattos who were descendants of French and Negro parents remained fertile. This, he argued, could be explained using the presence-and-absence model of Bateson. The latter model defined that "a dominant character is the condition due to the *presence* of a definite factor, while the corresponding recessive owes its condition to the *absence* of the same factor."[23] Moszkowski postulated that all whites had a hereditary factor, denoted with the letter A, which damaged the development of the female egg when exposed to continuous conditions of extreme heat. Among the Southern Europeans, who seemed more accustomed to living in tropical

[21] Max Marcuse, "Über die christlich-jüdische Mischehe," *Sexual-Probleme* 8, no. 10 (1912): 691–749; Paul Kaznelson, "Kritische Besprechungen und Referate," *ARGB* 10, no. 5 (1913): 685–686; Hans Fehlinger, "Kreuzung beim Menschen," *ARGB* 8, no. 4 (1911): 447–457; Fehlinger, "Koloniale Mischehen in biologischer Beziehung."

[22] Max Moszkowski, "Klima, Rasse und Nationalität in ihrer Bedeutung für die Ehe," in C[arl] Noorden and Siegfried Kaminer (eds.), *Krankheiten und Ehe: Darstellung der Beziehungen zwischen Gesundheitsstörungen und Ehegemeinschaft*, 2nd ed. (Munich: J. F. Lehmanns, 1916), 100–156. See his reproduction of Fischer's table from the Rehoboth study on p. 145, and his lengthy quotes of Fischer's teaching, almost word for word, in pp. 142–148.

[23] Bateson, *Mendel's Principles of Heredity*, 53.

conditions, this factor was neutralized by another hereditary factor, which he called B. Negroes had none of these, an absence denoted with ab. Moszkowski described the crossing of Northern Europeans and Negroes using the following scheme (Figure 3.1):

$$\frac{\overbrace{A + a}^{}}{\underbrace{Aa}_{}} \quad P$$

$$\overline{AA + Aa + aA + aa} \quad F_2$$

with F_1.

Figure 3.1 A 1916 model for crossing Northern Europeans with Negroes. A represents a factor damaging the female egg in extreme conditions of heat.
Source: Max Moszkowski, "Klima, Rasse und Nationalität in ihrer Bedeutung für die Ehe," in C[arl] Noorden and Siegfried Kaminer (eds.), *Krankheiten und Ehe: Darstellung der Beziehungen zwischen Gesundheitsstörungen und Ehegemeinschaft*, 2nd ed. (Munich: J. F. Lehmanns, 1916), 151

According to this Mendelian formula, three-quarters of the descendants of Northern European–Negro marriages were expected to carry the damaging element. This could explain why in subsequent generations of such crossing, Europeans eventually died out and only the Negroes survived. When crossing a Southern European with a Negro, on the other hand, the relevant formula was as follows (Figure 3.2):

$$\frac{AB + ab}{\overline{ABab}} \quad \begin{matrix} P \\ F_1 \end{matrix}$$

$$\overline{ABAB + abAB + ABab + abab} \quad F_2$$

Figure 3.2 A 1916 model for crossing Southern Europeans with Negroes. B represents a factor blocking the expression of the factor A.
Source: Ibid.

According to this formula, which conveniently assumes a coupling of the A and B factors, all of the descendants were expected to be viable: whenever the damaging factor existed, it was always counterbalanced by a blocking factor. This explained why the crossings of Southern Europeans and Negroes were fertile.[24] Moszkowski did not fail to mention that not enough data were yet available to corroborate his hypotheses;

[24] Moszkowski, "Klima, Rasse und Nationalität," 150–152. In a footnote, Moszkowski further explained that even if coupling was not assumed and A and B were independently

these were put forward, he explained, mainly to demonstrate how Mendelian reasoning could shed light on the causes of various biological
phenomena and help analyze and even quantify them.

Whether light was really shed on the question of infertility due to
Moszkowski's model will be left for the reader to judge. As time went
by, persistent inability to provide genuine biological support for hybrid
infertility pressed scientists to the dim sphere of suggestive examples and
analogies. The most frequently used one mobilized the example of the
mule. A booklet on family history and heredity from 1924 stated that "if
traits are represented in a heterozygous state ... how will this find expression in the manifested form (the phenotype)? [...] There might come
about a middle form, just like the mule whose appearance is between a
horse and a donkey, or like Mulattos who stand between Europeans and
Negroes (intermediate form)."[25] The mule analogy was loaded with
implicit connotations: built into the syntax of the sentence was a correlation between human races and the mentioned animals, wherein Europeans paralleled horses, Negroes were like donkeys, and Mulattos were
analogous to mules, who were themselves sterile. The similarity in the
sound of the words themselves (*Maultier*, Mulatto) also helped to
strengthen their alleged biological relationship.[26] Nonetheless, with very
little evidence for the damaging effects of hybridization, the same analogy
could also be interpreted in the reverse manner. In 1916 anthropologist
Felix von Luschan criticized the common association of mules with
Mulattos; in 1922 he explained that the mule was much more suited
for a wide range of physical labors than both of its progenitors. "Horse
and donkey are undoubtedly much further apart from each other, systematically and genetically, than any [two] human groups, but we are all
aware of the excellent properties of the mules, and the new German
breeding literature is particularly full of their praise; there is also little
doubt about the fact that the mule is disproportionally more suitable and
valuable for many tasks, especially in agriculture, than the horse." Luschan nevertheless hastened to clarify that "it is not implied that a Mulatto

inherited, of every 16 offspring 11 would survive and 5 wouldn't, which would still be
compatible with the earlier mentioned observations. The 5 non-viable children
Moszkowski referred to are in all probability those who carried more damaging than
blocking factors, i.e., Aabb, aAbb, AABb, AAbB and AAbb – an interesting and
distinctly accumulative perception of the work of hereditary factors.

[25] Sandor Kaestner, *Was muss der Familiengeschichtsforscher von der Vererbungswissenschaft
wissen?* Praktikum für Familienforscher 5 (Leipzig: Degener & Co., 1924), 5.

[26] See similarly Ottokar Lorenz, *Lehrbuch der gesamten wissenschaftlichen Genealogie* (Berlin:
Wilhelm Hertz, 1898), 371–372. The similarity in the sound of the words was also noted
by contemporaries; see W[ilhelm] J[ulius] Ruttmann, *Erblichkeitslehre und Pädagogik*
(Leipzig: Schulwissenschaftlicher Verlag A. Haase, 1917), 67.

is *per se* more valuable than a white person, but this should put in the proper light the foolish tale of the inferiority of the hybrids, which is induced through the physiological 'disparity' of the parents."[27]

Hybrid infertility was therefore difficult to rule out, but also hard to support from the Mendelian perspective. Another potential harm traditionally attributed to racial hybridization was the creation of disharmonious or monstrous combinations – the "tale of the physiological disparity" mentioned by Luschan. Anthropological literature, especially in the United States, was replete with discussions of such disharmonies – from "crowded jaws" and "irregular dentation" of hybridized Americans, to long feet and short arms of Jamaicans, which "put them at a disadvantage when picking up things from the ground."[28] The actual proof provided for such disharmonies was scant and contested, and, perhaps as a result, the alleged harm caused by hybridization gradually shifted to the much less accessible mental domain, where to refute a hypothesis was at least as difficult as to corroborate it. In 1913 Fischer discussed the possibility that a kernel of truth existed in these widely held beliefs, and that a combination of courage and energy from one parent and lack of intelligence and self-control from the other could create criminal tendencies (a view already expressed in 1846 by the English physician George Gardner).[29] Fischer nevertheless stressed that bastards' mental inferiority was usually the result of social circumstances, not biological ones: he could find no proof that crossing was biologically damaging. That being said, Fischer did not recommend intermixture with Negro populations due to their "obvious" cultural inferiority.[30] In a public lecture in 1924 he admitted that supposedly disharmonic combinations were viable in the physical sense: an arm is an arm and a nose is a nose, whatever the race

---

[27] Von Luschan and Struck, *Kriegsgefangene*, 7–10; von Luschan, *Völker, Rassen, Sprachen*, 26.

[28] For useful reviews of the topic, see Thomas Teo, "The Historical Problematization of Mixed Race in Psychological and Human-Scientific Discourse," in Andrew S. Winston (ed.), *Defining Difference: Race and Racism in the History of Psychology* (Washington, DC: American Psychological Association, 2004), 79–108; William H. Tucker, "'Inharmoniously Adapted to Each Other': Science and Racial Crosses," in ibid., 109–133. The most vocal proponent of the theory of disharmony in the United States was Charles Davenport; the last quote is taken from his "Race Crossing in Jamaica," *Scientific Monthly*, 27 (1928): 225–238, esp. 238. See, however, the comments in p. 16, fn. 23.

[29] George Gardner, *Travels in the Interior of Brazil* (London: Revee Brothers, 1846), 21: "the savage rapacity of the mixed race ... the worst of criminals spring from this class, who inherit in some degree the superior intellect of the white, while they retain much of the cunning and ferocity of the black."

[30] Fischer, *Die Rehobother Bastards*, 298–300. More on Fischer's position on this issue, see Hutton, *Race and the Third Reich*, 64–79.

may be, Fischer said. "That is to say, that despite all the subtle differences between race-arms and race-noses there is no principal opposing difference between these two organs [in different races]."[31] But, according to Fischer, racial crossing was nevertheless dangerous, because when it came to their mental endowment, "It is difficult to imagine anything more diametrically opposed than the typical Semitic and the typical Germanic soul, and nothing equivalent to it can be found in the physical-anthropological field."[32]

Egon von Eickstedt had even more stringent views: the genes themselves, so he thought, strove to distinguish disharmonic combinations. "The mendeling-out of gene-complexes which had already become interconnected in previous genotypes," he wrote in 1920, "strengthens, in my opinion, the assumption of the striving of the genes towards harmonious union, and conversely the tendency to eliminate disharmonies (already visually disturbing in individuals from unbalanced population conglomerates)."[33] Any European individual was heterozygous, he admitted, but certain gene combinations simply did not hold together; natural selection favored organs that were intrinsically correlated, and this underlay the empirical observation that racial types were repeatedly detected in the highly mixed European population.[34] Genes of alien races simply did not

---

[31] See a report from Fischer's lecture in Adolf Stoll, "Über Familienforschung und Vererbung," *Zeitschrift für Kulturgeschichte und biologische Familienkunde* 1, no. 2 (1924): 62–67 (quote from p. 66). For a U.S. example of the same viewpoint, consider Herbert S. Jennings, *The Biological Basis of Human Nature* (New York, NY: W. W. Norton, 1930), 278: "The Negro and the white man each has 24 pairs of chromosomes. These work perfectly together in forming vigorous offspring, and in the much more delicate test of later uniting to form germ cells in these offspring. The same is true for crosses between any of the other races of man. ... We may dismiss from consideration, so far as the crosses of human races are concerned, the question of serious incompatibility of chromosomes or genes, such as we find in crosses between organisms standing far apart in their structure and physiology." William B. Provine defined the 1930s–1940s as a period of agnosticism, when, due to lack of evidence, anthropologists stopped claiming that racial crossing was harmful, but left the question of its effects scientifically undecided; only after World War II did scientists phrase their lack of knowledge as proof that racial crossing was harmless. See William B. Provine, "Geneticists and the Biology of Race Crossing," *Science* 182, no. 4114 (1973): 790–796.

[32] Stoll, "Über Familienforschung und Vererbung," 66.

[33] Egon von Eickstedt, "Rassenelemente der Sikh. Mit einem Anhang über biometrische Methoden," *Zeitschrift für Ethnologie* 52/53, no. 4/5 (1920): 317–394 (quotes from p. 353, 378–379).

[34] Ibid., 353. Similarly, see Edward M. East and Donald F. Jones, *Inbreeding and Outbreeding: Their Genetic and Sociological Significance* (Philadelphia, PA: J. B. Lippincott, 1919), 253: "The races differ by so many transmissible factors, factors which are probably linked in varied ways. ... The real result of such a wild racial cross, therefore, is to break apart those compatible physical and mental qualities which have established a smoothly operating whole in each race by hundreds of generations of natural selection."

match. The same issue was also the subject of a talk given by the Norwegian scholar Jon Alfred Mjøen at a 1921 meeting of the German Ethnological Society. Mjøen swiftly moved from discussing studies on disharmonic crosses of rabbit strains to addressing the effects of crossing Norwegians and Laplanders, whose descendants, so he argued, seemed to have had a less-balanced character and tended to lie, steal and drink.[35]

Yet to the great dismay of some thinkers, no sustainable proof for an inherent biological inferiority of racial hybrids had been found: neither were they less fertile than their parents, nor did they exhibit only the degenerative traits of their parental races, nor did their physical components create monstrosities – nor could real evidence be found to testify that they were intellectually damaged. Hybridization led to heterozygosity, and there was nothing extraordinary or hazardous about it. Furthermore, contrary to the claims (and hopes) of racial thinkers, humans did not exhibit instinctive aversion to coupling with other races, and some racial theorists – including Fischer – even claimed that the mixture of closely related races was beneficial and had created thriving cultures during the development of human civilization.[36] What therefore, wondered anthropologist Ernst Rodenwaldt in 1934, were the origins of the persistent sentiment against racial mixture, the wish to remain racially pure and the popular resentment against acts of *Rassenschande* (racial defilement)?

The disparaging judgment of the *Mischling* ... **is based not** on a **proven physical inferiority**, nor on **inferior biological prospects of the hybrid strains** and the new combinations of physical traits realized in them, not even on **defects of the elementary material of their intellectual property** ... [but on] **the inevitable consequences of the social isolation of the *Mischling*, his intermediate position and the peculiarity of his personality development, determined by this** [intermediate social position].[37]

---

[35] Jon Alfred Mjøen, "Harmonische und unharmonische Kreuzungen (Vortrag, ausserordentliche Sitzung vom 31. Mai 1921)," *Zeitschrift für Ethnologie* 53, no. 4/5 (1921): 470–479. Mjøen was central in promoting racial-hygiene in Norway and had close ties with prominent German racial-hygienists, such as Alfred Ploetz and Fritz Lenz. See Jon Røyne Kyllingstad, *Measuring the Master Race: Physical Anthropology in Norway, 1890–1945* (Cambridge: Open Book Publishers, 2016), ch. 5, 9.
[36] Eugen Fischer, "Zur Frage der 'Kreuzungen beim Menschen,'" *ARGB* 9, no. 1 (1912): 8–9. See also Fischer's speech given on February 1, 1933, and the furor that it created in Nazi circles, described in Niels Lösch, *Rasse als Konstrukt: Leben und Wirken Eugen Fischers* (Frankfurt a. M.: Peter Lang, 1997), 231–233. On the "distance theory" of racial crossing (closely related races can be crossed, those far apart should not be), see Teo, "Historical Problematization," 95–96.
[37] Ernst Rodenwaldt, "Vom Seelenkonflikt des Mischlings," *ZMA* 34 (1934): 364–375 (quote from p. 368, emphases in the original).

The feeling of estrangement that the young, racially mixed child experienced left its eternal mark on the soul of the *Mischling*, Rodenwaldt explained. A racially mixed child would return from school to ask his father – why am I different? This "inferiority complex" or "feeling of being a pariah" (*Minderwertigkeitskomplex, Pariagefühl*) was the root of the *Mischling* problem. Therefore, "Racial mixture is a risk for every human community from the family to the national state, a risk that is imposed on the next generation. Since no one can foresee its effects, it is irresponsible to take it."[38]

From the ideological perspective, Mendelian thought influenced the notions of racial purity and racial mixture in almost diametrically opposing ways. It toughened the prerequisites for racial purity by boosting the requirement for uniformity both down to the level of the genes and up to the level of the population. At the same time, Mendelization helped in neutralizing the imagined effects of racial crossing: one way or another, everyone was a hybrid. Before the rise of Mendelism it had already been agreed that all European peoples were mixed; now it was realized that all European people – indeed, all people in general – were mixed as well. The damages of racial crossing could therefore no longer lurk in the process of hybridization itself; only specific combinations were harmful. These alleged harms, however, were open to causal interpretations drawing from the social and cultural domains. It was against this backdrop that the 1930s saw the development of the concept of "the marginal man" in American scholarship, a concept that, quite similar to Rodenwaldt's phrasing above, shifted the peculiarities of the hybrids to the cultural and social spheres, explaining disharmony primarily through psychological terms (estrangement, ambivalence, emotional instability) and existing social conditions, not through biological inheritance.[39]

In the hands of liberally minded social reformers possessing political power, this understanding could have opened up possibilities for

---

[38] Ibid., 374.
[39] See Robert Ezra Park, "Human Migration and the Marginal Man," *American Journal of Sociology*, 33 (1928): 881–893; Everett V. Stonequist, "The Problem of the Marginal Man," *The American Journal of Sociology* 41, no. 1 (1935): 1–12; Stonequist, *The Marginal Man: A Study in Personality and Cultural Conflict* (New York, NY: Scribner, 1937). See in this respect also Melville J. Herskovits, "A Critical Discussion of the 'Mulatto Hypothesis,'" *The Journal of Negro Education* 3, no. 3, *The Physical and Mental Abilities of the American Negro* (1934): 389–402, as well as the psychologist (and racial segregationist) Raymond B. Cattell, *Psychology and Social Progress: Mankind and Destiny from the Standpoint of a Scientist* (London: C. W. Daniel, 1933), 63. Similar reconfigurations of the social problematique of racial hybridization were expressed elsewhere, too. For example, Norwegian anthropologist Halfdan Bryn analyzed jointly the damage of hybrids' "genetic chaos" together with their social alienation. See Kyllingstad, *Measuring the Master Race*, 128–129.

neutralizing the effects of social fragmentation and discrimination. In the hands of those who came to lead German society, it was taken in the opposite direction. Hitler's own perceptions regarding the impact of racial mixture were not bound by Mendelian reasoning; for him, racial mixture was against nature, detrimental to the higher culture and equivalent to blood poisoning.[40] For those who were more biologically informed, even if hybridization (or, heterozygosity) in and of itself was not deleterious, it was still the antithesis of purity (or, homozygosity); in that sense it could continue to be seen as harmful, or detrimental to the population or race. The Nazis had various reasons for insisting on the dreaded results of sexual mixture with racial aliens; but from the Mendelian perspective, the challenge that hybridization posed could be narrowed to their fear of the inability to achieve purity due to the "mendeling-out" of traits. Such mendeling-out was the result of a particular kind of genetic mechanism, which posed a special challenge to those seeking to regenerate and purify the nation: recessive traits.

The Dangerous Recessive Factors

In 1908, on the basis of initial empirical findings, American anthropologist Charles Davenport postulated that "where various [developmental] stages ... are found in individuals of the same race or species, the more progressive condition will often behave as a dominant toward the less progressive condition," which will therefore be recessive.[41] Plate severely criticized the circuitous nature of the argument, commenting that the definition of what was considered progressive in itself relied on what Davenport discovered to be dominant.[42] Within a few years, enough findings had accumulated for the eugenicist Fritz Lenz to assert that "the vast majority of hereditary diseases are grounded on recessive pathological dispositions." His examples came from the field of mental deficiencies and nervous disorders, all of which he considered recessive.[43] Plate again did not fully agree: the following year he claimed that "most hereditary diseases and deformities in man are dominant." This, however, did not conflict with the notion that "much more

[40] Adolf Hitler, *Mein Kampf*, 311ff., 441–448, 629.
[41] Charles B. Davenport, "Determination of Dominance in Mendelian Inheritance," *Proceedings of the American Philosophical Society* 47 (1908): 59–63 (quote from p. 61).
[42] Ludwig Plate, "Kritische Besprechungen und Referate," *ARGB* 6, no. 1 (1909): 101–102.
[43] Fritz Lenz, "Über die idioplasmatischen Ursachen der physiologischen und pathologischen Sexualcharaktere des Menschen," *ARGB* 9, no. 5 (1912): 545–603 (quote from p. 597).

difficult to control and therefore much more dangerous are the recessive diseases."[44]

Negative perceptions regarding recessiveness became more
entrenched as time went by. In 1927 Lenz noted that human pathologies
were usually caused by genetic mutations, which were themselves mostly
deleterious. These mutations were transmitted in either a simple dominant or a simple recessive manner; but most of them were recessive.[45] In
addition, when it came to specific deficient states, such as ataxia or
idiotism, "the rule applied that the most severe forms are inherited in a
recessive manner, the light forms in a dominant one."[46] If severe pathologies would not have been recessive, they would have quickly been
wiped out by natural selection; it was only their recessive state that
enabled them to continue spreading. Thus, both quantitatively and
qualitatively, recessive traits were the least favorable alternative. By the
late 1920s, equating recessiveness with pathological dispositions in fact
became so prevalent that Lenz had to clarify to his peers: "the assertion
that I (Lenz) considered blondness as a 'defective variant' is a misun-
derstanding." All he wrote, Lenz explained, was that due to the absence
of certain genes, blondness was recessive with respect to darker pigmentation; but "from this it does not follow that it was defective in the sense
of reduced life competence."[47]

In addition to the quantitative claim on the correlation between deleterious mutations and recessivity, and the qualitative claim on the severity of recessive pathologies, recessive factors were also less accessible to
human detection and, consequently, posed an unforeseeably greater
danger to the coming generations. As Plate's statement reveals, they were
"difficult to control and **therefore** much more dangerous" (my
emphasis). This third characterization mirrored the first one: a genetic
mutation, in itself an unplanned and undesirable process at the genomic
level, created a recessive disease, which then reproduced the state of
uncontrollability at the human and scientific levels.

Recessive traits thus became a focal point of what Zygmunt Bauman
has identified as one of the salient features of modernity: the obsession

---

[44] Ludwig Plate, *Vererbungslehre: mit besonderer Berücksichtigung des Menschen, für
Studierende, Ärzte und Züchter* (Leipzig: Wilhelm Engelmann, 1913), 392–3.
[45] Cf. Lenz, "Die krankhaften Erbanlagen," in *BFL* (1927), 397. For a discussion on the
possible reasons for mutations to be predominantly recessive as postulated throughout
the twentieth century, see Vidyanand Nanjundiah, "Why Are Most Mutations
Recessive?" *Journal of Genetics* 72, no. 2/3 (1993): 85–97.
[46] Lenz, "Die krankhaften Erbanlagen," in *BFL* (1927), 356.
[47] Fritz Lenz, "Kritische Besprechungen und Referate," *ARGB* 21, no. 2 (1929): 194–204
(quote from p. 202).

to control and ultimately remove uncertainties from the natural world.[48] The fact that recessive traits remained hidden for an undefined number of generations, only to ultimately unveil themselves, was catastrophic for any attempt to construct a homogenous, constant, uniform and lasting pure type. This was a problem for agricultural manufacturing, because recessive unpredictability frustrated the aspiration to produce and later distribute in large scale biological products – seeds, crops or even vaccines. Standardization of seed weight, sugarcane content, pest resistance or bacterial virulence was a necessary precondition for successful industrialization; without it, neither mass production, nor product transference, nor safety and quality control could be achieved.[49] In an article published in 1922, the zoologist Hans Nachtsheim accordingly addressed the issue of "Mendelism and Animal Breeding" and quoted excerpts from professional breeders' journals such as "the most desirable breeding objective must be to prevent as much as possible the effect of Mendelism; that is the yardstick for the high quality of a breed." Such remarks, explained Nachtsheim, represented "the horror of the segregation processes (*Aufspaltungen*)," a horror that was widespread in the community of breeders. The segregation of the second generation into the original distinct types of the parental generation, Nachtsheim wrote, ruined breeders' attempts to build a new constant race, since the desired combinations did not hold together. For the breeders, "*mendeln pendeln bedeutet, pendeln von einer Kombination zur anderen*" (to mendelize means to oscillate, oscillate from one combination to another).[50]

In the human domain, as we have seen above (p. 93), another element was added to these cravings for homogeneity, repeatedly frustrated by the reemergence of recessive traits. In view of the protagonists of human racial purity such as Günther, one could not formulate a proper ideal of

---

[48] Zygmunt Bauman, *Modernity and the Holocaust* (Ithaca, NY: Cornell University Press, 1989); Bauman, *Modernity and Ambivalence* (Ithaca, NY: Cornell University Press, 1991).

[49] For insightful analyses of the relation between industrialization, the manufacturing of purity and hereditary thinking, see Staffan Müller-Wille, "Hybrids, Pure Cultures, and Pure Lines: From Nineteenth-Century Biology to Twentieth-Century Genetics," *Studies in History and Philosophy of Science Part C: Studies in History and Philosophy of Biological and Biomedical Sciences* 38, no. 4 (2007): 796–806; Christophe Bonneuil, "Pure Lines as Industrial Simulacra: A Cultural History of Genetics from Darwin to Johannsen," in Staffan Müller-Wille and Christina Brandt (eds.), *Heredity Explored: Between Public Domain and Experimental Science, 1850–1930* (Cambridge, MA: MIT Press, 2016), 213–242; Andrew Mendelsohn, "Message in a Bottle: Vaccines and the Nature of Heredity after 1880," in ibid., 243–264.

[50] Hans Nachtsheim, "Mendelismus und Tierzucht," *Die Naturwissenschaften* 29 (1922): 635–640 (quote from p. 636). See similarly Carl Kronacher, *Grundzüge der Züchtungsbiologie* (Berlin: Paul Parey, 1912), 275.

beauty – something he considered culturally essential – as long as children did not resemble their parents, and such deviations were inevitable when phenotypes and genotypes did not collide – that is, when the parents contained within them hidden recessive traits.[51] Notably, the very words for dominant and recessive dispositions in the German language – *überdeckend* and *überdeckt*, literally "covering" and "covered" – implied that recessive genes were playing a sinister game of hide-and-seek with humans.

And there was yet another negative aspect of recessive traits. It quickly became agreed that it was due to the recombination of recessive genes that pathologies became so common in inbred populations. In 1912, the Swedish racial anthropologist Herman Lundborg could therefore declare that "Mendelism throws a clear light on the essence and the biological effect of consanguinity."[52] In 1919 Lenz provided a simple mathematical proof substantiating the correlation between rare recessive diseases and close kind marriages. For any theoretical recessive factor dispersed in the population with a given frequency, Lenz showed that the chances of manifesting the respective disease were considerably higher if there had been consanguineous marriages among the families of the parents.[53] Consequently, it became customary to assume that the opposite was true as well: if a certain disease was more common among inbred populations, a recessive hereditary factor must have underpinned it. In 1927, Lenz even claimed that "for the detection of recessive inheritance, one does not at all have to rely on the proof of theoretical [Mendelian] numerical ratios among siblings. The determination of comparative frequencies of cousin marriages among parents is more important in this respect."[54]

Lenz' reasoning quickly spread to his colleagues. Referring to the work of Lenz, one scholar explained in 1932 that

modern hereditary study supplied the scientific explanation for the old experience: we know today that the danger of consanguinity lies in the increased probability of the recombination of recessive dispositions – and most of the pathological traits are inherited in a recessive manner. And vice versa: if we find out that the parents of individuals with certain attributes are more frequently

[51] Günther, *Rassenkunde des deutschen Volkes*, 211–212.
[52] Herman Lundborg, "Über die Erblichkeitsverhältnisse der konstitutionellen (hereditären) Taubstummheit, und einige Worte über die Bedeutung der Erblichkeitsforschung für die Krankheitslehre," *ARGB* 9, no. 2 (1912): 133–149 (quote from p. 149).
[53] Fritz Lenz, "Die Bedeutung der statistisch ermittelten Belastung mit Blutsverwandtschaft der Eltern," *Münchener medizinische Wochenschrift* 66 (1919): 1340–1342.
[54] Fritz Lenz, "Die Methoden menschlicher Erblichkeitsforschung," in *BFL, Band I: Menschliche Erblichkeitslehre* (1927), 435–440.

relatives than the percentage of relative marriages in the entire population, we need to assume that the respective attribute is inherited, and namely inherited recessively. The more rarely the given attribute can be found in the population, the more frequently one finds blood relations among the parents."[55]

Pedigrees of highly inbred families displaying mental and physical deformities were a long-cherished visual genre of racial-hygienists. They strengthened the notion that disease was a familial feature, characterizing hermetically closed social groups (or "degenerate clans") whose marital practices belonged to foregone ages or expressed deviant sexual desires. More generally, such pedigrees, packed with criminals, alcoholics, vagabonds, the feeble-minded and mental deviants, helped to reformulate social problems as alleged biological ones.[56] Now the same pedigrees had finally found their scientific legitimization as depictions of the consequences of recessive diseases. As illustrations of a concrete Mendelian mechanism, these charts would continue to embellish both professional and popular publications of racial-hygienists.

In 1933, the racial-hygienist Wilhelm Weitz challenged the reasoning underlying this common practice. According to Weitz, the almost automatic attribution of multiple expressions of a rare disease in families displaying close-kin marriages to recessive inheritance was mathematically untenable.[57] But by 1933, Weitz' argument made very little impact. The cultural and scientific connotations of recessive diseases were already well established, including their prevalence, severity, elusiveness and inbreeding peril. All these unwelcome virtues often converged, as a public lecture by psychiatrist Hans Luxenburger may illustrate. Luxenburger explained that cured schizophrenic patients presented a particularly great danger to society: while they were capable of marrying and bearing children, their value in the marriage market was still somewhat diminished compared with that of normal persons, and so they tended to marry their own kind. This, "by the recessive nature of the disease, must lead to a catastrophic contamination of the descendants." Could schizophrenics then be coupled with healthy people? Of course not: "even if a

---

[55] Herbert Orel, "Die Verwandtenehen in der Erzdiözese Wien," *ARGB* 26, no. 3 (1932): 249–278 (esp. 249). See similarly Max Bigler, "Beitrag zur Vererbung und Klinik der sporadischen Taubstummheit," *Archiv für Ohren-, Nasen- und Kehlkopfheilkunde* 120 (1929): 81–92; Gustav Wulz, "Ein Beitrag zur Statistik der Verwandtenehen," *ARGB* 17, no. 1 (1927): 82–94.

[56] Nicole Hahn Rafter (ed.), *White Trash: The Eugenic Family Studies, 1877–1919* (Boston, MA: Northeastern University Press, 1988).

[57] Wilhelm Weitz, "Über die Häufigkeit des Vorkommens des gleichen Leidens bei den Verwandten eines an einem einfach rezessiven Leiden Erkrankten," *ARGB* 27, no. 1 (1933): 12–24 (quote from p. 24). Weitz' sister was Lenz' first wife; she passed away four years earlier, and Lenz was quick to marry another woman.

cross with hereditary healthy individuals takes place, the degenerate tendency is inherited as latent among the descendants."[58]

The notion of a "recessive Mendelian trait" was therefore burdened with medical and scientific anxieties, which had to do with its elusive character and the difficulty of both recognizing its existence and controlling its expression. These anxieties had a certain empirical basis, related to the nature of mutations and to inbreeding in families. But the negative perceptions regarding recessivity did not depend on, or stem from, these scientific findings. The source of the cultural connotations of recessive traits seems to lie not within genetics, but outside of it; above all, its roots point to another transformation in medical thinking that occurred just prior to the rise of Mendelian thinking. The 1870s and 1880s saw the emergence of modern medicine, in what was later to be called the bacteriological revolution. The studies of Robert Koch and Louis Pasteur tangibly altered the landscape of medical understanding, bringing into the world new mechanisms for the causes of disease and new methods for checking their spread. Germs, bacteria and, later, viruses were identified and characterized, and images of their microscopic structures became widely publicized. Through multiple media, citizens of the late nineteenth century learned of methods to reduce the danger of infection by disinfecting clothes, ventilating rooms and washing hands that came into contact with ill individuals. They also learned to fear non-human and human carriers of diseases – lice, mosquitos, rodents or even persons, who were immune to the detrimental effects of certain germs but harbored and later spread them to others. In different states, immigrants, racial others and the poor were identified as carriers of pathogens of multiple kinds; they needed to be inspected by state medical authorities, or quarantined, before contact with them could be permitted.

It was against this public and scientific preoccupation with germs and diseases that Mendelism rose. Unlike plant and animal genetics, human genetics did not make its first steps as a study of the mechanisms governing hereditary transmission or gene expression of normal traits. In its earlier days, Mendelian genetics was associated above all with the generation of human pathologies. This was noted in 1927 by Lenz. Pathological traits, Lenz explained, obeyed the same hereditary mechanisms as normal ones, since biologically speaking there was no essential distinction between health and disease. Pathologies nevertheless had two

---

[58] Hans Luxenburger, "Die wichtigsten Ergebnisse der psychiatrischen Erbforschung und ihre Bedeutung für die eugenische Praxis (Vortrag, gehalten in der Sitzung der Münchner gynäkologischen Gesellschaft, am 23. 1. 1930)," *Archiv für Gynäkologie* 141 (1930): 237–254 (quote from p. 245).

unique qualities: "Since many pathological dispositions distinguish their carriers very clearly from the rest of the population, it is possible to track their mode of inheritance particularly well." But they were not only easier to detect: "It so happens that precisely with regard to pathological dispositions one could show best the validity of Mendel's laws for humans."[59] The reason for this was a biological one: "pathological hereditary states are usually determined by single genes (monomer), normal features by many (polymer)."[60] Thus, hereditary pathologies were both inherited in a simple fashion as well as easier to detect and follow, unlike normal traits, which were both less conspicuous as well as governed by more complex mechanisms, involving delicate interactions between different genes and between them and various environmental signals. As a result of this scientific reality, human geneticists made their first and most impressive advances by studying human disorders. In its initial phases, human genetics was therefore primarily medical. Soon enough, the fears of contamination, impurity and degeneration that were widespread in the prevailing medical discourse became localized in Mendelian reasoning – and, in particular, in the concept of recessive genes.

In many senses, recessive genes thus turned into the germs of heredity: invisible, delusive, harmful biological entities, they hid in the bodies of others who needed to be avoided – or sterilized. Along with the idea of recessivity came the notion of disease carriers, which was adopted and adapted to describe vertical, not only horizontal, transmission of illness. Old anxieties now became imbued in the specific mechanisms of transmission and manifestation of recessive diseases; fears of contamination went through a process of "Mendelization" and became modernized, concrete and grounded in hereditary knowledge and terminology. As noted above, when it came to germs, certain social groups – immigrants, racial others and the poor – were often identified as carriers of diseases. In Germany, one segment of the population simultaneously represented those three social groups. This was the *Ostjuden*, or East European Jews. Long seen as carrying and spreading germs and disease, Jews were now also identified as carriers of recessive malignant dispositions.[61]

---

[59] Lenz, "Die krankhaften Erbanlagen," in *BFL* (1927), 176.

[60] Ibid., 177. The entire sentence is emphasized in the original.

[61] On the identification of Jews with diseases and bacteria, see Howard Markel, *Quarantine! East European Jewish Immigrants and the New York City Epidemics of 1892* (Baltimore, MD/London: Johns Hopkins University Press, 1997); Paul Julian Weindling, *Epidemics and Genocide in Eastern Europe, 1890–1945* (Oxford/New York, NY: Oxford University Press, 2000); Barbara Lüthi, "Germs of Anarchy, Crime, Disease and Degeneracy: Jewish Migration to the United States and the Medicalization of European Borders

Mendelizing Jews

When eugenic thinking became popularized during the late nineteenth century, Jews became one of its most favored targets. Sociologists, anthropologists and doctors on both sides of the Atlantic found in Jews a glaring example for the working of social Darwinian mechanisms in human society. Scholars began to consider the Mosaic code, and Jewish adherence to strict marriage laws, as representing an ancient form of eugenic consciousness. The historical, continual persecution of the Jews and their forced adaptation to Ghetto life were reconceptualized as forces that facilitated natural selection or toughened the Jewish type, its immunity to disease or its propensity to mental fragility. Jewish scholars were prominent in the ensuing debate over the nature and uniqueness of Jews in Europe, and they readily adopted eugenic schemes to account for whatever qualities they found in Jews, most notably their racial tenacity.[62]

With the spread of Mendelian teaching, Jews became an apt object also for the application of Mendelian notions to racial thought; and here, too, Jewish scholars were at the front line. In 1910, Jewish racial theorist Leo Sofer still doubted that Mendelism was at all applicable to the human domain. But, he thought, "the only phenomenon from the area of the Mendelian laws that appears also among humans, albeit in an impure form, is the return of earlier generations, and in a more striking manner the occasional occurrence of distant ancestors amid a mixed race. One example is the Negro type ... which sometimes occurs among the Jews of today." It was well known that Negroes had contributed their share to the Jewish bloodline, explained Sofer. The reappearance of the Negro components among today's Jews seemed to him to be "the single point among humans close to the Mendelian laws."[63] In another article from

around 1900," in Tobias Brinkmann (ed.), *Points of Passage: Jewish Migrants from Eastern Europe in Scandinavia, Germany and Britain 1800–1914* (New York, NY: Berghan Books, 2013), 27–46. See also Berger, *Bakterien in Krieg und Frieden*.

[62] John Efron, *Defenders of the Race: Jewish Doctors and Race Science in Fin-de-Siècle Europe* (New Haven, CT: Yale University Press, 1994); Mitchell B. Hart, "Racial Science, Social Science, and the Politics of Jewish Assimilation," *Isis* 90, no. 2 (1999): 268–97; Hart, *Social Science and the Politics of Modern Jewish Identity* (Stanford, CA: Stanford University Press, 2000); Hart, *The Healthy Jew: The Symbiosis of Judaism and Modern Medicine* (Cambridge: Cambridge University Press, 2007); Veronika Lipphardt, *Biologie der Juden. Jüdische Wissenschaftler über "Rasse" und Vererbung 1900–1935* (Göttingen: Vandenhoeck & Ruprecht, 2008); Dafna Hirsch, "Zionist Eugenics, Mixed Marriage, and the Creation of a 'New Jewish Type,'" *Journal of the Royal Anthropological Institute* (*JRAI*) 15 (2009): 592–609.

[63] Leo Sofer, "Auf den Spuren der mendelschen Gesetze," *Politisch-Anthropologische Revue* 7, no. 7 (1908): 345–351.

the same year, Sofer explained that racial crossing did not lead to the creation of an intermediate race, "but according to Mendel each feature may be inherited independently. We see among the Jews that they came into being mainly through the crossing of a long headed race with a broad headed one. The result was not a mesocephaly (medium skull) – which may indeed appear sporadically – but brachycephaly [wide skull] ... and along with it, as a correlative modification, the hooked nose displaced the straight Semitic nose."[64]

Mendelism also proved useful to account for the origin of blond hair and blue eyes among Jews, which several prominent scholars attributed to past racial intermixture. In 1913, based on the Hardy-Weinberg equilibrium and the fact that blue eyes were recessive, Paul Kaznelson calculated the amount of foreign racial intrusion into the Jewish nation that was required to account for the prevalence of blue eyes among Jews.[65] His result greatly exceeded what historical sources could corroborate: there were 40% to 45% of bright eyes among the Jews, which meant that the ratio between dark and bright eyes in the founder population had to be the square root of that, that is, 65% – and this was a modest estimation, Kaznelson explained. This, in his view, ruled out the theories that posited the intermixture between Jews and host nations as the prime or exclusive source for their bright eyes. The theory of von Luschan, who postulated that blond components had been intrinsic to the Jewish racial composition (the "Amorites theory"), therefore became more plausible.[66]

In another article from the same year, Kaznelson went even further in his application of Mendelian logic to Jewish racial history. Kaznelson challenged the claims of Austrian Jewish physician Ignaz Zollschan, regarding the racial purity of the Jews. Zollschan had argued that since the Kohanim (priests) were direct, unconverted descendants of the biblical Aaron, they were surely not mixed with other races; furthermore,

---

[64] Leo Sofer, "Über die Plastizität der menschlichen Rassen," *ARGB* 5, no. 5/6 (1908): 660–668.

[65] The Hardy-Weinberg equilibrium states that in conditions of random mating and without selection pressures, the proportions of different alleles in a population are not expected to change from one generation to the next. It also makes it possible to infer, under certain assumptions, from the frequencies of certain traits in a population to the frequencies of the alleles determining those traits.

[66] Paul Kaznelson, "Über einige 'Rassenmerkmale' des jüdischen Volkes," *ARGB* 10, no. 5 (1913): 490, 492–494. To finalize his thesis, Kaznelson claimed that statistics had shown that modern Samaritans were around 20%–25% blue eyed, which, again, according to Hardy-Weinberg, required a prevalence of 50% in the original population. To bridge the gap between these 50% and the required 65%–70%, Kaznelson assumed that some intermixture with local populations did take place, and that selection processes were in effect.

because their cephalic index and physiognomy were similar to those of the general Jewish population, one could deduce that Jews in general were not mixed.[67] Kaznelson found these assertions untenable. First, he explained that the Kohanim married daughters of proselytes and therefore introduced foreign blood into the Jewish nation. He then evaluated the effects of such introduction of foreign blood by denoting the Jews and the Kohanim as AA and the proselytes as BB. Assuming that there were a Jews, b Kohanim and c proselytes, it was now possible to calculate the rate of foreign intrusion in the F1 generation. Using Mendelian suppositions, Kaznelson went on to compute the amount of foreign gametes among the priests in any n-th generation and showed that the genetic difference between the intermixed Kohanim and the non-intermixed Jews was negligible. One could therefore not deduce that the Jews were racially pure.[68]

However, for many Jews and non-Jews alike, the burning question was not that of ancient intermixtures but of contemporary ones. What happened when Jews and Germans interbred? The prevailing view held that the Jewish type was predominant over the German or European type. The Semitic components – brachycephalic skull, prominent nose, darker skin color – had a stronger impact on the offspring of racial mixture than the European components, so it was assumed. Presumably, because of years of strict inbreeding in conditions of social isolation, the Jewish blood became more condensed and hereditarily effective; its infusion into German families was accordingly considered to have had an everlasting influence on subsequent generations. As the U.S. eugenicist Madison Grant put it, "The cross between a white man and an Indian is an Indian; the cross between a white man and a Negro is a Negro ... and the cross between any of the three European races and a Jew is a Jew."[69]

But in 1911, Redcliffe N. Salaman, a Jewish student of Bateson, challenged this view in a short paper in the *British Journal of Genetics*. "Impressed with the great frequency and the distinctiveness of the Jewish type of face, it occurred to me that this character might form excellent

---

[67] See similar reasoning in Redcliffe N. Salaman, "Heredity and the Jew," *Journal of Genetics* 1, no. 3 (1911): 273–292 (esp. 279).

[68] Paul Kaznelson, "Kritische Besprechungen und Referate," *ARGB* 10, no. 6 (1913): 797–798.

[69] Madison Grant, *The Passing of the Great Race; or, The Racial Basis of European History* (New York, NY: Charles Scribner's Sons, 1916), 16. In various versions, these views were shared by scholars in and outside of Germany, such as Friedrich von Hellwald, Richard Andree and Carl Heinrich Stratz. See Maurice Fishberg, *Die Rassenmerkmale der Juden* (Munich: Ernst Reinhardt, 1913), 178–179.

material for research on Mendelian lines," Salaman wrote.[70] He therefore examined the children of English-Jewish intermarriages and asked a
group of Jewish informants to determine whether these children
appeared Jewish or Gentile. The results seemed to favor the hypothesis
that it was the Gentile type, not the Jewish type, which was dominant. It
did appear to be an incomplete dominance, Salaman conceded, but that
might have been a result of the bias of his informants to judge in favor of
Jewish classification in questionable cases. Furthermore, the "recessive
Jewish facial expression" occasionally came to the surface even in Gentile
faces:

I have in some cases found that observers not specially acquainted with the
subject, although agreeing that a given individual of the first generation is of
Gentile appearance have yet felt that there was somewhere lurking in the face an
expression which suggested "Jewishness" and there is very little doubt that such
opinion may often be well founded.

This "lurking" Jewishness could surface when the individual matured.
Salaman went on to consider families wherein the father was a Jew and
the mother a hybrid: of 25 children in 9 such families, 13 were "undoubtedly Gentile" and 12 "unequivocally Jewish" – in perfect agreement with
the Mendelian expectation in crossing RR with DR. Gentile-hybrid
crossings yielded strictly gentile children – again, in agreement with the
recessive supposition of the Jewish character.[71]

Salaman further tested his new hypothesis by examining data on
different communities around the world where Jews resided. His survey
strengthened his view that the Jewish facial expression was recessive with
respect to the Northern European Teutonic race, the native Indian, the
Chinese and also the Negro type.[72] Salaman therefore concluded: "complex as the origin of the Jew may be, close inbreeding for at least two
thousand years, has resulted in certain stable or homozygous combinations of factors which react in accordance with the laws of Mendel and
which may explain the occurrence of the peculiar facial expression recognized as Jewish."[73]

Salaman's thesis quickly found its way into German literature. It was
quoted and discussed even in general works of biology, such as Plate's
1913 *Heredity Science: With a Special Consideration of Man.*[74] In Fischer's

---

[70] Salaman, "Heredity and the Jew," 280.   [71] Ibid., 282–283.   [72] Ibid., 285–288.
[73] Ibid., 290.
[74] Kaznelson, "Über einige 'Rassenmerkmale' des jüdischen Volkes," 490; Valentin Haecker,
"Einige Ergebnisse der Erblichkeitsforschung," *Deutsche medizinische Wochenschrift* 27
(1912): 1292–1294; Plate, *Vererbungslehre: Mit besonderer Berücksichtigung des Menschen,*
327.

Rehoboth study, published the same year, Salaman's conclusions were addressed several times; the fact that in faces that did not seem Jewish there was sometimes "a small detail which gave away the Jewish origin – what is that if not the independent heredity of each trait," explained Fischer, adding another level of Mendelian interpretation (i.e., independence) to Salaman's material.[75] Also in the same year, American anthropologist Maurice Fishberg published in German an extensive study of the Jewish race. With regard to Salaman's thesis, Fishberg claimed that the face could not be considered as a Mendelian unit, but as a complex composition of many factors. Attempting to fend off Jewish stereotypes, Fishberg further argued that most of the proclaimed Jewish features were not at all apparent among the majority of Jews.[76] The recessive hypothesis, however, was not so easy to refute: the fact that Jewish facial features were not apparent among Jews, explained Swiss anthropologist Otto Schlaginhaufen in the *Journal for Morphology and Anthropology*, did not make those same features less "Jewish," and could be fully accounted for by "the hereditary recessive state of parts of the Jewish racial traits."[77] Just because the Jews didn't have it, it didn't mean that it wasn't Jewish; it was, after all, recessive.

Salaman's thesis was developed further by Kaznelson, who agreed that there was a gene for Jewish physiognomy. As Salaman had shown, "The Jewish face is inherited as a unit."[78] This particular face was instinctively considered by Jews throughout history as a marker signifying their own kind and was therefore favored by sexual selection. When Jewish facial features were apparent, they testified to the purity of their carrier, in the sense of homozygosity (RR). By marrying persons with a distinctively Jewish face, there was no danger of reversion to other racial components that had formerly mixed into the Jewish nation, as could have happened if sexual selection operated on dominant features, where DR and DD are indistinguishable. Kaznelson was thus able to combine the Mendelian recessive nature of Jewish features with evolutionary theory, the discussion on the ancient Jewish racial components and the purity of the Jewish race.[79]

To an extent, the thesis on Jewish recessiveness created interest precisely because it contradicted the more customary attribution of dominance to Jewish (Semitic, or Negro) racial components. In most circles,

[75] Fischer, *Die Rehobother Bastards*, 136–137, 165, 213 (quote from p. 213).
[76] Fishberg, *Die Rassenmerkmale der Juden*, 183, 256–262.
[77] Otto Schlaginhaufen, "Die Rassenmerkmale der Juden by Maurice Fishberg," *ZMA* 19, no. 1 (1915): 265–269 (quote from p. 267).
[78] Kaznelson, "Über einige 'Rassenmerkmale' des jüdischen Volkes," 489–490.
[79] Ibid.

this latter view continued to prevail. The editors of the 1914 *Semi-Gotha* – an antisemitic publication devoted to exposing the infiltration of Jewish blood into German aristocracy – described the following phenomenon: when an Aryan man married a full-blooded Jewess, their children approached the Jewish type. If their own children kept on marrying Aryans for ten generations, the Jewish blood, which was "always predominant" (*immer vorherrschend*), became more and more isolated and diminished. "Nevertheless, numerous experiences have taught us that [...] in male lineages the Jewish type – the Jewish characteristic – continues to rule both outwardly and inertly, while the mixed-blooded descendants of the daughters have long been totally Aryanized or seem that way." They summed up the description of this curious phenomenon by declaring that Jews "therefore contradict the Mendelian law."[80]

The claim that Jews were able to break through normal scientific boundaries was dismissed as untenable by professional anthropologists, some of whom insisted on the dominant nature of Jewish traits.[81] In 1923, in one of the very few anthropological studies actually conducted on Jews, anatomist Ferdinand Wagenseil explained with relation to Salaman's study that it was "not the physiognomy which was inherited but single and separate anatomical traits." On the basis of his own and others' observations, Wagenseil wrote,

In what way the so-called hereditary prepotence, which plays such a great role especially in the Jewish physiognomy, may be explained, that has been shown by E. Fischer in his theoretical hereditary analysis of his Rehoboth Bastards. A whole series of such prepotent, that is homozygous or dominant and therefore racial traits, were documented by various authors for the Jews.

Among such dominant features, Wagenseil named "fleshy nose with a convex back and strongly flared wings (Jacob's 'nostrility'), strongly

---

[80] The Almenach de Gotha was a series of books on Europe's higher nobility and royal courts. Its publication began in 1763 and its content varied along with Europe's major political transformations. Between 1912 and 1919, a series of complementary publications was launched, titled "Semi-Gotha Genealogical Handbook of A̱ri(st) ocratic-Jewish Marriages" – the title itself pointing the reader's attention to the underlying racial issue of Aryan-Jewish mixture. See *Semigothaisches genealogisches Taschenbuch a̱ri(st)o̱kratisch-jüdischer Heiraten* (Munich: F. Bruckmann, 1914) (quote from p. xxv). Two decades after its publication, the book proved valuable to the Nazis for identifying Jewish lineages. See Gregor Hufenreuter, "Der 'Semi-Gotha' (1912–1919). Entstehung und Geschichte eines antisemitischen Adelshandbuches," *Herold-Jahrbuch*, Neue Folge 9 (2004): 71–88.

[81] See Lenz' comment: "Ref. möchte aber glauben, daß dieser Widerspruch nur scheinbar ist" in Fritz Lenz, "Kritische Besprechungen und Referate," *ARGB* 11, no. 6 (1914): 811.

protruding cheekbones, thick lips, pointed chin, 'giving the beaked look' (Bean), deep wrinkles of the face, long upper eyelids (Stratz)."[82]

Nevertheless, associating Jews with recessive traits had a special appeal for two main reasons. The first had to do with the high rate of consanguineous marriages among Jews. As mentioned above, even prior to the spread of Mendelian teaching, Jews were considered as the prime example for a nation built on the principle of inbreeding. "The Jews," stated an Austrian doctor in 1897, "are the only people that set up their inbreeding principle on a spiritual basis." Another 1904 article stated, "The entire national state and the law system of the Jews was essentially constructed around the principle of inbreeding. ... After returning from Babylonian captivity ... the community took upon itself the duty not to allow under any circumstances marriages with those who were not part of it."[83] Such historical characterization turned Jews into an exemplary population, and in fact the only European population that conformed to the required prerequisites for the consideration of biological hereditary processes. Their presumed isolation and strict inbreeding regime turned them into "pure lines" in the hereditary sense.[84] Under Mendelian suppositions, such inbreeding led not only to homozygosity and racial purity, but also to an increased rate of manifestations of malignant recessive traits. Both characterizations fit very well in describing the Jews as simultaneously the most pure race and the one most replete with genetic diseases.

Associating Jews with inbreeding, recessive Mendelian factors and degenerative phenomena thus became a recurring theme in anthropological and medical literature. In exceptional cases, it could even help lessen the burden of the pathological status of the Jewish race. If a certain disease was more common among Jews, explained Wilhelm Reutlinger in

---

[82] Wagenseil, "Beiträge zur physischen Anthropologie," 88. For a review of the German anthropological discussions on the Jewish race, see Annegret Kiefer, *Das Problem einer "jüdischen Rasse." Eine Diskussion zwischen Wissenschaft und Ideologie (1870–1930)* (Frankfurt a. M.: Peter Lang, 1991). An overview of the anthropological studies of the Jews in Germany also after the Nazis' rise to power is offered by Georg Lilienthal, "Die jüdischen 'Rassenmerkmale': Zur Geschichte der Anthropologie der Juden," *Medizinhistorisches Journal* 28, no. 2/3 (1993): 173–198. See also Lipphardt, *Biologie der Juden*, 133–186 and Doron Avraham, "The 'Racialization' of Jewish Self-Identity: The Response to Exclusion in Nazi Germany, 1933–1938," *Nationalism and Ethnic Politics* 19 (2013): 354–374.

[83] Albert Reibmayr, *Inzucht und Vermischung beim Menschen* (Leipzig/Vienna, 1987), 175; F. Kraus, "Blutsverwandtschaft in der Ehe und deren Folgen für die Nachkommenschaft," in Hermann Senator and Siegfried Kaminer (eds.), *Krankheiten und Ehe: Darstellung der Beziehungen zwischen Gesundheitsstörungen und Ehegemeinschaft* (Munich: J. F. Lehmanns, 1904), 56–88 (quote from p. 64).

[84] Lipphardt, *Biologie der Juden*.

1922, this did not necessarily imply that these diseases were particularly Jewish or that Jews were more prone to becoming affected by them, but simply that the Jewish rates of inner kin marriage were higher; had other populations married the same way, the same degenerative factors would probably appear among them as well.[85] The blame for degeneration could therefore be shifted from the ontologically irreversible Jewish racial components onto the practices of marriage, which by themselves could be altered.

The second reason that contributed to the identification of Jews with recessivity was related to the host of characteristics associated with recessive traits – most notably their malignancy, deceitfulness and troubling undetectability. All of these features corresponded neatly to popular antisemitic stereotypes of the Jewish deformed physique, their uncanny nature, and their incessant attempts to penetrate and subvert the German national body by camouflaging their true nature. These stereotypes were popularized through antisemitic propaganda, in pamphlets, newspapers, talks, popular tales and, later, movies.[86] The correspondence between such images of Jews and the notion of recessivity as hidden malignancy established the status of recessivity as particularly relevant for the analysis of the intermixture between Jews and Germans. Building on the long tradition that identified Jews as carriers of diseases, Jewish bodies now turned into carriers not only of malignant germs but also of deceitful genes, which needed to be eradicated, or at least avoided, to salvage the health of the national body. In Chapter 5, we will see how the concept of Jewish recessivity was translated, under the Nazis, into actual anti-Jewish policies. To heal and rebuild the nation, however, combating racial degeneration was not enough; a positive vision of regeneration was also necessary. And here, too, Mendelian thinking took center stage.

Peasants and the Mendelian Rebirth of the Nation

Facing the disturbing results of modernization, urbanization, dramatic demographic changes and all-too-quick alterations in the morals of society, many opinion makers in the early twentieth century turned to the peasants to constitute the ideal type for nascent political movements. This was true throughout the European continent, where conservative,

---

[85] Wilhelm Reutlinger, "Über die Häufigkeit der Verwandtenehen bei den Juden in Hohenzollern und über Untersuchungen bei Deszendenten aus jüdischen Verwandtenehen," *ARGB* 14, no. 3 (1922): 301–305.

[86] See Chapter 5 and Jeffrey Herf, *The Jewish Enemy: Nazi Propaganda during World War II and the Holocaust* (Cambridge, MA: Harvard University Press, 2006).

nationalistic and reactionary thinkers, building on romantic legacies, glorified peasanthood as representing the authentic national spirit. Peasant life was natural, intimately connected to the soil, distant from and therefore uncontaminated by the moral and physical decay caused by city life, and owing to geographical isolation, racially pure.[87] After World War I, peasants also became significant politically: it was the farmers who were safeguarding national survival by working the land, and it was they who were mobilized by German politicians as international pawns in the struggle to challenge the detestable Versailles treaty and to re-annex territories that had been "torn from Germany" in the east. The Nazi party realized the potential in recruiting support in rural areas in the late 1920s, and started emphasizing the idea that peasants still preserved the blood of the Germanic ancestors and would one day become the fountain of life of the Nordic race.[88] Yet admiration for the naturalness of village life was not singular to the political right; it found expression across the political spectrum and public life in Germany – in the architectural design of green-belt cities, the youth movements' long hikes to the mountains, and the artistic attraction toward rural landscapes (or, in contrast, the grim view of the sooty city).[89]

For genealogists, villages, more than cities, were also the source of invaluable genealogical data. It was in village churches that genealogists found their richest early sources for their research: parish registers. These registers documented the birth, baptism, marriage and death dates of the members of rural communities throughout the course of hundreds of years. Genealogists considered these records less prone to casual mistakes than state-organized, bureaucratic documentation, since such

---

[87] Hans-Ulrich Wehler, *Deutsche Gesellschaftsgeschichte*, vol. 4, *Vom Beginn des Ersten Weltkrieges bis zur Gründung der beiden deutschen Staaten 1914–1949* (Munich: C. H. Beck, 2003), 339–342, 347; Bernhard Dietz, "Countryside-versus-City in European Thought: German and British Anti-Urbanism between the Wars," *European Legacy* 1, no. 7 (2008): 801–814; Ulrich Herbert, *Geschichte Deutschlands im 20. Jahrhundert* (Munich: C. H. Beck, 2014), 251–254.

[88] Gustavo Corni, *Hitler and the Peasants: Agrarian Policy of the Third Reich, 1930-1939*, trans. David Kerr (New York/Oxford/Munich: Berg, 1990), 22, 27; Gesine Gerhard, "Das Ende der deutschen Bauernfrage – Ländliche Gesellschaft im Umbruch," in Daniela Münkel (ed.), *Der lange Abschied vom Agrarland. Agrarpolitik, Landwirtschaft und ländliche Gesellschaft zwischen Weimar und Bonn* (Göttingen: Wallstein, 2000), 124–142, esp. 134–135.

[89] Peter D. Stachura, *The German Youth Movement 1900–1945: An Interpretative and Documentary History* (London/Basingstoke: Macmillan, 1981), 14–15, 17, 94; Barbara Miller Lane, *Architecture and Politics in Germany 1918–1945* (Cambridge, MA/London: Harvard University Press, 1985), 44–45; Jill Lloyd, *German Expressionism. Primitivism and Modernity* (New Haven, CT/London: Yale University Press, 1991), 133–39; Ashley Bassie, "The Metropolis and Modernity," in Ashley Bassie, *Expressionism* (New York, NY: Parkstone International, 2012), 71–86.

records relied on the personal acquaintance of a community pastor with his spiritual flock and were usually closely monitored and maintained in strict order. With the help of these parish registers, which were systematic, relatively accessible and easy to handle (compared, for example, with diaries or gravestones), genealogists could map familial ties of entire communities over long periods of time and throughout many generations. At the same time, by working on parish registers genealogists could study rural societies, not just aristocratic dynasties. This reinforced genealogists' self-perception as active participants in a grand national project, supporting the (re)formation of the German *Volk*.[90]

Genealogists therefore devoted considerable resources to improving the retrieval, documentation and dissemination of data from such parish registers. The result was not only an elevation of the status of peasants in genealogical works but also, as a by-product, a diminished genealogical visibility of Jewish individuals, families and names. By definition, Jews appeared on such church records only if they resided in rural areas (many didn't) and if they abandoned their faith by converting to Christianity. Parish register–based genealogical work was therefore not so much antisemitic as much as it was a-semitic (or, a-Jewish). Partly as a response to the growing reliance of genealogists on church books, and the resulting extraction of Jews from the mapping of national roots and of Jewish genealogy from German genealogy, Crzellitzer chose in 1924 to establish an independent journal for Jewish familial studies.[91] Inevitably, however, Jews cropped up occasionally in German genealogical trees. When this happened, antisemitic sentiments quickly rose to the surface: "I readily admit that even the sight of another person's family lineage totality fascinates me," conceded the head of the Würzburg archive, August Sprel, in 1924, "if it doesn't mistakenly bring out the ascendancy of an

---

[90] Eduard Heydenreich (ed.), *Handbuch der praktischen Genealogie* (Leipzig: Degener, 1913); Adolf Förster, "Ahnenlistenaustausch," *FB* 19, no. 2 (1921): 33; Ernst Devrient, "Die Kirchenbücher und die Staatsarchive," *Mitteilungen der Zentralstelle für deutsche Personen- und Familiengeschichte* 6 (1910): 20–26; Friedrich von Klocke, "Deutsche Ahnentafeln," *FB* 19, no. 9 (1921): 257–262; [Emil] Bree, "Wie können die Kirchenbücher für die Familienforschung nutzbar gemacht werden?" *FB* 26, no. 7/8 (1926): 199–204, esp. 199; Walter Scheidt, "Erbbiologische und bevölkerungsbiologische Aufgaben der Familienforschung," *Archiv für Sippenforschung und alle verwandten Gebiete* 9 (1928): 289. For an analysis, see Eric Ehrenreich, *The Nazi Ancestral Proof* (Bloomington: Indiana University Press, 2007), ch. 1–3.

[91] Amir Teicher, "'Ahnenforschung macht frei': On the Correlation between Research Strategies and Socio-Political Bias in German Genealogy, 1898–1935," *Historische Anthropologie* 22, no. 1 (2014): 67–90.

enriched Hebrew from the depth of a Germanic farmer's family. For that is inappropriate (*stilwidrig*)."[92]

Like their peers in the field of familial research, anthropologists were similarly attracted to the study of rural areas. Anthropological investigations into the physical characteristics of peasants in lower Saxony began during the mid-1920s, and in 1928 Fischer launched a national project of an anthropological survey of rural populations, whose results were to appear as part of a new publication series, "German Racial Study."[93] In this context, Mendelian reasoning proved a significant obstacle to the rendering of peasants responsible for national renewal: isolated villages often exhibited high rates of consanguinity, which, as expected, led to an abundance of pathological or degenerative phenomena.

Fischer acknowledged the challenge and hoped to obtain more data on the distribution of these rural pathologies through his surveys. The problem, however, could not be easily brushed aside. Even much later, in 1942 and under the Nazi regime, a textbook on the dangers of heredity would describe to its readers a case where in one family "both parents originate from the same small village and are afflicted with the recessive disposition to hereditary muscular dystrophy. Maybe they are blood-related? If one searches far back enough, one sometimes finds among the inhabitants of small villages common ancestors for several families. The villagers have married each other again and again. Recessive inheritance is especially dangerous, because it involves persons that look outwardly healthy."[94]

Thus, the Mendelian schemes applied to Jews to account for the consolidation of their racial character and for the prevalence of pathologies among them could just as well be used to describe rural populations: isolation, inbreeding and consanguineous marriages were common in remote villages just as they were in the Ghetto. Nevertheless, when it came to peasants, scholars tended to downplay the negative elements

---

[92] August Sprel, "Wie wir zur Genealogie gekommen," *FB* 22, no. 1/2 (1924): 5–12 (quote from p. 12). As noted above (p. 114, fn. 80), the "semi-Gotha" was similarly motivated by strong antisemitic agenda.

[93] Eugen Fischer (ed.), *Deutsche Rassenkunde. Forschungen über Rassen und Stämme, Volkstum und Familien im Deutschen Volk*, vols. 1–4 (Jena: Gustav Fischer, 1929–1930); Paul Weindling, *Health, Race, and German Politics between National Unification and Nazism: 1870–1945* (Cambridge: Cambridge University Press, 1989), 466–467; Hans-Walter Schmuhl, *The Kaiser Wilhelm Institute for Anthropology, Human Heredity, and Eugenics, 1927–1945: Crossing Boundaries*, Boston Studies in the Philosophy of Science, vol. 259 (Dordrecht: Springer, 2008), 88.

[94] Karl Tornow and Herbert Weinert, *Erbe und Schicksal: Von geschädigten Menschen, Erbkrankheiten und deren Bekämpfung* (Berlin: Alfred Metzner, 1942), 67–68.

stemming from consanguinity. For example, a study of "Burkhards and Kaulstoss, Two Upper Hessian Villages," published in 1936 as part of the German Racial Study series, showed that in both villages there was a considerable rate of kin-marriage; in one family, for example, half of the ancestors in the eighth ancestral generation appeared more than once in the *Ahnentafel* (they were great-grandparents from at least two different sides of the ancestral chart). More generally, a substantial portion of the families was related through multiple blood ties. Yet "direct damaging effects of inbreeding can hardly be found in the two locations. The population's overall state of health is generally good. However, there is a strong tendency to goiter (endemic goiter), which nowadays can be effectively fought against with iodine preparations. ... Rachitic curvatures of the legs is extremely common, which is probably due to the fact that water in this area is poor with calcium."[95] Curiously, when it came to rural populations, use of medical treatment and improvement of environmental conditions seemed to solve even hereditary problems.

In general, the presumed impact of inbreeding and consanguineous marriages depended mainly on the cultural evaluation of the traits of the examined populations. Accordingly, Max Marcuse, a German-Jewish sexologist, explained in his 1926 *Handbook of Sexology* that kin-marriage could be good or bad depending on the hereditary constitution of the family. If both parents were talented, or if their traits constitutionally matched, inbreeding could by all means be beneficial. Inbreeding could be precarious; but consanguineous marriages "should not be automatically advised against, because they can also lead to the homozygotization of *desirable* hereditary traits." Marcuse stressed that "**what** is [genetically] combined – that is the essential thing," and not the process of combination itself.[96] Since the evaluation of the significance of certain personal attributes, or of the severity of certain diseases, depended on social and cultural norms, Mendelian theory could be accommodated and applied differentially according to the population in question. Among the Jews, inbreeding could explain racial tenacity as well as high rates of genetic

---

[95] Brigitte Richter, *Burkhards und Kaulstoß: Zwei oberhessische Dörfer*, Deutsche Rassenkunde, vol. 14 (Jena: Gustav Fischer, 1936), 24–26.

[96] Max Marcuse, "Inzucht und Verwantenehe," in Marcuse (ed.), *Handwörterbuch der Sexualwissenschaft* (Bonn: A. Marcus & E. Webers Verlag, 1926), 312–313. On the longer tradition of discussions on the ultimate beneficial/harmful impact of inbreeding, especially with relation to cattle improvement, see Vítězlav Orel, "The Spectre of Inbreeding in the Early Investigation of Heredity," *History and Philosophy of the Life Sciences* 19, no. 3 (1997): 315–330. These discussions, according to Orel, also motivated Mendel's own study.

pathologies; for rural populations, the same mechanism had to be acknowledged, but its overall importance for creating pathologies could be minimized.

One thing nevertheless became common to the understanding of the characteristics of both Jews and German peasants: their racial traits were considered to behave as independent Mendelian units. Thus, when Brigitte Richter, the author of the above-mentioned study on two upper-Hessian villages, turned to assessing the population's racial composition, she did so by calculating the proportion of independent traits, not of entire racial types. Among half of the individuals in Richter's sample, these traits were intermixed in a way that did not allow Richter to determine a "type" at all. Among the other half, 18% were primarily Nordic, 20% primarily Phalic, 8% mostly Alpine and 7% Dinaric. But "[t]he significant crossbreeding of the two darker races, the Dinaric and the Alpine, makes the number of bright individuals especially meaningful," Richter explained, "because the bright colors are transferred in a recessive hereditary mode. There is therefore much more Nordic-Phalic blood in the population than the visible colors betray."[97] Just as the recessiveness of the Jewish face could allow for the Jewishness of non-Jewish-looking persons, so could the recessiveness of blond hair support the assumption that non-Nordic-looking peasants were actually Nordic.

The disintegration of races into separate traits did not characterize only the studies of professional anthropologists, but informed the works of popular racial ideologues as well. Günther estimated in his widely read racial study that the Nordic component mounted to roughly 50% to 60% of the German *Volk*. These numbers, Günther clarified, represented not the proportion of Nordic "types," but that of Nordic "blood," which stood for inherited traits. The percentage of "pure Nordic persons" was much lower: 6% to 8%. Revealingly, Günther referred to pure Nordic men as "persons, who unite in them **all** the features of the Nordic race." Within such a definition, the whole (Nordic race) was defined as the sum of its parts (independent racial features), at least as far as the contemporary German population was concerned.[98]

---

[97] Richter, *Burkhards und Kaulstoß*, 78–79.
[98] Günther, *Rassenkunde des deutschen Volkes*, 208. The same held true for Jews. As Fischer's heir as the head of the Institute for Anthropology, Otmarr Freiherr von Verschuer, explained in 1938, "It goes without saying that one should not expect to be able to sort every Jew to one of these [Ashkenazi or Sephardi] types; these [types] would match only a minority of them. ... 'Pure' types... are to be found only rarely among them." Otmar Frhr. v. Verschuer, "Rassenbiologie der Juden," in *Forschungen zur Judenfrage*, vol. 3 (Hamburg: Hanseatische Verlagsanstalt, 1938), 137–151 (quote from pp. 142–143).

The legitimacy of this dissection of races into particular traits was a contested issue in anthropological scholarship. Proponents of the holistic approach claimed that race was something that needed to be perceived and seen, and that it could not be decomposed into its constituent parts, each measured or studied on its own.[99] Günther was one of the most vocal proponents of such a holistic approach to the study of race – although he, too, partly surrendered to the competing view, as his quote above reveals. Clearly, Mendelized racial-anthropology strengthened the opposite, atomistic camp, a trend not everybody looked favorably upon, especially not under the Nazis. In 1942, in a polemical article bearing the informative title "Trait Counting or Racial Research?" the Breslau anthropologist Ilse Schwidetzky fiercely criticized two leading German researchers – SS officer Wilhelm Gieseler, who was then the chairman of the German Society for Racial Study (*Deutsche Gesellschaft für Rassenforschung*), and the racial-hygienist Lenz. Recent works by Gieseler and Lenz, wrote Schwidetzky, gave the impression that the main task of racial scientists was not to study racial types, but to count individual traits. Gieseler and Lenz's justification for what Schwidetzky described as their "downright paradoxical attitude" was that it was not entire racial types that were geographically distributed but only individual racial attributes. Schwidetzky unequivocally rejected this view. For her, it was crucial that questions regarding racial types remain at the core of anthropological work. As Schwidetzky saw it, "Giving up a real race-research, i.e. racial-types research, and settling for pure feature-counts, would deprive anthropology of its most fruitful questions and its most vivid impacts."[100]

Given that the Mendelization of anthropology was accompanied by such an atomistic decomposition of the notion of race, how can we account for the relatively smooth adoption of Mendelian theory by racial anthropologists – and not only by medical practitioners – both before and after 1933? The elevated paradigmatic status of Mendelian theory in biology provides only part of the answer; as we have already seen, genealogists resisted the temptation to "mendelize" their field because

---

[99] The debate among anthropologists was as old as the anthropological discipline itself, and evident in the very first issue of the *Archiv für Anthropologie*: Wilhelm Hiss, "Beschreibung einiger Schädel altschweizerischer Bevölkerung nebst Bemerkungen über die Aufstellung von Schädeltypen," *Archiv für Anthropologie* 1, no. 1 (1866): 61–74, esp. 69. On holism vs. atomism in Germany, see Mitchell G. Ash, *Gestalt Psychology in German Culture, 1890–1967: Holism and the Quest for Objectivity* (Cambridge: Cambridge University Press, 1995); Anne Harrington, *Reenchanted Science: Holism in German Culture from Wilhelm II to Hitler* (Princeton, NJ: Princeton University Press, 1996).

[100] Ilse Schwidetzky, "Merkmalszählung oder Rassenforschung?" *Zeitschrift für Rassenkunde und die gesamte Forschung am Menschen* 13 (1942): 177–182 (quote from p. 178).

such a Mendelization undermined too great a portion of their professional identity. Schwidetzky's pronouncements show that potentially, a similar rejection of Mendelian assumptions could have happened in the anthropological community as well. Anti-atomistic views were certainly prominent in the writings of racial popularizers such as Günther and Ludwig Ferdinand Clauss.[101] Why did racially minded anthropologists nevertheless cling to Mendelian theory?

The answer seems to lie in another element of Mendelian thinking that had significant appeal for racial anthropologists – namely, its ability to meet a fundamental challenge of racial thought: the ambition to correlate, on the basis of physical affinities, present-day peoples with ancient ones. Pre-Mendelian concepts of hereditary transmission allowed for the hereditary factors to be influenced by external stimuli (as in Lamarck's work), to lose their impact with the progression of generations (as in Galton's "Law of Ancestral Heredity"), or to merge into each other ("blending inheritance").[102] Mendelian theory, in contrast, insisted that the gametes remained pure and the inherited factors unaltered from one generation to the next. This idea of gametic purity complemented August Weismann's germ plasm theory, which insisted that somatic elements had no influence on the hereditary material. Taken together, Mendelism and Weissmannism meant that – disregarding rare, random and mostly fatal mutations – the genes of the present were essentially identical to those of ancient times. Processes of selection modified the prevalence and distribution of human attributes, but they did not alter the genes themselves in any fundamental way.[103] As a result, the thought that contemporary Germans – especially those in remote rural areas – still carried the blood of their ancient forefathers could be taken seriously.

If the gametes of past races existed in the population in their pure, original form, then, by properly directing reproduction, these races from the past could literally be resurrected. Such resurrection was essentially what influential Nordic racial-hygienists strived toward. In 1930, the future Reich Minister of Food and Agriculture and senior officer of the SS, Richard Walter Darré, emphasized that although the German *Volk*

---

[101] On Clauss, see p. 184 below.

[102] In Galton's model, the individual's features were determined by cumulative effects of an infinite series of the individual's ancestors, the effect decreasing exponentially the more generations separated an ancestor from the proband. See Michael Bulmer, "Galton's Law of Ancestral Heredity," *Heredity* 81 (1998): 579–585.

[103] Critically considered, this view was built on a very narrow understanding of the meaning of genes and their dynamics of interaction, and overlooked the impact of selection pressures, genetic drift and other evolutionary mechanisms on the content of the hereditary material itself.

was racially mixed, this mixture, "unlike café-au-lait or raspberry lemon-ade," was simply a grouping of independent hereditary factors that were inherited unchanged from ancestors to progeny, in accordance with Mendel's laws. Cultivating the German *Volk* meant "deliberately and with a planned utilization of the available aids, to generate a progeny whose value will stand at least not below that of their begetters, but if possible, will ascend over time to the value of the original generations." For that purpose, a proper breeding goal, a model for directing selection processes, was essential. The Nordic race, explained Darré, was precisely that.[104] The biological goal propagated by Nazi ideologues like Darré was to recollect the scattered genes of formerly existing races, reunite them and reinstitute the character of the culturally glorified ancient ancestors.

In other words, what was at stake in the Mendelized version of Nordic racial thought was not only Social Darwinian eugenics, which aimed at increasing the propagation of those deemed fit and checking the spread of those of lesser value. When Mendelian logic was implanted into racial thinking, one could promote the screening and selection of ancient genes so as to reconstitute – gradually, but steadily – the Nordic race. For that end, the German public had to be educated on the proper way of choosing mates, and actions needed to be taken to eradicate malignant or alien genes wherever these surfaced. This was the essence of racial-hygiene. As part of this "biologisation of national belonging" (to borrow a term from Marius Turda),[105] national rebirth, so crucial for the fascist vision of the future, could take shape. On the basis of new, planned births, the racial composition of the national body could be refashioned until it would reach the ideal, past-and-future Nordic racial type. This was a vision that could easily find supporters among racial anthropologists. Since the peasants harbored uncontaminated, desirable racial traits, they could provide the necessary hereditary material to make such a national renewal possible.

Frustratingly, there remained several obstacles on the path toward this yearned-for Nordic renewal. It is to the removal of these obstacles that we will now turn.

---

[104] Richard Walter Darré, *Neuadel aus Blut und Boden* (Munich: Lehmanns, 1930), 144–146, 174–200 (quotes from pp. 144, 181).
[105] Marius Turda, *Modernism and Eugenics* (New York, NY: Palgrave Macmillan, 2010).

4   Annihilating Defective Genes: Mendelian Consciousness and the Sterilization Campaign

Beginning in 1933, the educational play *Erbstrom* (literally: hereditary stream) began its tour of Germany, attracting audiences of tens of thousands. Many of them were schoolchildren, shepherded to the local theater by their teachers; many others were factory workers whose workplace purchased tickets after being pressured to do so by the local government. "In fact, this is a piece that brings closer to every member of the *Volk* the exceptional meaning of hereditary health for the race and for the future of the nation," explained a circular from the Health Department of the Reich Ministry of Interior in 1935. The circular further noted that in the field of eugenic policies, "many things are not properly understood and are still being combated in a non-factual manner." The play, which was designed to "speak to the hearts of the people," was intended to help in "promoting the understanding of the biological population policies of the National-Socialist state."[1]

Erbstrom was an overtly didactic play in three acts, whose plot takes place between 1931 and 1933. The struggle between the biological, morally superior world view of the Nazis and the individualistic, greed-driven, capitalistic conduct of the Weimar years is represented through several conflicts between the play's characters, including a doctor and his patients, a town mayor, and a couple about to get married. Questions on the nature of heredity and its significance both for the individual as well as for the entire nation are debated throughout the play, leading to tensions and struggles that are only resolved with the triumphant rise of the Nazis to power. Hereditary

---

The argument and some of the passages in the first part of this chapter have been previously published in Amir Teicher, "Why Did the Nazis Sterilize the Blind? Genetics and the Shaping of the Sterilization Law of 1933," *Central European History* 52, no. 2(2019): 289–309.

[1] See Bayrisches Hauptstaatsarchiv [hereafter BayHStA] MInn 79480 and Hans-Christian Harten, Uwe Neirich and Matthias Schwerendt, *Rassenhygiene als Erziehungsideologie des Dritten Reichs* (Berlin: De Gruyter, 2006), 43.

mechanisms, and in particular dominance and recessivity, are explained at length in the play as reasons for the couple's eventual decision to refrain from marrying and as an indication for one of the characters' (denied) fatherhood of a child with six fingers. During the first act, one character says to another that in the second generation of crosses, the blue eyes of 25% of the progeny would be out-mendeled. – "What did you say?" – "Out-mendeled – m-e-n-d-e-l-e-d – after the famous Augustinian pastor Gregor Mendel."[2]

These explications, however, might have seemed insufficient to convey such complex scientific ideas to the audience; thus, the playwright supplemented them with visual demonstrations. While the stage managers were working to rearrange the set between the first act (which took place at the rural doctor's clinic) and the second act (at the local welfare office), a colorful slide, titled "Dominant heredity," was projected onto the curtain. The slide showed the Mendelian inheritance of brown and blue eyes; it was displayed for two full minutes. Between the second and third acts, the audience stared for another two minutes at a chart illustrating the "recessive inheritance of myoclonus epilepsy" (Figure 4.1).[3]

Throughout the years of the Third Reich, Germans of all ranks would continue to encounter Mendelian representations as explanations for eugenic policies. Mendelian charts, terminologies and mechanisms would be conveyed to the public through multiple channels: popular lectures by doctors, anthropologists and psychiatrists; official governmental announcements and laws; newspaper articles, leaflets and pamphlets; and films and plays. Medical personnel, SS officers, those suspected of carrying a hereditary disease and their families and schoolchildren were particularly exposed to Mendelian propaganda. In the universities, too, students studying biology could be asked to watch a seven-minute film, produced especially for this purpose in 1934 by the Reich Office for Educational Films, on hereditarily diseased rabbits. The film explained the Mendelian mechanisms that led to the manifestation of shaking palsy (Parkinson disease), defined to be "a simple recessive hereditary disease." Two of the seven minutes of the film were devoted to animated illustrations of Mendelian mechanisms; the remaining five were composed of long scenes of miserable, helpless rabbits, unable to control the incessant shaking of their own limbs, as photographed in the laboratory of heredity researcher Hans Nachtsheim (Figure 4.2). The

---

[2] Konrad Dürre, *Erbstrom: Schauspiel in drei Akten* (Berlin: Bühnenverlag Ahn & Simrock, 1933) (quote from p. 9). I thank Dr. Jenny Hestermann for her assistance with locating this play.
[3] Dürre, *Erbstrom*, 17, 31.

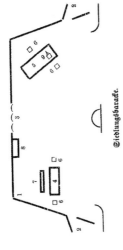
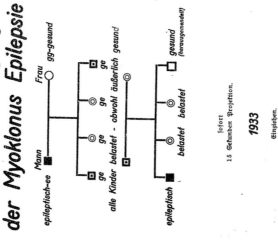

Figure 4.1 Instructions for the projection of a slide explaining the recessive Mendelian inheritance of myoclonus epilepsy for two full minutes, while changing of the set of the 1933 play *Erbstrom* (hereditary stream).
Source: Konrad Dürre, *Erbstrom: Schauspiel in drei Akten* (Berlin: Bühnenverlag Ahn & Simrock, 1933), 30–31.

Figure 4.2 Four scenes from a 1934 educational-scientific film on the terrifying results of allowing hares carrying recessive deleterious genes to reproduce.
Source: Reichstelle für den Unterrichtsfilm, *Erbkranke Kaninchen: Schüttellähmung, eine erbliche Nervenkrankheit (1938)*. MPG Archive, VII. Abt., Rep. 1, F 116

message was clear. Nervous malfunctions resulted from simple, scientifically identified Mendelian genes; for the sake of those afflicted, their continuous reproduction must be arrested. Any other policy would be inhumane.[4]

The campaign against the mentally ill was officially launched with the passing of the compulsory Sterilization Law on July 14, 1933. The law led to the establishment of a hierarchical system of Hereditary Courts, where individuals with "damaged" hereditary endowment were sentenced to sterilization. According to the most recent estimations, these courts discussed the cases of approximately 436,000 individuals between

[4] *Erbkranke Kaninchen: Schüttellähmung, eine erbliche Nervenkrankheit* (Reichstelle für den Unterrichtsfilm, 1934), available in MPG Archive, III Abt. Rep. 0020A, Nr. 158. As Alexander von Schwerin has shown, although Nachtsheim's work was on rabbits, the eugenic motivations and implications of his studies were quite evident, to himself as well as to others. See Alexander von Schwerin, *Experimentalisierung des Menschen. Der Genetiker Hans Nachtsheim und die Vergleichende Erbpathologie 1920–1945* (Göttingen: Wallstein, 2004), 193–197.

1934 and 1945, and led to the sterilization of roughly 300,000 persons.[5] Sterilization was especially intensive in the first three years, after which the rate was somewhat reduced, partly because a substantial portion of the potential human reservoir had already been processed: either sterilized or officially exempt. From 1939 began the murder of children with severe congenital deficiencies and of inmates of mental asylums – first by starvation and lethal injections, later also in gas chambers. In two years, more than 70,000 Germans with mental or physical disabilities were executed in the Nazi "euthanasia" program. Until the end of the war, additional tens of thousands of non-Jewish Germans deemed hereditarily or socially undesirable were similarly exterminated.[6]

Mendelian thinking played a key role in the formulation and implementation of the sterilization campaign that, in certain important respects, also paved the way toward the subsequent murder of the mentally ill. It became central to the way Germans learned to think about medical dangers, their meaning and their implications for their own lives. It influenced the choice of categories of diseases targeted by the law, and in some cases discussed by Hereditary Courts it became part of the decision-making processes that determined the fates of individuals. More crucially, the reasoning that it enshrined enabled the development of a fanatic eugenic state of mind, which made it easier for the Nazi state to take actions against those deemed inferior.[7] The first section in this

---

[5] See the analysis offered in Udo Benzenhöfer and Hanns Ackermann, *Die Zahl der Verfahren und der Sterilisationen nach dem Gesetz zur Verhütung erbkranken Nachwuchses* (Münster: Kontur-Verlag, 2015).

[6] Robert Proctor, *Racial Hygiene: Medicine under the Nazis* (Cambridge, MA: Harvard University Press, 1988), 177–178; Michael Burleigh, *Death and Deliverance: "Euthanasia" in Germany, 1900–1945* (Cambridge: Cambridge University Press, 1994); Ernst Klee, *"Euthanasie" im NS-Staat. Die "Vernichtung lebensunwerten Lebens,"* 2nd ed. (Frankfurt a. M.: S. Fischer, 1983); Hans-Walter Schmuhl, *Rassenhygiene, Nationalsozialismus, Euthanasie* (Göttingen: Vandenhoeck & Ruprecht, 1987); Henry Friedlander, *The Origins of the Nazi Genocide: From Euthanasia to the Final Solution* (Chapel Hill: University of North Carolina Press, 1995).

[7] No claim is made here, however, as to a direct path leading from the sterilization to the annihilation of the mentally ill. The "euthanasia" program was implemented in circumstances of war and was greatly influenced by economic considerations relating to the victims' "productivity" and "utility" for the community, often regardless of the presumed hereditary nature of their illness. See the results of the research by Maike Rotzoll, Gerrit Hohendorf, Petra Fuchs, Paul Richter, Wolfgang U. Eckart and Christoph Mundt (eds.), *Die nationalsozialistische "Euthanasie" – Aktion "T4" und ihre Opfer: Geschichte und ethische Konsequenzen für die Gegenwart* (Paderborn: Schöningh, 2010); Maike Rotzoll, Petra Fuchs, Paul Richter and Gerrit Hohendorf, "Die nationalsozialistische 'Euthanasie-Aktion T4.' Historische Forschung, individuelle Lebensgeschichten und Erinnerungskultur," *Nervenarzt* 81 (2010): 1326–1332. Still, a certain type of biological state of mind, in which Mendelian assumptions were central, set the groundwork for the later annihilation of the psychically infirm. See, in addition to

chapter will show how developments within Mendelian research led to the inclusion of certain disease categories in the sterilization law. Once included, the sterilization campaign received a Mendelian "edge," which was essential for its overall legitimization. We will then examine the way Mendel's laws, and the need for eugenic measures, were taught as inseparable themes in German high schools from 1932 through 1936. With the help of apprentice teachers' detailed reports on their actual experiences in the classroom, we will get a glimpse into what really happened when educators tried to explain the rules of Mendelian heredity to their pupils and also retrieve the voices and reactions of students to Nazi biological propaganda. The third section will look into the implementation of the sterilization policy and the way it was informed by Mendelian theory. Taken together, we will see how from 1933 onward, Mendelian concepts were propagated to the German public as integral parts of eugenic reasoning, shaping, legitimizing and radicalizing Nazi policies toward the mentally ill.

Mendelizing the Sterilization Law

In July 1932, the Prussian Health Council drew up a draft for a sterilization law. According to this draft, it would become legal to perform sterilizations on any "person, that suffers from hereditary mental illness, hereditary mental deficiency, hereditary epilepsy or any other hereditary disease … if it is voluntary and if according to the teaching of medical science, severe physical or mental hereditary damage is to be expected with high probability in his progeny."[8]

Because of the political turmoil of the time, the draft did not materialize into an actual law. But a year later, and six months after Hitler became chancellor, a newly formulated sterilization law was passed. It was compulsory, not voluntary, and was officially titled "The Law for the Prevention of Hereditarily Diseased Offspring." The first clause of the law stated that

---

what follows, Maike Rotzoll and Gerrit Hohendorf, "Murdering the Sick in the Name of Progress? The Heidelberg Psychiatrist Carl Schneider as a Brain Researcher and 'Therapeutic Idealist,'" in Paul Weindling (ed.), *From Clinic to Concentration Camp. Reassessing Nazi Medical and Racial Research, 1933–1945* (London: Routledge, 2017), 163–182.

[8] Heinrich Schopohl (ed.), *Die Eugenik im Dienste der Volkswohlfahrt: Bericht über die Verhandlungen eines zusammengesetzten Ausschusses des preussischen Landesgesundheitsrats vom 2. Juli 1932.* Veröffentlichungen aus dem Gebiete der Medizinalverwaltung, vol. 38, no. 5 (Berlin: Richard Schoetz, 1932), 107 (735).

(1) Anyone who is hereditarily ill can be sterilized by a surgical operation if, according to the experience of medical science, it is to be expected with high probability that his offspring will suffer from severe physical or mental hereditary damages.

(2) Hereditarily ill in the meaning of this law is anyone who suffers from any of the following diseases:

   1. Congenital feeble-mindedness,
   2. Schizophrenia,
   3. Manic-depressive insanity,
   4. Hereditary epilepsy,
   5. Huntington's chorea,
   6. Hereditary blindness,
   7. Hereditary deafness,
   8. Severe hereditary physical deformity.

(3) Furthermore, anyone suffering from severe alcoholism can be made infertile.[9]

The list of diseases specified in the law was a novelty: no other sterilization law prior to the German law went into such detail in defining concrete medical conditions requiring sterilization.[10] Evaluated against previous legislation proposals both in Germany and abroad, the inclusion in the list of Huntington's chorea (5), blindness (6) and deafness (7) stands out. With few exceptions, eugenic campaigns to prevent the socially or medically "inferior" from propagating centered on the feeble-minded, the mentally ill, the "asocial" and the criminals. As heirs to the nineteenth-century degeneration anxiety, eugenicists were principally preoccupied with mental aberrations – feeble-mindedness and insanity – not with physical disabilities. As members of the educated middle class, eugenicists were concerned about the uncontrolled propagation of lower social strata – the poor, wayward, criminals and alcoholics – not with those with visual or hearing impairments.[11] Blindness, deafness and Huntington's chorea were not entirely absent from eugenic

[9] Arthur Gütt, Ernst Rüdin and Falk Ruttke, *Zur Verhütung erbkranken Nachwuchses. Gesetz und Erläuterungen* (Munich: J. F. Lehmanns, 1933), 56.

[10] An exception is Harry Laughlin's 1922 model law, which was never legislated and which will be analyzed below.

[11] This was emphasized particularly with respect to eugenics in Britain, where eugenicists openly aimed at combating pauperism. See Pauline Mazumdar, *Eugenics, Human Genetics and Human Failings: The Eugenics Society, Its Sources and Its Critics in Britain* (London: Routledge, 1992), and the analysis (later to become a topic of debate in the "Science Wars") by Donald Mackenzie, "Statistical Theory and Social Interests: A Case-Study," *Social Studies of Science* 8, no. 1, theme issue: *Sociology of Mathematics* (1978), 35–83.

discourse; one can find them mentioned in discussions on hereditary and economic burdens. Notably, however, when it came to taking actions against the reproduction of certain groups of people, these categories were usually exempt, and in none of the sterilization laws actually passed prior to the German law do they appear. How did they make their way into the Nazi law?

Before we attempt to answer this question, it is useful to examine closely two notable exceptions to the overriding pattern delineated above. One of them was the sterilization proposal of a district physician from Saxony by the name of Gustav Boeters. An avid supporter of legal measures to hinder the propagation of the mentally and physically inferior, in 1923 Boeters published his own draft sterilization law ("Lex Zwickau"). Boeters called for the sterilization of children who, at the time of entering elementary school, were blind, deaf-mute or idiotic (blödsinnig), congenital deficiencies that apparently prevented them from taking part in the normal course of studies. This was the year of German hyper-inflation, and Boeters' explicit motivation was an economic one: without schooling, these children could not become productive citizens, rendering them a burden to the rest of society.

Boeters propagated his law incessantly in any forum he found access to, and his suggestions attracted considerable academic and public attention.[12] But the responses to the content of his proposed law, even among his like-minded colleagues, were mostly negative. In an article published in 1926, surveying the scientific arguments for and against sterilization, Erna Kohls called Boeters' proposals "very far-reaching," because "inborn blindness and deaf-mutism do not imply mental inferiority" (the latter, obviously, did provide reasonable grounds for sterilization in Kohls' eyes). Without exception, all the other plans for sterilization bills mentioned in Kohls' extensive review excluded the blind and the deaf.[13] Two years later, in 1928, Boeters omitted these latter categories from a revised draft for his law ("Lex Zwickau III"). The new draft targeted the "mentally ill, mentally deficient, epileptic

---

[12] Boeters' letters to the Ministry of Interior can be found in BArch R1501/126248, fol. 142. For his public campaign, see BArch R 86/2374.

[13] Erna Kohls, "Über die Sterilisation zur Verhütung geistig minderwertiger Nachkommen," Archiv für Psychiatrie 77 (1926): 285–302 (quote from p. 290). See also the articles in Ärztliches Vereinsblatt 51 (1924) as well as Jenny Blasbalg, "Ausländische und deutsche Gesetze und Gesetzentwürfe über Unfruchtbarmachung," Zeitschrift für die gesamte Strafrechtswissenschaft 52, no. 1 (1932): 477–496 (esp. 492).

and morally unrestrained" as well as the general category of "hereditarily inferior."[14]

Another exception to the pattern described above was the so-called Model Eugenic Sterilization Law, presented in 1922 by U.S. eugenicist Harry Laughlin. The Model Law listed the following "socially inadequate classes" as candidates for sterilization:

(1) Feeble-minded; (2) Insane (including the psychopathic); (3) Criminalistic (including the delinquent and wayward); (4) Epileptic; (5) Inebriate (including drug habitués); (6) Diseased (including the tuberculous, the syphilitic, the leprous and others with chronic, infectious and legally segregable diseases); (7) Blind (including those with seriously impaired vision); (8) Deaf (including those with seriously impaired hearing); (9) Deformed (including the crippled); and (10) Dependent (including orphans, ne'er-do-wells, the homeless, tramps and paupers).[15]

Including the blind and deaf (Huntington's chorea, again, left unmentioned) in this Model Law was a novelty for Laughlin himself; eight years earlier, when discussing "the best practical means of cutting off the defective germ-plasm in the American population," he explained that "[t]he crippled, the blind, the deaf, and the tubercular are thus not subject to the provisions of this act, because, unlike the classes enumerated in the statute, they are capable of education, and consequently, eugenical training rather than enforced sterilization should apply to them."[16] By 1922 he seems to have changed his mind. Importantly, however, despite the inclusion of the blind and deaf in his updated model law, and in spite of Laughlin's leading position in American eugenics, none of the sterilization bills passed by any of the legislative bodies in the United States named these categories.[17] As he formulated it, the Model Law remained a dead letter, and no actual law was modeled upon it.

---

[14] See E[rich] Hesse, "Die Unfruchtbarmachung aus eugenischen Gründen," *Beiheft zum Reichs-Gesundheitsblatt* 15 (April 12, 1933), 19. On the discussions surrounding Boeters' draft and the omission of the blind and mute, see also Fritz Engelmann and August Mayer, *Sterilität und Sterilisation. Bedeutung der Konstitution für die Frauenheilkunde* (Munich: J. F. Bergmann, 1927), 223–224.

[15] Harry. H. Laughlin, *Eugenical Sterilization in the United States* (Chicago, IL: Psychopathic Laboratory of the Municipal Court of Chicago, 1922), 446–447.

[16] Harry H. Laughlin, *Bulletin No. 10B: Report of the Committee to Study and to Report on the Best Practical Means of Cutting Off the Defective Germ-Plasm in the American Population. II The Legal, Legislative and Administrative Aspects of Sterilization* (New York, NY: Eugenics Record Office, 1914), 125.

[17] See Edwin Black, *War against the Weak: Eugenics and America's Campaign to Create a Master Race* (New York, NY: Four Walls Eight Windows, 2003); Harry Bruinius, *Better for All the World: The Secret History of Forced Sterilization and America's Quest for Racial Purity* (New York, NY: Vintage Books, 2007); Jonathan P. Spiro, *Defending the Master*

From the quantitative perspective, it should come as no surprise that blindness, deafness and Huntington's chorea were exempt from sterilization bills: all three conditions were of negligible overall significance, according to the calculations made by eugenicists themselves. In Germany, official estimates published before the enactment of the Sterilization Law suggested that the expected number of feeble-minded to be affected by the law in the entire German Reich was roughly 200,000; schizophrenics amounted to 80,000; those who suffered from manic-depressive insanity, 20,000, and epileptics, 60,000. These four categories, with feeble-mindedness taking the lead, were therefore not only among the most prominent in eugenic discourse, but also – and correlatively – the quantitative majority, eventually comprising around 95% of the actual sterilizations. It was also estimated that 20,000 Germans suffered from various kinds of severe physical deformities; the number of (hereditary) deaf was estimated to be 16,000; the number of (hereditary) blind, 4,000; and the number of those suffering from Huntington's chorea, a mere 600.[18] In competing assessments published by leading eugenicists, the numbers of the blind and deaf-mute were similarly smaller than those of the other categories, and Huntington's chorea was negligible and hardly even mentioned.[19]

Because of the relatively small numbers of the blind, deaf and Huntington's chorea sufferers, the economic burden attributed to them was similarly marginal, both when measured against the expenses allocated for caring for the mentally ill and feeble-minded, as well as when compared with the funds that were needed for supporting the non-hereditarily blind and deaf or other categories of "dependents" – costs

Race: Conservation, Eugenics, and the Legacy of Madison Grant (Burlington: University of Vermont Press, 2009).

[18] It was further estimated that 10,000 suffered from chronic alcoholism. See Ernst Klee, *Deutsche Medizin im Dritten Reich: Karrieren vor und nach 1945* (Frankfurt a. M.: S. Fischer, 2001), 62. This numerical estimation was often quoted and accepted as authoritative both in Germany and abroad; see for example Leon F. Whitney, *The Case for Sterilization* (New York, NY: Frederick A. Stokes, 1934), 137.

[19] See Otmar Frhr. von Verschuer, "Vom Umfang der erblichen Belastung im deutschen Volke," *ARGB*, 24 (1930): 238–268, and the reproduction of his final figures in Rainer Fetscher, "Die Sterilisierung aus eugenischen Gründen," *Zeitschrift für die gesamte Strafrechtswissenschaft* 52, no. 1 (1932): 404–424. Fritz Lenz estimated in 1931 that of the entire 65 million German citizens, 1 million were feeble-minded, another million mentally ill, 100,000 epileptic and several millions psychopathic; all in all, more than 6 million were mentally damaged (*geistig nicht vollwertig*) and a further 6 million physically weak (*körperlich schwach oder siech*); of these, only 10,000 were presumably blind and 15,000 deaf-mute. See Fritz Lenz, "Praktische Rassenhygiene," *BFL* (1931), 272–273.

that sterilization was not expected to reduce.[20] Supporters of eugenic sterilization were well aware of these figures. In the second edition of the official commentary to the Sterilization Law, the authors – who were also the ones responsible for drafting the law itself – admitted with respect to Huntington's chorea that "this disease plays almost no role in the average population."[21] Six days after the Sterilization Law became effective, the psychiatrist Robert Gaupp noted that "[i]n their prevalence and significance, 'hereditary blindness' and 'hereditary muteness' … lie well behind the psychic diseases named in [clauses] 1–4 and [behind] severe alcoholism."[22]

What was it, then, about these three medical conditions that made them attractive for Nazi legislators? What benefit did their specification harbor for the new Nazi regime? Why was it useful to name them and not be content with applying a generalized category of "severe physical or mental damage," as previous legislation bills had done?

Definite answers to these question are hard to pin down, because relevant archival sources are missing at a crucial point.[23] But a provisional answer may be achieved by noting the one thing that these three

---

[20] For example, those regarded as hereditary blind constituted no more than 10%–15% of the number of blind in general; and the financial cost they engendered lagged far behind the amounts spent on supporting the "mentally deprived." See the calculations published in 1937 and reproduced in Christine C. Makowski, *Eugenik, Sterilisationspolitik, "Euthanasie" und Bevölkerungspolitik in der nationalsozialistischen Parteipresse* (Husum: Matthiesen, 1996), 178, where it was estimated that 5,000 out of a total of 33,000 blind were hereditary blind, and that their yearly cost in 1937 was 5 million marks out of a total of 301 million marks spent on caring for the hereditary burdened. The condition of 20,000 of 40,000 deaf-mutes was deemed hereditary and cost 15 million marks a year; these sums did not include an additional 200 million marks spent on caring for alcoholics and "psychopaths." This was negligible compared with the purported total costs allocated for caring for the hereditary ill in general: 1.2 billion marks a year.

[21] Gütt et al., *Zur Verhütung*, 115.

[22] [Robert] Gaupp, "Das Gesetz zur Verhütung erbkranken Nachwuchses und die Psychiatrie," *Klinische Wochenschrift* 13, no. 1 (1934): 1–4 (quote from p. 1).

[23] We do not know, for example, if the legislators integrated into the law whatever advice they received from members of the second working group of the Expert Committee for Population and Racial policy (*Sachverständigenbeirat für Bevölkerungs- und Rassenpolitik*), convened precisely for that purpose by Interior Minister Wilhelm Frick. We also have little information on the discussions that the legislators held during the two weeks that separated the first meeting of the Expert Committee and the announcement of the law. For an attempt to reconstruct the processes leading to the formulation of the law based on detailed documentation and analysis of the available sources, see Udo Benzenhöfer, *Zur Genese des Gesetzes zur Verhütung erbkranken Nachwuchses* (Münster: Klemm u. Oelschläger, 2006). In English, cf. Proctor, *Racial Hygiene*; Gisela Bock, "Nazi Sterilization and Reproductive Policies," in Susan Bachrach and Dieter Kuntz (eds.), *Deadly Medicine: Creating the Master Race* (Washington, DC: United States Holocaust Memorial Museum, 2004), 61–87.

conditions had in common – aside from their quantitative marginality: the three of them were regarded as indubitable "Mendelian diseases." Long before 1933, blindness, deafness and Huntington's chorea became standardized as fundamental examples of Mendelian processes among humans and, in this capacity, were repeatedly referred to in both professional and popular writings on biology and heredity. They became pivotal to the promotion of Mendelian research in the scientific communities of geneticists, physicians and psychiatrists, and were therefore also regularly mobilized by psychiatrists and eugenicists to prove that diseases in general and, by implication, mental illnesses as well were genetically determined. Later, it was this special status in Mendelian research that made these disease categories useful for propagating eugenic causes and for presenting the sterilization campaign as grounded in biological knowledge, and therefore scientifically sound.

Eye diseases caught the attention of Mendelian scholars as early as 1905, when English physician Edward Nettleship began publishing his studies on cataract, glaucoma and retinitis pigmentosa.[24] Fascination with eye diseases stemmed from their unique hereditary character: healthy mothers gave birth to affected sons and healthy daughters, who again produced affected (grand)sons and healthy (grand)daughters. This curious hereditary pattern was soon recognized as the third basic category of Mendelian teaching after simple dominance and simple recessiveness, and one especially captivating for heredity researchers: sex-linked inheritance. Not before long, the relatively innocuous phenomenon of color-blindness became a favorite illustration of Mendelian mechanisms in humans.[25] Rüdin's 1911 treatise reveals how deep the impact of this last category was on Rüdin's own scientific thinking. A true science, Rüdin explained there, allowed

---

[24] Eduard Nettleship, "On Heredity in the Various Forms of Cataract," *The Royal London Ophthalmic Hospital Reports* 16 (1905): 1; William Bateson, "An Address on Mendelian Heredity and Its Application to Man (Delivered before the Neurological Society of London, on Thursday, February 1st, 1906)," *British Medical Journal* 2, no. 2376 (1906): 61–67; Gertrude C. Davenport and Charles B. Davenport, "Heredity of Eye Color in Man," *Science* 26 (1907): 589–592; Eduard Nettleship, "A History of Congenital Stationary Night Blindness in Nine Consecutive Generations," *Transactions of the Ophthalmological Societies of the United Kingdom* 27 (1907), 269–293; Nettleship, *The Bowman Lecture on Some Hereditary Diseases of the Eye* (London: Adlard, 1909).

[25] Adolf Steiger, "Über die Bedeutung von Augenuntersuchungen für die Vererbungsforschung," *ARGB* 5, no. 5/6 (1908): 623–634; Alfred Vogt and Richard Klainguti, "Weitere Untersuchungen über die Entstehung der Rotgrünblindheit beim Weibe," *ARGB* 14, no. 2 (1922): 129–140; Adolf Franceschetti, "Die Vererbung von Augenleiden," in Franz Schieck and Arthur Brückner (eds.), *Kurzes Handbuch der Ophthalmologie* (Berlin: Springer, 1930), 631–855; *BFL* (1927), 178, 212.

for no exceptions: superficial irregularities only reinforced a rule that was momentarily unknown to the investigator, but would eventually be revealed. Color-blindness was the example, or proof, for this assertion: at first it appeared to follow no hereditary pattern, certainly not a Mendelian one; but further research revealed it was inherited in a Mendelian sex-linked manner. Thus, a thorough acquaintance with the Mendelian rules and their implications could demonstrate how even exceptional phenomena were eventually determined by biological mechanisms. And vice versa: Mendelian laws formed the ultimate focal point from which all the yet-unknown mechanisms of heredity were destined to emanate.[26]

Huntington's chorea attracted the attention of William Bateson, the leading proponent of Mendelism in Britain, during 1906. Initially, Bateson did not think that Huntington's disease was Mendelian, but by 1908 he had changed his mind.[27] Before the first decade of the twentieth century came to a close, this disease became an internationally recognized, indisputable example of simple dominance in humans.[28] Deaf-mutism followed immediately, and was shown to conform to a simple recessive hereditary mode.[29] Together with polydactyly (the growth of additional fingers/toes) and hemophilia – both also proven to follow Mendelian regularities – this accumulation of pathologies strengthened the notion that Mendelism provided the ultimate key for studying, and combating, degenerative phenomena in humans.[30]

As soon as they became standardized as Mendelian diseases, color-blindness, deaf-mutism and Huntington's chorea (along with polydactyly and hemophilia) were mobilized repeatedly in eugenic and genetic text-books as unambiguous demonstrations of the applicability of Mendelian

[26] Ernst Rüdin, "Einige Wege und Ziele der Familienforschung, mit Rücksicht auf die Psychiatrie," *ZgNP* 7, no. 1 (1911): 487–585, here 508.
[27] See Alice Wexler, *The Woman Who Walked into the Sea: Huntington's and the Making of a Genetic Disease* (New Haven, CT: Yale University Press, 2008), 129–131.
[28] See R[eginald] C. Punnett, "Mendelism in Relation to Disease," *Proceedings, Royal Society of Medicine* 1 (1908): 135–168, 144; William Bateson, *Mendel's Principles of Heredity* (Cambridge: Cambridge University Press, 1909), 229; Charles B. Davenport, "Huntington's Chorea in Relation to Heredity and Eugenics," *Proceedings of the National Academy of Sciences of the United States of America* 1, no. 5 (1915): 283–285.
[29] Viktor Hammerschlag, "Zur Kenntnis der hereditär-degenerativen Taubstummheit," *Zeitschrift für Ohrenheilkunde* 61 (1910), 225–253; Hermann Lundborg, "Über die Erblichkeitsverhältnisse der konstitutionellen (hereditären) Taubstummheit und einige Worte über die Bedeutung der Erblichkeitsforschung für die Krankheitslehre," *ARGB* 9, no. 2 (1912): 133–149.
[30] On hemophilia's status in hereditary thinking, even prior to the rise of Mendelism, see Stephen Pemberton, *The Bleeding Disease: Hemophilia and the Unintended Consequences of Medical Progress* (Baltimore, MD: Johns Hopkins University Press, 2011), ch. 1.

theory to human traits in general.[31] The more psychiatrists struggled with finding Mendelian models for psychiatric conditions, the more they turned to these diseases to prove that their eugenic aspirations were sound. It was not the overall medical danger that these diseases represented, but their utility as "scientific exemplars" that made eugenicists refer to them in their talks – but not (yet) in their sterilization proposals. As Jena Professor O. Abel admitted when speaking at a meeting of the Reich's Health Council in 1920, "[s]pecific deviations from health, which have become so important for the study of hereditary relations in humans, such as hemophilia, night-blindness, polydactyly and many others, are probably occurrences too rare to require separate consideration in the present context."[32]

The meeting at which Abel spoke was convened to discuss a law that would require prospective marriage partners to exchange their health certificates before commencing marriage procedures. This was an early precursor of the later Sterilization Law, for which Mendelian diseases, at least in Abel's view, were not yet seen as relevant, despite their "importance for the study of human heredity." A year later, however, in a meeting held at the Prussian Ministry for People's Welfare to discuss the "eugenic indication" for sterilization, gynecologist Dr. Max Hirsch demonstrated how Mendelian diseases and psychic disorders could be welded together. He proudly announced that "we know that the most important psychic diseases such as dementia praecox, epilepsy, [and] feeble-mindedness are inherited recessively in the Mendelian sense. [...] Similarly with atrophy of the optic nerve, retinitis pigmentosa, deaf-mutism, familial juvenile glaucoma, [and] hemophilia."[33]

The fact that schizophrenia could be regarded as dihybrid recessive (see Chapter 2) was mobilized for the same rhetorical purposes. Speaking before the Bavarian County Assembly in 1931, Dr. Friedrich (Fritz) Ast, director of an institution for the mentally ill, described the efforts made by Rüdin to identify the hereditary mechanisms of mental illnesses, for example, of schizophrenia. "The results of his research seemed to provide the proof that schizophrenia, just as all other hereditarily determined

[31] See Erwin Baur, *Einführung in die experimentelle Vererbungslehre* (Berlin: Gebrüder Borntraeger, 1911), 154; Carl Chun and Wilhelm Johannsen, *Allgemeine Biologie* (Leipzig/Berlin: B. G. Teubner, 1915), 640; *BFL* (1927), 428; Hermann Werner Siemens, *Grundzüge der Vererbungslehre, Rassenhygiene und Bevölkerungspolitik* (Munich: J. F. Lehmanns, 1930), 28, 43.

[32] BArch R86/2372, fols. 184–216, esp. fol. 199 (p. 16 of the original document). See also Berliner Gesellschaft für Rassenhygiene, *Über den gesetzlichen Austausch von Gesundheitszeugnissen vor der Eheschließung und rassenhygienische Eheverbote* (Munich: J. F. Lehmanns, 1917).

[33] BArch R86/2371, fols. 153–164 (quote from fol. 160).

diseases and traits, is inherited according to the biological hereditary laws whose discovery we owe to the Augustinian priest Mendel." Ast gave a short exposition on Mendel's laws and clarified such key terms as homo- zygosity, heterozygosity, dominance and recessiveness. "All findings speak for the fact that the hereditary trait for schizophrenia is recessive," he claimed, and proceeded to survey other mental diseases and their hereditary dangers. At the same time, Ast also conceded that "we are not interested in the rare diseases that are studied the best, but in the frequent kinds [of diseases] that fill our asylums, that is, schizophrenia or dementia praecox … , manic-depressive insanity, epileptic diseases, feeble-mindedness and asocial and criminal psychopathy."[34]

Unlike Ast, the Nazi legislators opted to include in the law "the rare diseases that are studied best," and to use them to present "the frequent kinds that fill our asylums" as genetically grounded and scientifically understood. The official commentary to the Nazi Sterilization Law, written jointly by Rüdin, the Director at the Interior Ministry Arthur Gütt and the jurist Falk Ruttke, skillfully weaved together Mendelian diseases with the necessity of applying eugenic measures. With thousands of copies published in 1934 alone, the commentary begins with detailed explanations of Mendel's laws and their relevance for humans.[35] A handful of neatly devised charts comprise entire pages to illustrate in a clear and precise manner the various mechanisms of Mendelian hered- ity. But these charts are not only hypothetical constructs: for every kind of hereditary pattern and for every theoretical chart, an actual pedigree of a real disease inherited in the same manner is attached. Huntington's chorea helps to exemplify dominant inheritance; deaf-mutism demon- strates recessive inheritance; and a pedigree of glaucoma is used for illustrating sex-linked inheritance. It is thereby not only shown that the rules of Mendelian heredity really apply to human diseases; a firm rela- tion between Mendel's laws and the Sterilization Law is also established.[36]

The introduction to the commentary does not end there. The authors move to discuss at great length the heritability patterns of traits deter- mined by two hereditary factors. They then explain that "there is much evidence that, for example, schizophrenia is based on a pair of factors, a dihybrid hereditary mode, although a satisfactory, comprehensive proof

[34] BArch R 86/5631 fols. 14–15.
[35] According to the contract between the authors and Lehmanns publishing house (accessible in MPIP-GDA 59), the first edition was to be printed in 3,000 copies, but it was professed that the eventual number of copies needed may reach 40,000. Already in 1936, a new and revised edition had been published.
[36] Gütt et al., *Zur Verhütung*, 13–35.

[for that] is difficult to obtain."[37] This latter difficulty is portrayed as stemming primarily from computational complexities and in no way undermines the validity of Mendelian suppositions. As the authors see it, even if one could not determine for certain whether a particular disorder was inherited, "simple recessive or sex-linked recessive, dominant or recessive monohybrid or polyhybrid, recessive and dominant mixed, or pure recessive," the law was still valid, "as long as it is generally clear that some kind of heredity must underlie it."[38] Having established that Mendelian factors always lie in the background, the authors now move to discuss epilepsy and manic-depressive insanity, both of which seem to follow some kind of Mendelian mechanisms, even if precisely which type is not clear yet.

Rüdin used an almost identical line of justification on many other occasions. For example, there is a striking resemblance, in both wording and argumentative structure, between the commentary's introduction and a draft Rüdin composed circa 1925, at the request of the Prussian Ministry of Public Welfare (*Preußisches Ministerium für Volkswohlfahrt*), on the "Eugenic Indication for the Termination of Pregnancy in the Field of Psychiatry."[39] The same pattern of justification repeated itself elsewhere. For example, in November 1933, a two-day professional training session was held for judges who were to become members of Hereditary Courts. It used a similar strategy: it opened with a lecture on Mendelian heredity and dominance and recessive mechanisms in humans, explained "in simple, easy-to-grasp modes of representation and in a succinct manner." Immediately after that came a lecture on "clinical manifestations and [their] meaning for hereditary factors of the diseases listed in the Law for the Prevention of Hereditarily Diseased Offspring."[40] This way, Mendelism was presented as the foundation on which eugenic policy stood. As we will see in the next section, indoctrination in high schools worked similarly, linking Mendelism to the law by making repeated references to the Mendelian diseases mentioned in it as if they were representative of the larger category of mental diseases. The slide projected to the audience of the play *Hereditary Stream* on the

---

[37] Ibid., 35–36.   [38] Ibid., 46.

[39] See MPIP-GDA 54. Among other similarities, the draft contained the statement, "In the field of psychiatry, a conclusive proof for most dispositions is however still missing, [regarding] which specific hereditary mode they follow, whether recessive or dominant, a sex-linked mode, recessive or dominant or mixed dihybrid, or any other polymeric [mode]." The second part of the sentence, specifying possible Mendelian patterns, was erased in the 1925 draft, only to resurface with slight modifications eight years later, in the text of the commentary.

[40] BayHStA MJU 18650 (GzVeN), and Staatsarchiv Munich Oberlandesgericht [hereafter OLG] 607.

recessive nature of myoclonus epilepsy (see Figure 4.1) and the film on rabbits' nervous malfunctions (see Figure 4.2) also used the same manner of legitimizing and justifying Nazi sterilization policy.

Was the list of diseases formulated intentionally to serve the propagandistic purposes of lending the law a Mendelian outlook? Because of lack of sources, we will probably never know.[41] But by 1933, "Mendelizing" the law by including classically recognized Mendelian diseases was a conventional rhetorical move, and might have emerged naturally during the exchange between Gütt, Rüdin and others. Obviously, these specific medical conditions would not have been chosen unless they represented some form of damaged hereditary endowment; but neither would they have been chosen as independent categories had they not brought with them a dowry in the form of Mendelian legitimization.

Some of the intentions of the Nazi legislators become apparent when we compare the Nazi law with Laughlin's aforementioned Model Law – the only one that contained, prior to 1933, a detailed list of medical conditions justifying sterilization. Notably, the Germans split Laughlin's category of "insanity" into two separate categories; they abstained from mentioning criminality, dependency or infectious diseases; and they presented alcoholism as an addition to, and not an integral part of, their list. They also added Huntington's chorea (Table 4.1). Revealingly, there is one common denominator to all of these modifications: they all testify to the Nazis' deliberate attempt to "medicalize" their law. The Nazi legislators avoided using what was seen by contemporaries as explicitly socially determined categories and favored naming illnesses that (by the standards of the time) could be regarded as hereditarily determined medical conditions. The legislators themselves revealed later that "the law intentionally limits itself to those diseases of which the hereditary mechanism is sufficiently scientifically researched."[42] Replying to critics who complained about the exclusion of criminality, they argued that "the relevant authorities refrained from including criminal dispositions so as not to depart from the pure medical foundation of the law."[43] A law against "habitual criminals," which legitimized their sterilization and

---

[41] According to Gisela Bock, Gütt, Rüdin and Ruttke worked simultaneously on the law and the commentary to it. If that is true, that could support the idea that the law was devised with constant reference to its method of justification. But I could find no corroboration for Bock's assertion. The contract between Gütt, Rüdin and Ruttke and Lehmanns publishing house was signed in November 1933, four months after the law had been legislated. Gisela Bock, *Zwangssterilisation im Nationalsozialismus* (Opladen: Westdeutscher Verlag, 1986), 84; Christian Ganssmüller, *Die Erbgesundheitspolitik des Dritten Reiches: Planung, Durchführung und Durchsetzung* (Cologne: Böhlau, 1987), 64; MPIP-GDA 59.

[42] Gütt et al., *Zur Verhütung*, 78.   [43] Ibid., 60.

Table 4.1 *A comparison of the categories designated for sterilization in Laughlin's 1922 Model Law and the Nazi Sterilization Law.*

Laughlin's 1922 "Model Law"	The 1933 Nazi Sterilization Law
1. Feeble-minded	1. Congenital feeble-mindedness
2. Insane (including the psychopathic)	2. Schizophrenia
	3. Manic-depressive insanity
3. Criminalistic (including the delinquent and wayward)	Omitted (but legislated separately)
4. Epileptic	4. Hereditary epilepsy
5. Inebriate (including drug habitués)	(Relegated to the end). Furthermore, anyone suffering from severe alcoholism can be made infertile.
6. Diseased (including the tuberculous, the syphilitic, the leprous and others with chronic, infectious and legally segregable diseases)	Omitted. Instead: 5. Huntington's chorea
7. Blind (including those with seriously impaired vision)	6. Hereditary blindness
8. Deaf (including those with seriously impaired hearing)	7. Hereditary deafness
9. Deformed (including the crippled)	8. Severe hereditary physical deformity
10. Dependent (including orphans, ne'er-do-wells, the homeless, tramps and paupers)	Omitted

castration, came into force together with the Sterilization Law – but was legislated separately.[44] Shaping a crucial building block of future eugenic policy, Nazi lawmakers wanted to ensure that their legislation would be as biologically grounded and as scientifically respectable as possible. With the help of several widely recognized Mendelian diseases, that goal could be achieved much more easily.

An episode that took place while the law was being prepared will serve as a final illustration for the eugenic standpoint on the entire affair. In April 1933 the jurist Rudolf Kraemer, himself blind, expressed in writing his fierce objection to the possibility of a compulsory sterilization law of blind persons. Kraemer's arguments against the possible targeting of the blind were based on a variety of moral, social, economic, historical and scientific grounds, which questioned the necessity and professed

---

[44] For more on the administrative and personal dynamics that led to the exclusion of criminals from the law, see Richard F. Wetzell, *Inventing the Criminal: A History of German Criminology 1880–1945* (Chapel Hill: University of North Carolina Press, 2000), 255ff.

utility of sterilization both within and outside the framework of eugenic reasoning.[45] Kraemer sent his reservations to the Reich Ministry of the Interior, where it was forwarded to the Director of the Hygiene Institute of the Friedrich-Wilhelm University of Berlin, Heinrich (Heinz) Zeiss. In his professional assessment of Kraemer's arguments, Zeiss wrote that "it is advisable to verify the statements of Kraemer in the drafting of the sterilization law and to use them as [raw] material" (probably referring to the statistical data that Kraemer provided on the distribution of hereditary and non-hereditary blindness).[46]

But Kraemer also published his small treatise, a publication that was promptly countered by Otmar Freiherr von Verschuer, who was by that time the Head of the Human Heredity Department in Eugen Fischer's Kaiser-Wilhlem Institute. Verschuer's response was very different from that of Zeiss. He did not bother to answer Kraemer's doubts regarding the alleged economic, social or eugenic burden imposed by the blind on society. For Verschuer, the crucial thing about blindness was not that it was – or wasn't – hampering society, but that it was hereditary; and it was to this aspect that he devoted most of his counterattack.[47] To strengthen his case, Verschuer provided two illustrations. These were neither pedigrees of families overburdened by blindness nor graphs visualizing the costs of blind welfare, but basic schemes of Mendelian heredity (Figure 4.3). To Kraemer's argument that blindness should not be seen as a disease, Verschuer's reply was simple: it is nevertheless Mendelian. Expressing doubt on the utility of sterilization when the mode of inheritance was perfectly understood was, in Verschuer's eyes, simply nonsensical.

Teaching Mendel, Explaining Eugenics:
A View from the Classroom

Public education provided the Nazi state an opportunity to propagate its worldview to a captive audience. Already during the Weimar years,

---

[45] Rudolf Kraemer, *Kritik der Eugenik, vom Standpunkt des Betroffenen* (Berlin: Reichsdeutscher Blindenverband, 1933).

[46] BArch R1501/126248, fol. 42 (Kraemer's booklet) and 263–265 (Zeiss' review).

[47] Otmar Frhr von Verschuer, *Blindheit und Eugenik* (Berlin: Reichsdeutscher Blindenverband, 1933). Kraemer was most likely unpopular among eugenicists after he published, several months earlier, a short satirical piece mocking the "world-known eugenicists, professors Lenzmann, Mucker and Fischbaur" (a pun on the names of Lenz, Muckermann, Fischer and Baur) who, purportedly for the benefit of the blind, managed to cross-breed a seeing-eye dog and a mare. See Rudolf Kraemer, "Neujahrs-Zauber, 2. Bild: Das Blindenpferd, der Triumph der angewandten Vererbungswissenschaft," *Die Blindenwelt* 21, no. 1 (1933): 7–11.

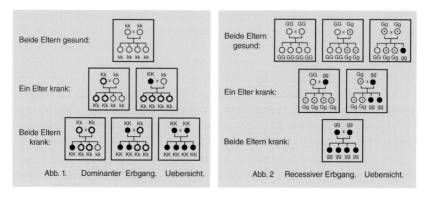

Figure 4.3 Mendelian heredity explained as a background/justification for sterilizing the blind (reproduction).
Source: Otmar Freiherr von Verschuer, *Blindheit und Eugenik* (Berlin: Reichsdeutscher Blindenverband, 1933), 6–7.

biology educators advanced the view that biology instruction was intimately linked with the cultivation of love for one's homeland, the strengthening of emotional ties between students and their natural surroundings and the development of students' consciousness that they were part of a larger, organic whole – their nation.[48] Heredity was part of the biology curriculum in the higher grades, and Mendelism was central to the teaching of heredity, as any textbook designated for primary or high school from the time makes plain.[49] To an extent, this centrality of Mendelism for teaching heredity was trivial and almost inevitable. What is less self-evident is how German teachers explained to their students the nature of the relation between the Mendelian understanding of heredity and Nazi eugenic policy. This section will reconstruct the significance of Mendelism for eugenic teaching not only by exploring textbooks and curricula but also by analyzing a unique

---

[48] Sheila F. Weiss, "Pedagogy, Professionalism and Politics: Biology Instruction during the Third Reich," in Monika Renneberg and Mark Walker (eds.), *Science, Technology and National Socialism* (Cambridge: Cambridge University Press, 1994), 184–196; see also Weiss, *The Nazi Symbiosis* (Chicago, IL: University of Chicago Press, 2010), ch. 6.

[49] For example, see Siemens, *Grundzüge der Vererbungslehre*; Otto Steche, *Leitfaden der Rassenkunde und Vererbungslehre, der Erbgesundheitspflege und Familienkunde für die Mittelstufe* (Leipzig: Quelle & Meyer, 1934); Hans Feldkampf, *Vererbungslehre, Rassenkunde, Volkspflege* (Münster: Aschendorff, 1935); Karl Bareth and Alfred Vogel, *Erblehre und Rassenkunde für die Grund- und Hauptschule* (Bühl-Baden: Konkordia, 1937). The number of textbooks and supplementary materials for biology instruction from the period is enormous; see Harten et al., *Rassenhygiene als Erziehungsideologie*.

collection of reports by prospective teachers on their experiences in high schools. These reports, which document both teachers' pedagogic intentions as well as students' reactions and objections, make it possible to paint a rich picture of the dynamics of biological indoctrination, and of the role that Mendelism acquired in the process.

The first years of the Third Reich saw an exponential growth in publications on the teaching of heredity, its methods and its aims. The demand for such publications was high, and school curricula had to be adjusted to the demands of the new regime. On June 19, 1933, the Interior Ministry circulated a preliminary draft regarding a plan of biological teaching in various German lands under the title "The Place of Biology in the New German school." On its second page, the circular clarified: "Hereditary teaching is never a goal in its own right in school. It is always only the preschool for understanding its larger areas of application in human society, [that is,] racial hygiene (eugenic)." Scientific teaching, the circular emphasized, was to be fully subjected to its practical use; it was not supposed to prepare students for biological research but give the next generation of statesmen, politicians, jurists, clerks, teachers and doctors, as well as those who would leave school at an early stage, tools for life.[50] In September, an official decree from the Prussian Ministry of Culture ordered that biology be taught two or three hours a week in secondary and primary schools and that genetics, eugenics and racial science be part of biological instruction.[51]

When it came to the teaching of heredity, one issue that had to be clarified from the outset was Mendel's national identity. Was he German? Austrian? Czechoslovakian? His name could even imply – god forbid – Jewish ancestry. In 1922, Lenz found it necessary to reassure his readers that both Mendel's outlook, as well as historical data on migrations in the area, substantiated the assumption of Mendel's primarily Nordic racial origins. Furthermore, "the name Mendel suggests Jewish origins, but this is certainly not the case. The name would be written in earlier generations as 'Mandel,' which in the Bavarian dialect designates a small man."[52] Hugo Iltis, Mendel's first biographer, also addressed this issue. Despite the fact that Czechs and Slavs resided for centuries where Mendel was born, Iltis explained that in the same area there was good reason to suppose that "continuous infusion of fresh German blood, together with occasional crossing with Slav neighbours, must have

[50] Yad-Vashem Archives [hereafter YVA] JM22411, files 0419–0427, originally BArch R1501/126852.
[51] Weiss, "Pedagogy, Professionalism and Politics," 188.
[52] Fritz Lenz, "Gregor Mendel," *Münchener medizinische Wochenschrift* 69 (1922): 1349–1350.

helped to save the [local inhabitants] from a racial degeneration." The adoption of the name Mendel by Jewish families, Iltis explained, anteceded its use by Gregor Mendel's forefathers and was accordingly irrelevant.[53]

These concerns also found their way into pedagogic materials. In a small sixty-page booklet published in 1934, Mendelian inheritance was explained to the young readers through an imaginary conversation between a father and his children. Toward the end of the booklet, the daughter, Magdalena, becomes concerned that Mendel was not German. The father reassures her that in fact he was: "My loved child, German are all the men whose hereditary mass stems from the German *Volk*," regardless of where they live; Gregor Mendel was therefore a son of the German *Volk*. The booklet ends with a poem glorifying Mendel and his work:

> Was Mendel als Forscher erdacht und gefunden,
> Die Wissenschaft setzt ihm den Ehrenstein;
> Doch wie als Mensch er gedacht und empfunden,
> das wissen nur, die ihn kannten, allein.
>
> Und wenn Eure Kinder Euch staunend fragen,
> Was ist's mit dem Manne, was hat er getan?
> Dann könnt Ihr ihnen herzensfroh sagen:
> Er war ein echter, ein deutscher, ein ganzer Mann![54]

Roughly:

> For what Mendel has conceived and discovered as a researcher,
> Science erects him a monument;
> But how he thought and felt as a man,
> only those who knew him can tell.
>
> And whenever your children ask you, amazed,
> What about this man, what has he done?
> Then you can say to them with delight:
> He was a genuine, a German, a real man![55]

Another strategy of claiming ownership of Mendel's teaching was by emphasizing the importance of Carl Correns, and marginalizing the role

---

[53] Hugo Iltis, *Gregor Johann Mendel: Leben, Werk und Wirkung* (Berlin: Springer, 1924), 3–4; in English: Hugo Iltis, *Life of Mendel*, trans. Eden and Cedar Paul (London: George Allen & Unwin, 1932), 22–23. Iltis himself was of Jewish origin, and in later years entered a fierce debate with Lenz on the scientific status of Nordic racial teaching. See below p. 225, fn. 55.

[54] Hans Bartmann, "Das Erbe der Väter: Eine kleine Erblehre," in Otto Rabes (ed.), *Mutter Natur*, 16 (Berlin/Leipzig: Julius Beltz, 1934), 59–60.

[55] I thank Dr. Joachim Warmbold for his helpful advice with this translation.

of Erich Tschermak and Hugo de Vries, in the discovery of Mendel's laws. Teachers made continuous references to Correns' experiments in the classroom, and corrected students who attributed these experiments to Mendel himself.[56] This strategy received its clearest expression in anthropologist Hans Weinert's 1943 *Biological Foundations for Racial Science and Racial Hygiene.* Carl Correns, Weinert explained, was the real founder of modern heredity science, and he was the man who was responsible for the name "Mendelian laws," which were in reality his – that is, Correns' – laws of heredity (this view was not unfounded: see Chapter 1). "And on these laws, our racial-hygiene is based!"[57]

Usually, however, it was Mendel himself who stood at the center of the teaching of heredity. Quite a few biology teachers consciously decided to teach heredity in the tenth grade (*Untersekunda*) by explaining Mendelian mechanisms without making any reference at all to chromosomes: as one of them saw it, "When teaching, the chromosome theory of heredity can be disposed with."[58] This pedagogic strategy was propagated by Philipp Depdolla, an esteemed writer on biology instruction, who recommended teaching Mendelian principles in the tenth grade but explicating cellular mechanisms only in the eleventh grade. Pupils could thereby appreciate the wonder of the fit between Mendel's projections and the later discoveries of chromosomal behavior; they could also understand the role of hypotheses in the development of scientific ideas.[59] However, explaining Mendelian theory as abstract regularities sometimes caused difficulties: one teacher-trainee claimed that students who did not learn about cells and chromosomes experienced the entire teaching section as "float[ing] in the empty space;" Mendelian laws appeared to her female students like an arbitrary calculation game. The teacher-trainee's evaluator was unsatisfied with this last remark and reminded her that Mendel's laws were discovered without any knowledge of chromosomes.[60]

One way or the other, teaching Mendelism required that teachers discuss with their students the emergence of statistical regularities. This challenging pedagogic task encouraged them to develop innovative

---

[56] See the biology exams in Landesarchiv Berlin [hereafter LAB], A Rep. 020-31 Nr. 60.

[57] Hans Weinert, *Biologische Grundlagen für Rassenkunde und Rassenhygiene* (Stuttgart: Ferdinand Enke, 1943), 83.

[58] Deutsches Institut für internationale pädagogische Forschung, Bibliothek für bildungsgeschichtliche Forschung [hereafter DIPF/BBF/Archiv]: Gut Ass 1464 (Vererbungslehre und Erbgesundheitspflege im Biologieunterricht der Untersekunda, 1935), 4.

[59] Philipp Depdolla, "Vererbungslehre und naturwissenschaftlicher Unterricht," in Günther Just (ed.), *Vererbung und Erziehung* (Berlin: Springer, 1930), 277–303.

[60] DIPF/BBF/Archiv Gut Ass 1201 (Die Vererbungslehre im Unterricht der Oberprima und Untersekunda. Oberlyzeum und Lyzem, 1934), 25–26.

methods, which often involved actively engaging the students with the demonstration of Mendelian mechanisms. Performing actual crossing experiments with flowers or fruit flies was highly recommended, but not always feasible. A popular substitute was drawing paper slips, or beans, or marbles, colored in black/red and white, from a bowl – the "drawing lots out of an urn," cherished by many statisticians – to represent different possibilities of gametic union during sexual reproduction. In one girls' high school classroom, the entire hereditary process turned into a small dramatic play, later described in detail by their teacher:

One student was called to come to the front. She sat on the table. She was to illustrate a red-flowering plant. What about her hereditary factors? – "What kind did you receive from your father?" – "for red coloring." The student received a red cardboard disk in her hand. – "And what kind of hereditary disposition from your mother?" – "for red coloring." The student received a red cardboard disk also in her other hand.

Another student now joined, representing the pure-bred white flower. "Now let's cross." A student "of smaller bodily size" stood in front of the table and received one cardboard disk from each of her classmates. A fourth student came forward to similarly represent the F1 generation. A stool was now laid in front of the standing students, on which a fifth student sat down, to represent the next hybrid generation, the F2. Mendelian hereditary transmission was now fully illustrated. The F1-generation student passed her red disk to the F2 student but then said: "But I can also pass the factor for white color."

In this particular classroom, Mendel's laws turned from abstract theory into a real performative act. The teacher summarized the experience, which aroused many laughs, by noting that "what is fun for the students, impresses itself better." After the exercise, the students testified that they better understood Mendelian regularities, and one student regretted only one thing: "it is a shame, that we didn't have a camera."[61]

Another teacher did have a camera, as well as some handicraft talent: Figure 4.4 shows the large wooden board, 60 cm high and 45 cm wide, decorated with cardboard cutouts, he prepared in order to illustrate Mendelian mechanisms to his classroom.[62] With the help of such methods, Mendelism could become palpable to the students – indeed, much more palpable than the hypothetical ideas of Darwinian evolution, which could not be demonstrated in class and for which no experimental corroborations yet existed.

[61] DIPF/BBF/Archiv Gut Ass 1400 (Erb- und Rassenkunde in der Untersekunda eines Oberlyzeums, 1935), 14–16.
[62] DIPF/BBF/Archiv Gut Ass 1464.

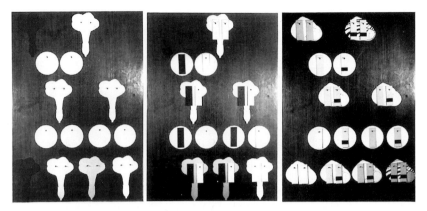

Figure 4.4 Three photos of a wooden board (60 × 45cm^2) with cardboard cutouts prepared especially for teaching Mendelian mechanisms in German high school by Rudolf Wagner, a creative teacher-trainee, 1935
Source: DIPF/BBF/Archiv: Gut Ass 1464, 18, 22, 25

Nevertheless, some more difficulties arose once Mendelian heredity was taught as a probabilistic game of chance. In one Berlin high school for girls, the teacher reported that after she charted on the blackboard several pedigrees of dominant and recessive heredity of diseases, "[t]he students were not entirely satisfied. They told me: 'what's the point of all this? One can never detect recessive inheritance before two ill hereditary factors come together. In the case of dominant inheritance the children can also be healthy, because the numerical relations are anyhow not exact. We can never say anything certain about the offspring!'"[63] The students, who were 16 years old (*Untersekunda*), found it difficult to perceive things that did not follow clear rules (or so thought the teacher); and since they did not master probabilistic computations, the teacher considered it was best to simply let them view as many pedigrees as possible. One thing was crucial, in her opinion: "The students must be convinced that the Mendelian laws apply also for human heredity and that their numbers will be reached on average."[64]

Interestingly, different reports on the teaching of heredity in German high schools indicate that at one point or another, the incessant propaganda and magnification of the dangers of heredity succeeded in creating fear among the students, who consequently tried to object to some of their teachers' lessons. Resignation and insecurity stemming from the

---

[63] DIPF/BBF/Archiv Gut Ass 1201, 33.   [64] Ibid, 45–46.

Figure 4.5 Nazi eugenic propaganda poster, demonstrating what would happen if the hereditarily inferior propagate at double the rate of those considered valuable, 1935.
Source: Staatsarchiv Bamberg, A 241 T 14 006

indeterministic elements of Mendelian calculations were one possible reaction. Full-fledged objection to the teacher's agenda was another. In one classroom, an apprentice teacher asked his students to calculate how differential reproduction rates would ultimately affect the population structure. Assuming that healthy couples produce two children per marriage, while the hereditarily inferior produce four children per marriage, and assuming that the population was composed of 50% valuable and 50% inferior, what would the proportions between the two groups be after three, four or five generations? This mathematical exercise was frequently used and visually portrayed in Nazi propaganda, in films, in textbooks and in newspaper articles. It was considered a simple and effective measure for making palpable the horrifying results of the exponential growth of those hereditarily burdened against the terrifyingly shrinking classes of valuable citizens – bold factory workers, muscular peasants and the educated middle class (Figure 4.5).[65]

---

[65] For example, see Otto Helmut, *Volk in Gefahr* (Munich: J. F. Lehmanns, 1933), 33; Ingo Kaul, *Das Wunder des Lebens: Ausstellung Berlin 1935, 25. März bis 5. Mai* (Berlin:

Initially, this mathematical exercise seemed to work well in the class-room, too: "the students were willingly engaged and followed [me] with eagerness," reported the apprentice teacher. "The last example, the gloomy prospect of the future, aroused an unusually lively participation, in which some doubts were expressed. This proved to me that the students were fully aware of the seriousness of the questions at hand." The students' doubts, however, were quite severe; the questions that they posed practically invalidated the entire scheme. The students began by raising a possibility that not only undercut the simplistic mathematical calculation, but was also heretical from the point of view of Nazi eugenic thinking: what would happen, they asked, if the hereditarily inferior married hereditarily healthy individuals? Further, the students pointed out that "if the calculation is correct, were not the German people supposed to have degenerated already long ago?" Third, they continued to inquire, with relation to the previous questions, "do the hereditarily ill exist only in recent times, or were there always hereditary illnesses, and where do they come from?"[66]

To answer these doubts, the teacher turned to Mendelian reasoning. He answered the first question with the help of Mendelian laws, showing the theoretical results expected from marriages between ill and healthy individuals – assuming, of course, that the illness was caused by a single Mendelian factor. The second and third questions led him to discuss lethal mutations in *Drosophila* (fruit fly), which again framed the entire discussion within the confines of Mendelian thinking. The teacher argued that it was modern human society and modern medicine that led to retaining such lethal mutations and to prioritizing individual welfare over the interests of the *Volk*. But this recent, modern, individu-alistic trend was now – at the end of 1935 – finally reversed.[67]

But simplifying things by referring back to Mendelian mechanisms, as this last apprentice teacher attempted to do, could also create pedagogic problems. In one classroom, the assertion that Mendelian laws also applied to humans was "constantly doubted by the girls. These regular-ities seemed to them to be too simple to be valid also for humans." The teacher reacted by insisting on the sway of Mendelian teaching, referring her students to certain human pedigrees that demonstrated Mendelian regularities.[68] Another teacher in a girls' high school was similarly puzzled: after she had analyzed pedigrees of human traits, gave examples

Meisenbach, 1935); Friedrich Burgdörfer, *Volk ohne Jugend* (Heidelberg/Berlin: Kurt Vowinckel, 1937), 76.
[66] DIPF/BBF/Archiv Gut Ass 1464, 34–35.   [67] Ibid.
[68] DIPF/BBF/Archiv Gut Ass 812 (Die Vererbungslehre in der Untersekunda, 1936), 35.

from the students' families on the inheritance of brown and blue eyes, and even discussed the results of twin studies, she wrote, "I assumed that it would now be taken for granted that the [Mendelian] hereditary regularities apply also to humans." But it wasn't: "doubts were expressed on the validity of heredity laws for humans." Frustrated, the teacher referred the students to a long list of traits whose mode of heredity was perfectly known. She seems to have done that with too much zeal: the students moved from expressing doubts to flatly and unanimously agreeing with everything she said. "Despite my kindly requests that they share with me calmly their thoughts, all the students claimed that they now see the validity of the hereditary laws for humans. Nevertheless I myself was unsatisfied: where were the difficulties [in comprehending that from the very beginning]? I could not extract the answer from the students" – who seemed to have realized the futility of challenging the teacher on this particular issue. The teacher resorted to a simple solution: "probably they didn't know themselves."[69]

But teaching teenagers that the mechanisms of human heredity worked so simply and deterministically, and that heredity was at the same time such a dangerous force, was also problematic for a regime that constantly propagated the need to marry and raise many children. "One has to be careful," wrote one teacher, "not to paint everything in black, because in many circles there already prevails a veritable heredity-anxiety (*Vererbungsfurcht*). ... [O]ne can point emphatically enough at the great dangers that many severe hereditary diseases present to our national health, but it is important to stress again and again that the layman cannot decide whether or not a hereditary disease is involved in a particular case."[70] Nevertheless, the same teacher thought that when it came to feeble-mindedness and mental illnesses, "I find it good if one can influence the girls also in a purely emotional way. Here, in my opinion, one can calmly describe the horrible impression that one gets when visiting an auxiliary school or when walking through an asylum."[71] The students did not have to rely solely on second-hand descriptions: from 1934 onward, schoolchildren throughout Germany regularly attended guided tours in institutions for the "hereditarily inferior" and viewed films that stressed the inhumanity of those who were "mentally defective."[72]

---

[69] DIPF/BBF/Archiv Gut Ass 1400, 26.   [70] DIPF/BBF/Archiv Gut Ass 812, 45–46.
[71] Ibid., 49.
[72] Burleigh, *Death and Deliverance*, ch. 2; Carol Poore, *Disability in Twentieth-Century German Culture* (Ann Arbor: University of Michigan Press, 2007), 96–97; Ulf Schmidt, *Medical Films, Ethics, and Euthanasia in Nazi Germany: The History of Medical Research*

The heredity anxiety described by the teacher was only partially related to the mechanisms of inheritance; in most respects, its cause lay in the actions of the state, not those of the genes. This found expression in one teacher's difficulties in extracting from her female students information about their family histories. She thought it would be wise to begin teaching heredity not with talk about plants and animals but by discussing students' familial traits. She soon realized that "it is not so simple to ask them to draw or talk about their families, because for some of them, a grandmother or granduncle maybe passed away in the past year or two, and the whole thing brings up lots of emotions." For others, she discovered, her assignment was annoying, because its completion required considerable time and effort. She therefore asked them to Simply write "What I Know about My Family," which she thought should not be too troublesome. Several students replied, "We know nothing about our families, we cannot say anything about it." Another student, clearly voicing her parents' frustration over the constant need to supply evidence on ancestral origins to one authority or the other, complained: "The procurement of the documents is too expensive. From the territories given to Poland [in the treaty of Versailles] it is generally fraught with difficulties."

Exasperated, the teacher told the class to refrain from using additional sources and simply supply whatever information was available to them. But the students continued to object, this time claiming that "writing these things in an essay is not nice." In her attempt to explain to herself the students' repeated reluctance to cooperate with their given assignment, the teacher suggested that "many students, precisely at this age, feel the disclosure of familial relations in a school assignment is an indiscretion." She recalled that when she was herself a student, she often felt that her teacher had special curiosity for anything personal from the students' own lives. Oblivious to the fact that her childhood experiences were very different from those of her students when it came to revealing familial information, she was sure that she had finally found the reason for the students' antagonism: "Some remarks in the essays themselves indicate that the pupils are concerned about whether their family will also stand before the judgment of those who read the essay. Otherwise I cannot explain to myself, why in so many essays – and in one of them even twice – it is repeatedly reassured with special emphasis: 'Hereditary

and Teaching Films of the Reich Office for Educational Films/Reich Institute for Films in Science and Education, 1933–1945 (Husum: Matthiesen, 2002).

diseases have never occurred in our family, and most ancestors died of old age.'"[73]

Grief over a dead granduncle, students' preference for non-demanding tasks, difficulty in obtaining information, indiscretion: the teacher could think of many reasons for students' reluctance to cooperate, aside from the most obvious one, to which her students (and, in all probability, their parents) were much more alert – that they feared the consequences. Their fears were not unjustified. The following year, the Health Bureau of Berchtesgaden asked that lists of mentally ill, criminals, alcoholics "and the like" be compiled, "because the longer we wait, the more the population becomes suspicious, and disguises and conceals all the hereditary defects in their families. It is best to start in schools and give in the final school year, as well as in all classes, ancestral- and kin-charts as school assignments. It is urgently desired that once the charts are completed, the teachers, on their own, but confidentially, mark with a pencil or ink of different color... what is known to them personally," regarding diseases, such as those detailed in the Sterilization Law. These marked charts were then to be transferred to the local health bureau to be collected and cataloged for further use.[74]

Fraught with various pedagogical and emotional challenges, teaching heredity was intrinsically linked with discussions of diseases considered to be inherited in a classic Mendelian fashion and with the sterilization policy that targeted some of these diseases. A 1934 guidebook for teachers recommended using the example of deaf-mutism to demonstrate recessive inheritance in humans. Blindness was also used for a similar purpose, to the extent that in 1936, one teacher reported that the moment she pronounced the word "heredity" in class, the students asked her about the inheritance of color-blindness.[75] In another classroom, the teacher chose short-fingeredness as an example of dominance, and illustrated recessive inheritance with the help of a pedigree of deafness. He also asked his students to chart the results of the following possible marriages: two healthy homozygous individuals; two ill homozygotes; a healthy homozygote with an ill homozygote; an ill heterozygote with a healthy homozygote; an ill heterozygote with an ill homozygote; and two

[73] DIPF/BBF/Archiv Gut Ass 1400, 7–9.
[74] Bock, *Zwangssterilisation*, 189. On the efforts to register all instances of inherited diseases in the German population, see Götz Aly and Karl Heinz Roth, *The Nazi Census. Identification and Control in the Third Reich* (Philadelphia, PA: Temple University Press, 2004), 104f.
[75] DIPF/BBF/Archiv Gut Ass 812, 28; Hermann Römpp (Gaureferent für Rasseforschung im NSLB), "Mendelismus," *Der praktische Schulmann* 10 (1934), 45. I thank Dr. Ina K. Uphoff for making this publication available to me.

ill heterozygotes. This gave his students the "opportunity to practice the Mendelian laws. ... [T]he students undertook this task with great zeal." In a subsequent lesson, deaf-mutism was referred to once again, and, with its help, "the dangers of recessive hereditary diseases were particularly emphasized."[76]

In this way, the "Mendelian diseases" listed in the Sterilization Law, along with several other classic examples of Mendelian inheritance, lent the authority of Mendelian research to the entire sterilization campaign. The following sentence appeared in one student's final exam in biology, written at the end of January 1934: "We know," wrote the student, "of a number of hereditary-diseases, that undoubtedly follow a simple hereditary mode (e.g., Huntington's chorea)." And he immediately added: "It is the blessing of the Law for the Prevention of Hereditarily Diseased Offspring, now in effect, that the hereditary mass that determines the quality of our *Volk* will not deteriorate further to such a horrible magnitude."[77] In the classroom, just like in the commentary to the Sterilization Law, Huntington's chorea, as a recognized example of Mendelian diseases, helped to associate Mendelism with the Sterilization Law and suggest that this law in its entirety – indeed, racial-hygiene in general – was grounded on Mendelian teaching. For high school students, it therefore became clear that Mendelism provided the foundation on which the Nazi regime's biological policy was built.

Mendelian Reasoning and the Implementation of the Sterilization Policy

"Please explain to me for the first time in clear German and then, I hope to be on the right track [to understand], where and since when is it proven that my case is hereditary???"

The sentence above appeared in a furious letter sent in October 1935 to a Hereditary Court in Berlin by Karl H., who was about to be sterilized after being diagnosed as schizophrenic. Karl was clearly outraged. The charges against him, he argued, resulted from a quarrel between himself and two officials from Brandenburg's employment office, who approached him following a false complaint made against him by a certain peasant. The whole affair escalated, he reacted angrily, and now he was reported as schizophrenic. "I readily admit that I suffer from all kinds of anxieties and also that I am very nervous, but that this

---

[76] DIPF/BBF/Archiv Gut Ass 1464, 31. For another example of the actual use of these diseases in the classroom, see DIPF/BBF/Archiv GUT ASS 1201.

[77] LAB A Rep. 020-25, No. 31.

proves that the whole story is hereditary and that it is inherited further, that I do not believe." Karl continued, "I can give you some good advice: You yourselves go once to assist such farmers for a long time ... and then you will see that even a healthy man cannot endure there a while. So, for me, a sterilization is out of the question. ... So now you know, not for me!!! Heil Hitler!"[78]

Karl's efforts to overturn his verdict proved successful: he was eventually diagnosed as schizoid, not schizophrenic, and his sterilization was revoked. Unfortunately, he was among the mere few who had the courage, brashness, intellectual ability or eloquence to effectively oppose the Nazi machinery, which from January 1, 1934, channeled its energy against those inner foes, the hereditarily burdened.

Werner A. presents another example. The legal procedures in his case began in early 1938 and ended in April 1939. Werner fought vehemently against the decision to treat his case as one of "genuine epilepsy." He used every imaginable stalling tactic: argued against his diagnosis on the basis of medical grounds, wrote letters protesting the way his case had been judged, showed that his doctors' reports failed to properly document his oral testimonies regarding his medical state, demonstrated that he has been healthy for a considerable time when treated with a homeopathic drug, delayed his replies to official inquiries, sent a personal letter to Hitler, supplied evidence that an operation could be life threatening for him, objected to further examinations on various grounds, and finally mobilized economic argumentation and the interests of the entire German *Volk* against the idea to relieve him from work for two weeks for medical examinations. In a letter he sent in February 1938 to a higher medical official he argued, "No one can say for certain that I am still ill, since my last seizure, which was relatively light, occurred over 3–4 years ago. ... The view that mine is a case of a congenital (genuine) disease or that I am a carrier of congenital epilepsy stands in contradiction to the fact that I had my first seizure when I was 27, whereas congenital cases appear during childhood; moreover, [it contradicts the fact] that my entire family [from myself] up until my grandparents was and is completely healthy."[79]

An eleven-page medical report compiled in June on Werner's case agreed that "the family history of the disease provides absolutely no indication, from which side a hereditary inferiority was transmitted to the proband"; but since the symptoms gave the clear impression of genuine epilepsy, and following the instructions in the official

---

[78] LAB A Rep. 042-08-01, Nr. 5486, Bl. 9.
[79] LAB A Rep. 042-08-01, Nr. 5499, esp. Bl. 13–14, 51–52, 73–74.

commentary to the law, Werner needed to be sterilized. Werner fought back: according to the law, he reminded the Hereditary Court, "one can (!) be sterilized if it is to be <u>expected</u> with <u>high</u> probability that the descendants would suffer from <u>severe</u> hereditary damage." None of these conditions, he argued, applied to him. "Even if it assumed that the children's children would suffer from instances [of the disease], in no way is it to be expected with <u>high</u> probability that in my special case they would suffer from <u>severe</u> hereditary damage." Implicating an entire family on the basis of cases like his, he maintained, would harm the efforts of a nation struggling to increase its population after the losses of the war; aside from that, he did not inherit this condition, it was caused by external stimulation. As mentioned above, Werner's efforts eventually succeeded; a final medical report determined that his disease was not genuine epilepsy, and his verdict was overturned.[80]

The decisions in both Karl's and Werner's cases demonstrate an important characteristic of the working procedure of Hereditary Courts: formally speaking, their rulings relied on clinical diagnoses, not on hereditary analyses. From the legal perspective, once the list of diseases requiring sterilization had been compiled and elaborated on in the official commentary to the Sterilization Law, Mendelian reasoning was no longer required in order to pass judgment on each individual case. Hereditary doctors and jurists were asked to determine whether the sterilization candidates suffered from one of the diseases defined by the law, not to verify actual heritability in the cases brought before them, something that was at any rate impossible to do.

The principal category targeted by the Sterilization Law was feeble-mindedness. Here, intelligence tests were used to determine the (purportedly biologically based) inferiority of the mentally weak. The proceedings against the feeble-minded demonstrate the degree of rhetorical flexibility that the perpetrators were willing to employ, in order to assure the sterilization of their most vulnerable and helpless victims. In one of the cases handled by a Berlin Hereditary Court, the father of the sterilization candidate appeared before the chamber and submitted a written objection to the actions against his son. "From my side, as the child's father, no hereditary disease (congenital mental deficiency) exists. I am totally healthy; nothing [negative] is known regarding my parents, four siblings, grandparents and relatives … I actively served three years as a soldier in the World War … my son, as a small child, was infected with severe rachitis … he is not hereditarily burdened with any diseases." The

---

[80] Ibid.

father further testified that his son worked in a train station, where he was given the task of signaling waiting passengers to step back before a train arrived; he liked this job and people liked him. He also noted that there was no danger that his son would father children, because the son's wife has already been sterilized.[81]

The father's appearance before the court "made a very good impression." His son's performance, however, did not, nor did the son's intelligence tests. The court therefore determined that the son was "undoubtedly mentally very retarded." The father's appearance before the court now backfired: the son's mental retardation was "all the more evident, when one compares him with his father, who according to the personal impression he made on the court is a capable man." Lacking a proof that the cause for the son's mental retardation was external, "the mental deficiency must be seen as congenital, even if he stems from a healthy family." An appeal against this decision was turned down: the higher court stated that "although no evidence exists for a familial burden," there are no grounds for **not** seeing the deficiency as a congenital one, and "the danger, that among his offspring the same or a similar case of hereditary burden would appear is highly probable."[82]

A healthy family and a competent parent could therefore become another reason for sterilization, highlighting the gap between the candidate's performance and his family. In a similar twist of logic, one "mentally feeble" woman managed to answer the intelligence test so well that her sterilization was revoked; but the head of her sanatorium objected, arguing that if she answered so well, she must have learned the answers by heart (something which indeed many attempted to do).[83] The initial decision was therefore reversed and she was sterilized.[84]

According to their own self-understanding, Hereditary Courts cared less about the actual health of individuals, which they saw as merely external or phenotypic; their aim was to annihilate malignant genes, not to cure diseases. In an article published in 1935, psychiatrist Hans Luxenburger stressed that "it is not the disease as such that burdens, but its hereditary basis, not the phenotype, but the genotype, not the timely, transitory, changeable, but the lasting, essentially constant, unchangeable. The trait, the disease, is only the indicator, [whereas] the substance

---

[81] LAB A Rep. 356 Microfilm A 5255, No. 7309, Bl. 9, 15, 20, 32.   [82] Ibid.

[83] The intelligence questionnaire was printed in the first edition of the official commentary to the Sterilization Law, thus making the questions known also to the general public. It was removed from the second edition and altered. For more on the intelligence tests, see Bock, *Zwangssterilisation*, 313–314.

[84] LAB A Rep. 356 Microfilm A 5255, No. 7311.

of the [hereditary] burden is [hereditary] disposition."[85] Continuing a similar line of thought, a detailed discussion of the case of a person with cleft lip eventually came down to deliberating, "Is a hereditary cleft lip a severe physical deformity or is it not? The decision, however, cannot be made solely according to the phenotype." What needed to be decided was whether the gene for this external deformity was such that it could also lead to severe deformities, even if it did not do so in the specific case of the person under discussion. If so, it should be eradicated. That is why "physical deformities that make their carriers appear like unfit natural life forms because of aesthetic reasons, are to be seen as *severe* physical deformities in the sense of the law … cosmetic operations cannot change that. For they do not eliminate the hereditary risk." Indeed, "precisely the light forms, from the standpoint of cultivating hereditary health, are the most dangerous ones, because their carriers reproduce more easily."[86] Potential genetic damage, not actual harm, was a sufficient reason for sterilization.

In a similar vein, in one case of an epileptic woman, the court acknowledged that the woman "stemmed from a healthy family," that her seizures had never been observed by a medical authority, and that these seizures completely ceased when the woman was treated with Luminal (a popular anti-epileptic drug). But "[p]recisely the latter circumstance is an indicator of great weight for the fact that we are dealing here with genuine epileptic seizures, because seizures [caused] by other sources are rarely influenced by Luminal." The conclusion was clear: "Since the examination in the [Berlin university hospital] Charité did not provide sufficient evidence for other kinds of seizures or for external causes for the [epileptic] seizures, the chamber had no other choice but to conclude that in the case of Mrs. N. we are dealing with a disease caused by [hereditary] disposition in the sense of genuine epilepsy, that is, a hereditary, that means, inherited epilepsy in the meaning of the [Sterilization] Law."[87]

---

[85] Hans Luxenburger, "Der Begriff der Belastung in der Eheberatungstätigkeit des Arztes," *Der Erbarzt* 1 (1935): 12–15.

[86] MPIP-GDA 140, a case from 10.2.1939 regarding Karl K. In a personal correspondence between Verschuer and Lenz on October 1939, Verschuer remarked that the fact that "otosclerosis does not have to lead in every case to hereditary deafness in the meaning of the law, seems to me to be irrelevant for deciding the present cases, for in [the cases of] many other hereditary diseases of the law, sterilization has been carried out, when we know that the relevant [hereditary] disposition does not have to lead in every case to the respective hereditary disease." MPG Archive Abt. III Rep. 86B Nr. 3, 12.10.1939, Verschuer to Lenz.

[87] LAB A. Rep. 356, 7312.

The same kind of reasoning was used in 1938 to advance charges against another person suffering from epilepsy. The Hereditary Court originally ruled against sterilization, but a district physician in Berlin appealed against the dismissal of that person's case and attempted to overturn all of the arguments previously used by the court. The fact that there had been no epileptic seizures for three years, claimed the district physician, resulted from the patient's abstention from drinking alcohol; it was only typical for epilepsy that it would manifest itself less under such conditions, he pointed out. That no evidence could be found of any familial burden, argued the physician, could be explained by the recessive nature of the disease. The higher court accepted the appeal and decided on sterilization.[88] Similarly, in a case laid before him regarding a woman who once suffered from a schizophrenic episode, the Higher Hereditary Court explained that "even if currently this episode has almost completely come to a cure, the pathological disposition must nevertheless be assumed to exist."[89]

Mendelian reasoning, even if it wasn't required for making decisions on sterilization, still found its way into legal proceedings. This was apparent first and foremost in some of the cases dealing with "Mendelian diseases" – visual impairments of all kinds, hemophilia, deformations of the fingers (polydactyly, syndactyly), deafness and Huntington's chorea.[90] For example, one medical assessment of the heritability of an individual's deaf-mutism referred to "the parents' blood-relations, which, through the potential recessive elements, together with the sporadic manifestation of the disease, is meaningful."[91] Similarly, the veracity of the clinical diagnosis "retinitis pigmentosa" for a man from a small village in the vicinity of Darmstadt was reinforced by the fact that the man had two siblings with the same diagnosis, as well as blood relations among his patrilineal grandparents. "The case is without doubt one of a recessive form, whereby the frequency of ancestral consanguinity is to be emphasized."[92]

Mendelian logic could also work to repel sterilization. For example, one case of questionable Huntington's chorea was debated (also) with

---

[88] LAB A Rep. 356 Microfilm A 5255, No. 7324.   [89] LAB 037-08-01 Nr. 5014.
[90] The following analysis is based on the examinations of hundreds of Hereditary Courts proceedings, most of which are in LAB A Rep. 042-08-01, 037-08-01, 356 Microfilm A 5255; MPIP-GDA 58, 63, 72, 81, 136-142; HHStD G29U; HHStW 473/4.
[91] LAB A Rep. 042-08-01, Nr. 5304, Letter from 18 April 1939.
[92] HStAD G29U/683, diagnosis from 11.01.1939 of N.E. In fact, given that no indications existed on blindness among the man's parents, blood relations among the grandparents could possibly explain why the man's father should have become blind (which he didn't), but not why the man himself became blind. Thus, even under genetic reductionist assumptions, Mendelian reasoning in this case was misapplied.

relation to the hereditary nature of the disease: "[T]he absence of a proof for dominant hereditary transmission, which is assumed without exception in the case of Huntington's chorea" strengthened the doctor's diagnosis that "we are dealing with a generalized tic on hysterical grounds" and not with Huntington's chorea proper.[93] Another medical evaluation, discussed by the Wiesbaden Hereditary Court, relied even more closely on Mendelian reasoning, probably owing to the fact that the individuals under discussion had unwittingly emulated the basic requirements of a Mendelian experiment. The case involved a deaf-mute couple who had five children, all of whom without apparent hearing difficulties. The deaf-mute husband had one deaf-mute brother, as did the deaf-mute wife. When discussing the heritability of the husband's condition, the examining doctors observed that the hearing loss initially gave the impression of an inherited defect. But medical theory acknowledged the existence of two principal forms of inherited deaf-mutism: the one recessive, the other dominant. If the husband and his wife's deaf-mutism were of the recessive type, then the same defect should have also manifested itself among their children (both parents being homozygous for the disease, thus necessarily transmitting the defective genes to their children). Yet as noted above, the couple's children were healthy, a fact that ruled out recessivity. On the other hand, if both the husband and the wife had the dominant type, then deaf-mutism should have also manifested itself among both partners' own parents; but the couple's parents, like the couple's children, were healthy. The inescapable conclusion was that at least one of the spouses had an acquired, not an inherited, deafness. According to clinical parameters, coupled with the fact that the wife's parents were blood related, the evaluating doctors concluded that it was the husband who had an acquired type, the wife an inherited one. Only she was therefore sterilized; he was acquitted.[94]

Explicit Mendelian reasoning was not confined, however, to discussions on Mendelian diseases, which were relatively few; on the contrary, it surfaced also with respect to schizophrenia, manic-depressive insanity, epilepsy and even feeble-mindedness. Most commonly it was deployed to justify the sterilization of those who were "outwardly healthy" but continued to threaten the public as carriers of pathological recessive dispositions. It was also used to justify bureaucratic lethargy. Replying

---

[93] LAB A Rep. 042-08-01, Nr. 5402.
[94] HHStAW 473/4 Nr. 87, 88. Dr. Schwarz, the head of the Ear-Nose-Throat Clinic in the University of Frankfurt, applied the same reasoning to help release other deaf-mute persons from sterilization; see for example a similar case of two deaf-mute parents with normal-hearing children in HStAD G29U/687.

to an inquiry about the implementation of the law, the Munich District Court reported in September 1935 that "since schizophrenia is inherited in a recessive manner, even a negative result of the assessment of parents' and siblings' [mental state] does not speak against the validity of the diagnosis." Consequently, "In general performing such an assessment can be dispensed with, unless there is doubt in the diagnosis itself."[95] In other words, there was no reason to bother examining the state of the health of family members of schizophrenics, because even if they were healthy, this proved nothing; only if one's diagnosis was uncertain could ancestral findings become meaningful.

One of the implications of this line of thinking was that pedigrees were charted in medical evaluations only when they aided in strengthening the case for sterilization and were dispensed with in virtually all other cases. Consequently, pedigree charts are often absent from the files of Hereditary Courts. A minority of cases do contain pedigrees that delineate the structure of the candidates' families but provide no additional medical data, and occasionally one may encounter pedigrees that document the medical status of family members but that contain little information of real value for hereditary research. More than anything, these latter pedigrees are reminiscent of late nineteenth- and early twentieth-century charts of mentally ill families, stressing overall burden rather than illuminating the actual mechanism of transmission, let alone a Mendelian one.

It is tempting to see this overall lack of family charts as proving that, with the exception of the above-mentioned cases, Mendelism was no longer relevant for Hereditary Courts, whose decisions were informed by medical, social, gender and ethnic biases much more than by genetic theories. Yet as we have seen, from the perspective of the participants themselves, the lack of pedigrees or Mendelian analysis in the files of sterilization courts did not indicate that these courts were uninformed by Mendelian reasoning. On the contrary, it was precisely the internalization of Mendelian logic that made it possible for the courts to try to eradicate hereditary diseases without even attempting to perform hereditary analyses in the cases they judged. It was none other than the Mendelian notion of recessivity that made the drawing of individual pedigrees redundant. Paradoxically, the lack of pedigrees in the majority of cases discussed by Hereditary Courts was "Mendelian" par excellence. In all of the cases described above – the cases of Kurt, Werner, the train station worker, the epileptic woman and others – individuals tried to fend off

---

[95] Staatsarchiv Munich OLG 605.

sterilization by showing that their family members were healthy and that their diseases were not hereditary. In most of these cases, what made such assertions meaningless was the underlying notion that these diseases stemmed from defective recessive genes.

Moreover, Mendelism was not the only source of the idea of recessive inheritance; Mendelian crossing experiments and even Mendel's own work also underlay the notion that hereditary patterns can be observed only when many breeding results are accumulated and analyzed statistically. In a Mendelian world, as Rüdin had already sensed correctly in his 1911 treatise, only state institutions, with ample resources and access to multiple types of medical records, could diagnose heritability. "[I]t is important to stress again and again that the layman cannot decide whether or not a hereditary disease is involved in a particular case," wrote one of the teachers mentioned in the previous section, fearing that her students would rush into negative conclusions regarding their own hereditary endowment.[96] The same message, but for opposite purposes, was reiterated over and over again by medical officers during Hereditary Court discussions: the genes of a person could be diagnosed only by the medical establishment. Mendelian theory therefore created the preconditions for exceptionally uneven power relations between individuals and the state, ones in which the ability of a person to say anything or know anything about his own "hereditary diseases" was close to nil.[97] Against this background, the involuntary nature of the Nazi sterilization campaign reflected not only the dictatorial character of the Nazi state but also a Mendelian-based scientific stance, wherein the hereditary doctor was the only one who was in the position to determine the genetic nature of medical conditions – along with the fates of those who harbored, and propagated, deviant genes.

Thus, despite the reliance of the sterilization campaign on the results of empirical hereditary prognosis, Mendelian thinking remained central to the understanding of hereditary illness and the threat that it posed. It was presented as, and understood to be, the basis of the sterilization

---

[96] DIPF/BBF/Archiv Gut Ass 812, 45–46.

[97] A brief thought experiment may be useful for highlighting the extent to which these uneven relationships relied on a Mendelian understanding of heredity. What would the operation of Hereditary Courts look like if it was not Mendel's laws, but Galton's "Law of Ancestral Heredity" that was the basis for launching a sterilization campaign? Certainly some of the parameters related to the strength of heritability of certain traits would still need to be determined by networks of scientists with access to aggregates of data. But in a hypothetical discussion between a chamber and a sterilization candidate, the candidate would be able to dispute his verdict, because his own family history would have been the most important criterion for determining his hereditary composition. Not so with Mendelian heredity.

policy. Even if in the majority of cases, Hereditary Courts did not refer directly to Mendelian suppositions, these were still assumed to be valid and were revisited whenever necessary. In rare instances, such as that of the questionable Huntington's chorea or the deaf-mute couple mentioned above, Mendelian thinking could work to "redeem" individuals from the accusation that they were posing a hereditary danger. But its overall impact was to radicalize eugenic thinking by annihilating the ability of individuals to object to court decisions and by strengthening the belief that hereditary danger continued to lurk in the genes of healthy and functional citizens. And, most crucially, through the concept of recessive inheritance, which was extended to practically all of the diseases named by the Sterilization Law (aside from Huntington's chorea), Mendelian theory was the first scientific theory of heredity in history that made it possible to accuse someone of carrying an inherited medical condition, even when no family member of that individual had ever been ill with the respective disease.

5   Mendelizing Racial Antisemitism

A little more than two months after Hitler became chancellor, on April 7, 1933, the "Law for the Restoration of the Professional Civil Service" was passed, purging civil servants who were not of Aryan descent. For the purposes of that law, a single Jewish grandparent was enough to make a person "non-Aryan." The need of proof of Aryan ancestry quickly spread to the army, the league of workers, youth organizations, women's societies, professional unions (physicians, lawyers, tax consultants, etc.) and all the organizations belonging to or affiliated with the Nazi party, where membership had already been confined to Aryans since 1920. In certain cases, the requirement to supply a document proving Aryan descent (*Ahnenpass*) was enshrined by law. In other cases, it was determined through a special decree or the internal regulations of the respective organizations.[1]

Proof of being Aryan relied heavily, almost exclusively, on ancestors' family names and religious affiliation, examined through parish registries and civil records. At last, the professional skills genealogists had developed in constructing ancestral charts became relevant for millions of citizens, as well as for the state authorities who wished to verify the authenticity of personal statements on ancestral origins. The unprecedented public demand for genealogical expertise harbored significant financial gains for genealogists.[2] Guidebooks on how to conduct genealogical investigations became popular and new editions of older publications were released, refurnished with quotes by Nazi leaders on the significance of the family for regenerating the nation.[3] Workbooks for

---

[1] Eric Ehrenreich, *The Nazi Ancestral Proof* (Bloomington: Indiana University Press, 2007), 58–77.

[2] Ibid., 58–77, 121–149. On the "registration mania" that followed, see esp. Karl-Heinz Roth and Götz Aly, *Die restlose Erfassung. Volkszählen, Identifizieren, Aussondern im Nationalsozialismus* (Berlin: Rothbuch Verlag, 1984), translated as *The Nazi Census* (Philadelphia, PA: Temple University Press, 2004).

[3] Compare the 1919, 1924, 1926 and 1935 editions of Ernst Reinstorf, *Wie erforsche und schreibe ich als Bürgerlicher meine Familiengeschichte? Eine kurze Anleitung dazu*

the "racially conscious youth" encouraged youngsters to perform their own family research. "The farther you pursue your studies on your ancestors and their descendants," explained the author of one of these workbooks to his young readers, "the larger your family will become before your very eyes. And if we imagine that we could learn everything about our ancestors until the remotest of times, then your family, your circle of relatives, would be nothing but your *Volk*."[4]

In 1934, Friedrich Wecken's booklet on "The *Ahnentafel* as a Proof of German Descent" was published in its seventh edition. Discussing the proper way of compiling an *Ahnentafel* and of recognizing Jewish forefathers, Wecken noted that family names were problematic indicators of descent because they were easily replaced with Christian names. But if a substantial gap between the date of birth and the date of baptism was observed in the records, Jewish descent could be assumed.[5] As we saw in Chapter 3, the motivation to collect information from parish records originally stemmed from their comprehensiveness and their ability to provide data on rural communities. Now the same sources became useful also as a means for uncovering hidden Jewish affiliations.

The following year, on September 15, 1935, during the seventh Nazi Party rally in the city of Nuremberg, the German Reichstag passed two infamous laws: the "Law for the Protection of German Blood and German Honor" and the "Reich Citizenship Law." These laws stripped Jews of their civil rights, denied them the privilege of displaying the Nazi flag, forbade Jews from marrying or having sexual relations with Germans, and made it illegal for Jews to employ female Germans under the age of 45 as domestic servants. The laws drew a clear separation line between Germans and Jews, preventing Jews from taking a full and equal part in German national life and criminalizing any sexual contact between Germans and Jews.

The Nuremberg legislation was acknowledged as a watershed in the relations between Jews and Germans.[6] Years later, a Jewish-Austrian

---

(Hamburg: Zentralstelle für niedersächsische Familiengeschichte, 1919), which are essentially the same, with the addition of quotes from Nazi authorities in the last edition.

[4] Emil Jörns, *Meine Sippe. Ein Arbeitsheft für rassebewusste deutsche Jugend* (Görlitz: Verlag für Sippenforschung u. Wappenkunde C. A. Starke, 1934), 33. At the same time, the author didn't fail to remind his readers that, when visiting their grandparents for the collection of information for familial research, "don't forget to inquire about diseases among the great-grandparents" (ibid., 18).

[5] Friedrich Wecken, *Die Ahnentafel als Nachweis deutscher Abstammung* (Leipzig: Degener & Co., 1934), 9, fn. 9.

[6] Raul Hilberg, *The Destruction of the European Jews*, 3rd ed. (New Haven, CT: Yale University Press, 2003), ch. 4.

writer would recall his memories from that period: "I was sitting over a newspaper in a Vienna coffeehouse and was studying the Nuremberg Laws, which had just been enacted across the border in Germany. [...] After I had read the Nuremberg Laws I was no more Jewish than a half hour before. My features had not become more Mediterranean-Semitic, my frame of reference had not suddenly been filled by magic power with Hebrew allusions, the Christmas tree had not wondrously transformed itself into the seven-armed candelabra. If the sentence that society had passed on me had a tangible meaning, it could only be that henceforth I was a quarry of Death."[7]

Then, on November 14, a supplemental decree was issued, altering the hitherto accepted legal definition of Jewish racial identity. The decree determined: "A Jew is anyone who descended from at least three grand-parents who were fully Jewish by race. A Jew is also anyone who des-cended from two fully Jewish grandparents, if: (a) he belonged to the Jewish religious community at the time this law was issued or joined the community later; (b) he was married to a Jewish person at the time the law was issued or married one subsequently." In addition: "A Jewish *Mischling* is one who is descended from one or two grandparents who were fully Jewish by race. ... One grandparent shall be considered as full-blooded if he or she belonged to the Jewish religious community."[8]

Within the framework of these genealogically bound racial definitions, it may seem surprising that Mendelism acquired any role at all. Never-theless, as we will now see, Mendelian thinking became a key factor in how the Nazis perceived the meaning of their Jewish identification prac-tices and in the establishment of policies to deal with the Jewish *Mischlinge*. In the previous chapter, we saw that the sterilization policy was carried out on the basis of clinical and mental diagnosis, but never-theless relied on Mendelian suppositions. In a parallel manner, racial policy was executed with the help of genealogical methods, but was informed, at certain key junctures, by Mendelian assumptions. To uncover these assumptions and reveal their influences, we will begin by

---

[7] Jean Améry, "On the Necessity and Impossibility of Being a Jew," *New German Critique* 20, special issue: Germans and Jews (1980): 15–29 (quote from p. 17).

[8] Text and translation adapted with minor modifications from German History in Documents and Images, http://germanhistorydocs.ghi-dc.org/sub_document.cfm?document_id=1523. Source of English translations: The Reich Citizenship Law of September 15, 1935, and the First Regulation to the Reich Citizenship Law of November 14, 1935, in United States Chief Counsel for the Prosecution of Axis Criminality, *Nazi Conspiracy and Aggression*, volume IV (Washington, DC: United States Government Printing Office, 1946), Documents 1416-PS and 1417-PS, 7-10 (English translation attributed to Nuremberg staff; edited by GHI staff).

exploring the way Mendelian reasoning entered the legal debates on the racial definition of Jews, debates that culminated in the formulation of the Nuremberg decree. Then, we will turn to the teaching of racial science in schools, and analyze the role Mendelian thinking acquired in the racial indoctrination of young Germans between 1933 and 1936. Finally, we will see how Mendelian notions informed the actual practices of racial determination in Nazi Germany and became central to discussions on the fate of the Jewish *Mischlinge*. Although the anti-Jewish measures of the Nazi regime never confined themselves to any single framework of thought, Mendelian notions will be shown to have constituted an important part of Nazi reasoning, successfully merging with prior and even competing modes of antisemitic thinking.

The Mendelian Route to the Nuremberg Laws

The fourth point of the Nazi Party's 1920 program stated that only individuals of German blood could be members of the nation and that, accordingly, no Jew could be a member of the nation.[9] The earliest documented attempt to translate these declarations into actual state policy came in late 1932 and early 1933, in a working paper prepared by the Nazi genealogist Achim Gercke. Gercke, who with the Nazis' rise to power became the Interior Ministry's expert on genealogical-racial identification, addressed the future civil status of German-Jewish *Mischlinge*, and referred to Mendel's laws to substantiate his views on the inability of Jewish racial components to progressively wane. "When one 'computes' ... with the decreasing series of 50 percent, 25 percent, 12.5 percent, 6.25 percent of a mixture etc., then one shows that one did not grasp the Mendelian laws at all. It is known that in addition to mixed types, the pure types also appear, which, however, cannot say much about humans, since body, mind and soul and the many hereditary traits cannot mendelize together." The intrusion of Jewish blood into the German nation was therefore unlimited, explained Gercke, always capable of "mendeling out."[10]

---

[9] See the full program of the German Worker's Party (the 25 points) in http://germanhistorydocs.ghi-dc.org/sub_document.cfm?document_id=3910
[10] Gercke's memorandum is quoted at length in Cornelia Essner, *Die "Nürnberger Gesetze" oder die Verwaltung des Rassenwahns 1933–1945* (Paderborn: Ferdinand Schöningh, 2002), 77–78. On Gercke and his Genealogical Office, see Diana Schulle, *Das Reichssippenamt: Eine Institution nationalsozialistischer Rassenpolitik* (Berlin: Logos, 2001). Gercke uses the verb "mendelize" in the sense of being "inherited according to Mendel's laws," and "mendel-out" in the sense of "reappear (as homozygous) after being hidden as recessive for several generations." For a discussion on the multiple meanings of the verb, see Chapter 1.

After the Nazi seizure of power, the attempt to separate Germans from Jews in the public and judicial spheres continued on various fronts, some of which were already mentioned earlier. During the first half of 1933, discussions began on an entirely new citizenship law. The racial popularizer Hans F. K. Günther was among the experts who took part in these discussions. Like Gercke before him, he insisted that, according to the principles of racial science, any person with even a remote ancestral element of Jewish blood should be denied citizenship. He acknowledged that this may be difficult to implement and even cause political hardships; but these were not his problems to solve. As he saw it, it was the responsibility of racial scientists (like himself) to present the scientific viewpoint, not to adapt it to practical or political needs.[11] A working paper of the Interior Ministry from July 24, 1933, critically reviewed Günther's position as well as other suggestions on how to deal with Jews and Jewish *Mischlinge*. It determined that in order to put an end to "the decomposition (*Zersetzung*) [of the national body] through Jewish blood," it would be practically sufficient – contra Günther – if from that point onward, further mixture with Jewish blood would be restricted to only those with remote Jewish ancestors. It reasoned, "Should an especially Jewish *Mischling* once 'mendel,'" his civil rights could always be denied so that he would be distanced from taking active participation in state life. "But the danger that descendants with one remote Jewish ancestor would mendel [with] particularly strong Jewish [influence], decreases with every generation, if the further influx of Jewish blood is terminated."[12]

The issue of the intrusion of Jewish blood was revisited a year later, as part of deliberations that took place in June and July 1934 in the German Interior Ministry. Conducted in the presence of the Minister of Justice, Franz Gürtner, the Nazi judge, Ronald Freisler and senior officials from the Justice and Interior Ministries, the discussion bore the title "Attacks on the [German] Race" and centered on the legal possibilities of preventing racial intermixture between Germans and alien races.[13] Among other issues, the discussants debated how to prevent extramarital births of racially mixed children. Was there a

[11] Günther's position appears in the working paper of the Interior Ministry mentioned subsequently, available in Institut für Zeitgeschichte München [hereafter IfZ Munich], F 71/1, document from July 24, 1933.

[12] Ibid.

[13] BArch R3001/20852, Bl. 75–322. The transcript of this meeting also appeared in Jürgen Regge and Werner Schubert (eds.), *Quellen zur Reform des Straf- und Strafprozeßrechts*, vol. 2.2 (Berlin: de Gruyter, 1989), 223–348.

way, one of the participants inquired, for racial experts to determine
on the basis of objective criteria that a child was racially mixed?
Bernhard Lösener, the Interior Ministry's expert on Jewish matters,
answered:

An effective means of determining for a given person, according to his behavior or
his blood or the like, whether he has Jewish infiltration, does not exist, or at least
for the time being has not yet been found. ... Our [state] expert for racial study
[Gercke] is in fact an expert on family research; that is, when he is given the task
to determine for a certain person whether he has Jewish blood, he will trace the
family tree and in this manner, that is by genealogical research, find out whether
Jewish ancestry is present.[14]

Thus, Lösener openly admitted that racial affiliation was determined
through genealogical practices and not by anthropological, anthropometric, biological or serological means. Continuing the discussion a week
later, the participants addressed this understanding. Ministerial Director
Schäfer said that when it came to racial differentiation, a distinction had
to be made between a precise scientific solution, which could be difficult
to obtain, and a practical solution, which needed to be simple.[15] Kurt
Möbius from Department IV (Public Health) of the Ministry of the
Interior reminded the discussants that it was not the race that was
inherited, but racial markers, which could be dominant or recessive.[16]
These remarks only strengthened Minister Gürtner's view that for the
differentiation of races, no solution other than the genealogical was
possible.

I'll take your statement as a confirmation of the idea that, as things are, in
delineating the term [race] we cannot work with anything else but genealogy.
From your point of view, that is cruelly primitive, because you say that racial
features, which are inherited in a dominant and recessive manner, skip
generations. We cannot deal with that here. We must do crude carpentry work,
and that is the family tree.[17]

Lacking the ability to biologically recognize, define or delineate the
Jewish race, the Nazi regime found no other alternative but to digress
from a scientific solution to a practical solution, namely, ancestral
research. Under the assumption that Jews and non-Jews remained
socially and, therefore, also reproductively isolated up until Jewish
emancipation in the early nineteenth century, checking the religious
affiliation of one's grandparents seemed like a reasonable, even if

[14] BArch R3001/20852, Bl. 98.   [15] BArch R3001/20852, Bl. 198–206.
[16] BArch R3001/20852, Bl. 206–207.   [17] BArch R3001/20852, Bl. 209.

imperfect, way to evaluate one's racial origins.[18] However, making such a scientific compromise could undermine the regime's pretensions of legitimacy for its racial conduct. It placed the most basic concepts of heredity science – Mendelian mechanisms and recessive and dominant inheritance – in opposition to, or as interfering with, the practice of racial identification. This was an unfavorable standpoint from the regime's perspective. As we will now see, future discussants managed to brilliantly reverse these relations, and present the same Mendelian principles as supportive of, and not contradictory to, Nazi racial conduct.

Hitler's proclamation of the Nuremberg Laws on September 15, 1935, revived intensive discussions at the Justice and Interior Ministries on where exactly the line should be drawn between German and Jewish racial identity. The goal was now to reach an acceptable definition of Jews and half-Jews, one that would satisfy the antisemitic aspirations of the Nazis, but also be practically applicable for jurisdiction and bureaucracy. State officials, bureaucrats, racial experts and Nazi fanatics took part in these deliberations, which led to the final decree of November 14.[19] The first cue for positively integrating Mendelian reasoning into the legislative process was given ten days after Hitler's announcement, in a short memorandum on the treatment of half-Jews compiled by Arthur Gütt. As of April 7, 1933, Gütt was the head of Department IV in the Interior Ministry and therefore in charge of national health policy; as we saw in the previous chapter, he was one of the driving forces behind the Sterilization Law and a co-author of its official commentary.

Gütt began by proclaiming his position as a scientific expert, titling his memorandum "Which demands must the hereditary biologist raise regarding the solution of the Jewish question?" The first page of his paper was then devoted to Mendel's three laws. In a later version of the same memo, he changed the title and simply started with "Principles of Mendel's Hereditary Law."[20] The second page visualized the basic mechanisms of Mendelian heredity with the help of an illustration of the results of crossing white rough-coat with black smooth-coat guinea pigs (Figure 5.1). This was an iconic example of Mendelian mechanisms,

---

[18] As one of the advisory bodies on racial legislation stated in July 1933, "since according to experience, in the great-grandparental generation in Germany mixed marriages with those belonging to the Jewish people did not yet take place, the council laid the boundary between the grandparents and the great-grandparents," IfZ Munich F71/1, 7.

[19] The most detailed description of these processes is provided by Essner, *Die "Nürnberger Gesetze."*

[20] BArch R 18/5246 and YVA JM 3579, Bl. 33–39 (probably the final draft) and Bl. 40–45 (probably the first draft).

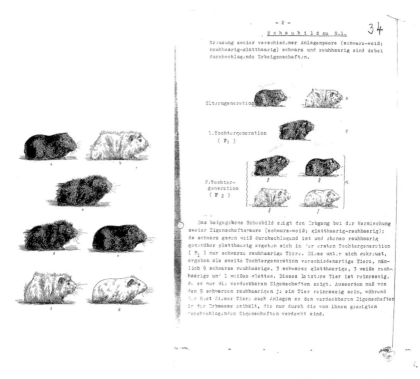

Figure 5.1 Left: A 1912 Mendelian chart describing the results of crossing guinea pigs with two independent traits. Right: The same chart integrated into Gütt's 1935 memo for the discussants on the Nuremberg Law.

Source: Ernst Teichmann, *Handwörterbuch der Naturwissenschaften* (Jena: Gustav Fischer, 1912), 860; BArch R 18/5246

commonly reproduced in biology textbooks for didactic purposes.[21] Now it was brought forward by Gütt as an analogy for what human racial crossing harbored.

With the visual imprint of this illustration still resonating, Gütt began his discussion of the "Applicability of Mendel's Laws for Humans."

---

[21] The genealogy of this particular chart demonstrates how research results can gradually morph into paradigmatic scientific exemplars. At the end of his 1905 "Heredity of Coat Characters in Guinea-Pigs and Rabbits," William E. Castle attached photographs of albino/red and smooth-/rough-coated guinea pigs. Those pictures were reproduced in Erwin Baur's 1911 "Introduction to Experimental Genetics," but only after they underwent two significant transformations: first, they were replaced by drawings; second, these drawings were positioned in the Mendelian P-F1-F2 (two parents–one

When Negroes and Whites were crossed, Gütt explained, the result was a Mulatto. Further crosses would never yield a pure White, and the colored blood will forever be visible. Fischer's studies on the Rehoboth Bastards, Gütt recalled, proved that human traits followed Mendel's laws. Since every trait mendelized independently, the chances that a combination of features would again produce a pure White were practically zero.[22] The guinea pigs, which Gütt had allegedly used to demonstrate the independent heredity of coat texture and coat color, conveniently actualized his discussion on the persistence of human "colored blood" as well. Finally turning to the "Jewish Problem," Gütt maintained that the mixture of quarter- and half-Jews with Germans could never yield pure-German blood. Nevertheless, continuous crossing would result in repeated division of the Jewish attributes, so that, as long as no back-crossing took place, a gradual dilution of Jewish traits might come about. After many generations, the German type would be at least approximated, although Jewish intrusion would still make itself apparent every now and then. But, Gütt explained, if one wished to avoid the eventual creation of an independent German-Jewish *Mischling* race, there was no alternative other than allowing some intermixture of the Jewish *Mischlinge* and the German-blooded.

Gütt continued: If such racial crossing between Jewish *Mischlinge* and the German-blooded was indeed allowed – a solution that was unsatisfactory from a biological perspective, since it could never lead to a complete annihilation of Jewish features – there were nevertheless some principal guidelines that he considered essential. Quarter-Jews should only be allowed to marry the German-blooded, but not Jewish *Mischlinge* or full Jews. Half-Jews, on the other hand, could marry German-blooded only after going through a selection process. Those who would not pass

---

hybrid–four descendants) format. In this format, the illustration was then reproduced in Baur's entry on Bastardization in the *Handbook of Natural Sciences*, published the following year. From that point onward, this chart spread throughout hereditary literature, to end up being used by Nazi officials for discussing the fate of the Jews. See William E. Castle, *Heredity of Coat Characters in Guinea-Pigs and Rabbits* (Washington, DC: Carnegie Institution, 1905); Erwin Baur, *Einführung in die experimentelle Vererbungslehre* (Berlin: Gebrüder Borntraeger, 1911), 85; Erwin Baur, "Bastardierung," in Ernst Teichmann et al. (eds.), *Handwörterbuch der Naturwissenschaften*, vol. 1 (Jena: Gustav Fischer, 1912), 850–873, 860, and compare Haecker's use of the original photos, still in 1911, to the later usages of the standardized and Mendelian-structured drawings in later publications: Valentin Haecker, *Allgemeine Vererbungslehre* (Brunswick: Friedrich Vieweg & Sohn, 1911), 229; Hermann Werner Siemens, *Grundzüge der Vererbungslehre, Rassenhygiene und Bevölkerungspolitik* (Munich: J. F. Lehmanns, 1930), 31; Wilhelm Gelfius, *Die gesetzmäßige Vererbung des Menschen* (Lübeck: self-pub., 1932), 62.

[22] BArch R 18/5246 and YVA JM 3579, Bl. 34, 40.

such a selection should be considered part of the Jewish people. The same held true for those half-Jews who married Jews: they should also be legally regarded and treated as Jews. State authorities would nonetheless hold the right to consider some of them as belonging to the German nation. Decisions on giving citizenship to half-Jews would be made on the basis of external appearance, personal character and mental capabilities. Descendants of East European Jews (*Ostjuden*), those with pronounced Jewish looks, the hereditarily ill and those with criminal tendencies or bad reputation should not be considered. When the time came and public opinion in Germany and abroad was calmer, voluntary sterilization should be offered to those half-Jews. At the same time, Gütt concluded, efforts should be made to ensure that on the German side, half-Jews remain unmarried and without children.[23]

Gütt's memo had significant impact on later proceedings, both in the short term as well as in the long term; we will later see that throughout the following decade, many of his assumptions and propositions would continue to surface in the discussion on the fate of the Jewish *Mischlinge*.[24] In the short term, Gütt influenced not only the content of later discussions but also their discursive framework by setting an example of how to construct an argument that would have a legitimate biological façade. In doing so, he followed closely the model of his earlier commentary on the Sterilization Law. It is probably no coincidence that both texts began with a salient reference to Mendel's laws, made especially conspicuous with the attachment of a diagram (in the case of the commentary, quite a few of them) showing the dynamics of Mendelian mechanisms. Thereafter, both the memorandum and the commentary moved swiftly to arguments that were less Mendelian bound – East European origins, "pronounced Jewish looks" and "criminal tendencies" in one case, the results of empirical hereditary prognosis in the other. In certain respects, these latter arguments were more significant in directing the suggested policies than the Mendelian background that preceded them. But the Mendelian exposition gave the impression that the detailed propositions in general were based on firm biological grounds, thus legitimizing the entire legislative act.

Gütt's example was followed by Lösener. In a memo dated October 11, Lösener made an effort to convince the other discussants that the decisions on the status of the half-Jews must not be left to the vagaries of state officials but needed to be done according to fixed, predetermined rules. There were many reasons why Lösener considered the possibility of

[23] BArch R 18/5246 and YVA JM 3579, Bl. 35–39, 41–45.
[24] This point was also made by Essner, *Die "Nürnberger Gesetze,"* 158.

individualized assessment problematic, including legal and bureaucratic difficulties and his fear of creating fertile ground for corruption. In the first page of his memo, however, under the category "Sorting the half-Jews by decision of a public authority" (which he opposed), the first words were these: "The following reasons speak against this solution: A. Mendel's Laws themselves."[25] Lösener explained that each half-Jew contained thousands of German and Jewish characteristics, which "mendelize separately from each other." The most important were the internal qualities, which evaded direct inspection; and since any examined half-Jew would also transmit to his descendants his recessive hereditary material, it would remain "unknown how his children would mendelize." Lösener therefore argued that personalized sorting would lead to countless misguided decisions from the racial-biological perspective, and would aid the efforts of foreign countries to undermine the scientific basis of the Nazi regime.[26]

Lösener thus relied on Mendelian reasoning to support his stance. Moreover, he made sure to position Mendel's laws in a prominent place, on the very first page of his memo, with a separate, underlined heading, and to use Mendelian notions to substantiate his propositions. Gütt's rhetoric was therefore followed almost to the letter: when arguing about racial policy, one should begin with Mendel and show how one's proposals were informed by Mendelian teaching. Better still, if possible, one should show how one's propositions were in fact the direct and inevitable consequences of Mendelian theory.[27]

References to Mendel's laws were used to support other agendas as well. Advising the discussants on the future decree, the radical antisemite Karl Astel, Head of Jena University's Institute for Human Heredity and Racial Politics, expressed his fierce opposition to allowing half-Jews to marry Germans, explaining that each independently inherited trait was genetically indestructible and, once introduced, doomed to reappear in the German nation in ages to come. Like Gercke and Günther before him, he argued that any degree of intermixture demanded that the racially mixed be negated the right to marry the German-blooded.[28]

---

[25] BArch R 18/5246 Bl. 141–155. Bl. 141: "Material zur Lösung der Halbjudenfrage."
[26] Ibid.
[27] Notably, in an earlier draft, which might have been written before Gütt presented his memo, the discussion on the implications of Mendelian teaching received a less prominent position. See BArch R 18/5246 Bl. 107–117. "Bemerkungen zur Reichsbürgerrechtsverordnung und zur Blutschutzverordnung," esp. Bl. 108–109.
[28] BArch DS (ehem. BDC) B0026, Karl Astel, Bl. 1014–1016. See also ibid., 162–163. Responding to a formal inquiry on his scientific education, Astel wrote that his knowledge was acquired through Rüdin, Ploetz, Lenz, Darré, Günther "and most of all through personal multifaceted animal breeding from early youth." See BArch RS A0131 Bl. 1872.

A week before the final announcement on the definition of Jews and Jewish *Mischlinge*, minister Stuckart also adopted the prevailing rhetorical convention. "The half-Jew is neither identical in nature to the German, nor to the Jew... in the half-Jew there are, according to Mendelian laws, 50% Jewish and 50% German hereditary mass, wherein maybe the German hereditary mass is not the best." In earlier drafts, prepared four days earlier, Stuckart made no mention of Mendel. In the revised version, he adopted the customary format, not only mentioning Mendel's name but also positioning him up front.[29]

Hence, during the discussions that led to the formation of the supplemental decree to the Nuremberg Laws, Mendelian theory became a constant point of reference. It was used to legitimize the legislative procedure in general as well as to support conflicting stances on the preferred direction that the legislation should follow. These references may seem superficial, attesting more to the wish to clothe policy-making procedures in biological garb. But they might just as well testify to the opposite, that is, that no practical stance on the Jewish question could be considered legitimate unless it could find support in Mendelian principles and that, consequently, all those involved had to become acquainted with Mendelian teaching and find ways to integrate their belief system and suggestions with it. Later, we will see further evidence that Mendelian thinking indeed became part of the reasoning apparatus of various Nazi leaders, up to Hitler himself. Moreover, the reference to Mendel did not stop at the inner circles of Nazi bureaucracy, but was propagated to the general public. Official commentaries on the Nuremberg decree made explicit reference to Mendel as background to the new legislation, and posters and articles explaining the law to the German populace made visual reference to biological thinking in order to convey, compellingly, that the law was biologically grounded.[30] Indeed, for the non-expert, the Nuremberg decree must have seemed entirely grounded in Mendelian reasoning: visually speaking, the poster explaining the Nuremberg legislation to the lay public was very much akin to common textbook

---

[29] IfZ Munich F71–2, Bl. 128, 137. "Diktat Stuckart im Verbindungstab am 6.11.1935."

[30] Wilhelm Stuckart and Hans Globke, *Kommentare zur deutschen Rassengesetzgebung* (Munich: C. H. Beck, 1936), 4; Bernhard Lösener and Friedrich August Knost, *Die Nürnberger Gesetze über das Reichsbürgerrecht und den Schutz des deutschen Blutes und der deutschen Ehre* (Berlin: Vahlen, 1936), 21; See also the articles by Frick and by Stuckart that appeared in the German daily press, in BArch R 8034 II 1497, fol. 55, 87.

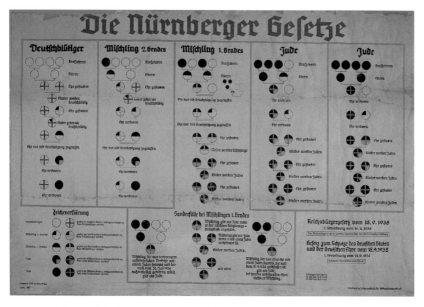

Figure 5.2 A 1936 poster explaining the Nuremberg Laws, using the conventions of Mendelian charts, that was hung on bulletin boards throughout Germany.
Source: Unites States Holocaust Memorial Museum, Item #1991.5.3

charts of Mendelian mechanisms (compare Figure 5.2 with Figure 5.3).[31] The relations between Mendel's laws and Nazi racial policy were made even more explicit for those who attended high school, as we will now see.

Teaching Mendelian Racism

Nazi educational materials, testimonies of those who were students during the Nazi period, and pictures taken in high schools during the first years of Nazi rule all provide ample evidence for the centrality of

---

[31] Strictly speaking, since the method for determining racial identity ultimately relied on ancestral charts, the poster shown in Figure 5.2 does not really represent a Mendelian mechanism. Nevertheless, as is quite apparent, there are notable similarities between its mode of representation and standard Mendelian schemes - from the use of black/white circles, through the magnified importance of unraveling inherited components by inspecting the **second** generation from the proband (of ancestors, in the Nuremberg decree; of descendants, in Mendelian schemes), to the affinity of the number of ideographic elements, which contain **four** grandparents, **two** parents and **one** offspring in one case, and **two** parents, **one** hybrid and **four** descendants in the other.

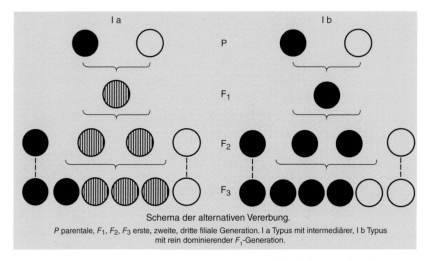

Figure 5.3 Mendelian mechanisms, as explained in a textbook on heredity, 1911.
Source: Valentin Haecker, *Allgemeine Vererbungslehre* (Brunswick: Friedrich Vieweg & Sohn, 1911), 220

Hans F. K. Günther's writings to the teaching of racial science.[32] Large tables showing the diversity of European racial types à-la-Günther were hung in classrooms, and students' own facial characteristics were compared with the depicted racial norms (see Figures 5.4 and 5.5). One teacher-trainee in a high school for girls reported how, after obtaining and preparing several measuring instruments,

[t]he method of measuring and the establishment of skull, head and facial indices were explained. Then, several relatively racially pure students were selected, measured, the indices calculated and these written on a table. In the tenth grade (*Untersekunda*) with which I worked, fairly good distinct racial types were available:

> One [girl] almost pure Nordic,
> One almost pure Phalic,
> One almost pure Eastern,
> One not-so-pure distinct Dinaric student, and
> One student, that exhibited the visible markers of the
>   East Baltic race.

---

[32] In addition to the examples below, see the detailed bibliographic analysis of Hans-Christian Harten, Uwe Neirich and Matthias Schwerendt, *Rassenhygiene als Erziehungsideologie* (Berlin: De Gruyter, 2006) and Gregory Paul Wegner, *Anti-Semitism and Schooling under the Third Reich* (New York, NY: Routledge, 2002).

Figure 5.4 The teaching of racial science in a German classroom, as shown in a 1934 article. Note that behind the chart showing facial types, there is another chart that is dedicated to the explication of Mendelian mechanisms.

Source: "Rassenkunde in der Volksschule," *Neues Volk* 7 (1934), 9

Figure 5.5 The teaching of racial science in a German classroom, as shown in a 1935 *Die Woche* article. Two of the five photos show students drawing charts of Mendelian heredity.
Source: Arbeitsgruppe Pädagogisches Museum (ed.), *Heil Hitler, Herr Lehrer: Volksschule 1933–1945, Das Beispiel Berlin* (Hamburg: Rowohlt, 1983), 104–105, reproduced from *Die Woche*, Berlin, October 1935

According to the teacher-trainee, the measurements taken from these five students indeed corresponded "in part completely, in part to a high degree" with the racial features of the five respective races.[33]

As stimulating as such class activities may have been, they also posed a risk, threatening to dissolve, rather than consolidate, national integrity. An article in the Nazi illustrated journal *Neues Volk* therefore stressed that, while the pride of being a member of the great Aryan race must be encouraged via children's activities, it was "up to the teacher to make sure that the seriousness of these exercises [in measuring heads] does not deteriorate into silly games (*Spielereien*) … the comparison of an almost documentary-good Nordic skull with a classmate of pure Phalic influence or with another, such as an Eastern, Western or Dinaric basic type,

---

[33] DIPF/BBF/Archiv Gut Ass 552a, "Über den Einbau der Vererbungslehre, Rassenkunde und Rassenhygiene in den Biologieunterricht der Untersekunda eines Oberlyzeums der Oberrealschulrichtung, 1933/1934," 28.

aims only at the planned working-out of similarities [and not of contra-distinctions], since all the blood and all the ancestors are German. The striking visible differences by no way mean inferiority or contradiction."[34]

The concerns of the article writers were not unfounded. One teacher-trainee with overt Nazi inclinations observed that "students at the ages of 15–16 make a big effort to play the inner-German races against each other. Under all circumstances, this needs to be avoided."[35] In his own classroom, he therefore intentionally refrained from performing racial analyses of the students themselves. First of all, he explained, doing so can mislead the young students to think that they can easily recognize racial affiliation, which can do more harm than good. Second, as said, he "wanted to avoid creating fissures within the school community, which [could] be determined by the different racial affiliations of each student. Thank god we overcame class antagonism, and we should do everything not to let racial antagonism take its place."[36] The teacher stressed the problematics of Günther's racial classification, the difficulty in identify-ing the mental features of different races, and the importance of avoiding overemphasizing the supremacy of the Nordic race. Similarly, in another high school for girls, where "Jewish and non-Aryan schoolgirls, as far as I could tell, were not in class," the students expressed their wish to be divided according to their race. The teacher refused, this time on the grounds that racial categorization cannot be done solely on the basis of appearance, but only with the help of extensive familial research.[37]

The importance of Günther's teaching notwithstanding, we will now see that the Nazis regarded Mendelian theory as supportive of, and integral and complementary to, their racial theories. Its centrality was expressed in several different areas. First, any attempt to teach about heredity without connecting it to racial science was denounced. Thus, one teacher who taught heredity in high school in 1935 without making explicit reference to its racial implications was reprimanded by his evalu-ator: "Important findings are withheld from the students. One gets the impression that heredity teaching was only handled as long as it found application in the teaching of hereditary health. [But] then racial teaching

---

[34] "Rassenkunde in der Volksschule," *Neues Volk* 7 (1934): 7–11.

[35] DIPF/BBF/Archiv Gut Ass 597, 31. See similarly a report by a student: "pupils continued during the break: 'you are totally Eastern, but I am Nordic.'" Arbeitsgruppe Pädagogisches Museum (ed.), *Heil Hitler, Herr Lehrer: Volksschule 1933–1945, Das Beispiel Berlin* (Hamburg: Rowohlt, 1983), 103.

[36] DIPF/BBF/Archiv Gut Ass 597, 28.   [37] DIPF/BBF/Archiv Gut Ass 1400, 4–5.

should have followed as a further development of heredity teaching. On this, nothing is reported."[38]

But this was probably a rare occurrence; more commonly, the teaching of racial science and the teaching of heredity science were intimately connected. A visual indication of this close connection can be seen in the same photographs referred to earlier. In the first picture (Figure 5.4), taken at a primary school classroom in Berlin-Wedding, students' physiognomy is compared with the facial types of European races according to Günther's scheme. Behind the table of racial types, however, another large chart may be observed. It shows the Mendelian mechanisms governing the crossing of the four o'clock flower (*Mirabilis Jalapa*) – a popular demonstration of Mendelian heredity in plants. Obviously, this chart was not left there by mere accident but was perceived by the teacher, the photographer and the journal editor alike to be an integral part of racial indoctrination. The 1934 article containing this photograph, entitled "Racial Study in Elementary School," gave considerable space to describing the centrality of Mendel's laws of heredity to the teaching of racial science. "The Mendelian law helped us to gain a glimpse into the secrets of the most sublime and most refined laws of nature, and all of a sudden it became clear to us what is meant by the concepts of racial-hygiene and Nordification (*Aufnordung*)."[39]

The second picture (Figure 5.5) was part of another article on "Racial Study in School" that appeared the following year in the illustrated weekly *Die Woche*. The article described how racial science was taught in a small school in a village in the vicinity of Lübeck. Two of the five pictures show the comparison of students' facial types with Günther's racial ideals; one centers on the striking similarity between two students who were identical twins; the remaining two photos show students drawing colorful representations of Mendelian regularities in flowers and snails. In the case of the latter, it is a crossing between a snail with a dark shell and one with a bright shell that "sharpens the [students'] feeling for the necessity of racial purity."[40]

In fact, however, the relations between Mendelian and racial teaching were more complex than what these pictures convey. First, Mendel's laws acquired several important roles in relieving the tension between the omnipresent idealization of the Nordic race, and the desire not to

---

[38] DIPF/BBF/Archiv Gut Ass 1464, "Vererbungslehre und Erbgesundheitspflege im Biologieunterricht der Untersekunda, 1935," evaluator's comments.
[39] "Rassenkunde in der Volksschule," 8.
[40] Arbeitsgruppe Pädagogisches Museum, *Heil Hitler, Herr Lehrer*, 104–105, reproduced from *Die Woche*, Berlin, October 1935.

alienate the majority of students, many of whom did not live up to the Nordic ideal. As the article in *Neues Volk* explained, "the wish, expressed spontaneously at the beginning by some boys, 'to want to look totally Nordic,' could easily and thoroughly be refuted by Mendel's law." This could be done by noting the fact that "not all hereditary material is externally visible. For example, an externally dark-looking type could be more spiritually Nordic than a man endowed with all Nordic markers."[41] The crucial point, then, was not the invisibility of some of the hereditary material, but the independent heredity of different racial features. Adopting this line of thought, one teacher referred in her classroom to the independence rule that was "once again important for racial study; for there it must indeed be acknowledged that the traits of a race do not necessarily remain together when inherited; one or even several traits alone cannot give away the essence of a man."[42]

Added to this notion on the independent heredity of different traits was the popular assumption, again following Günther, that the German *Volk* had 50% Nordic blood. When independent heredity and such presumed proportion of Nordic blood were taken together, it was easier to attribute Nordic traits also to non-Nordic-looking students. A 1934 textbook on *Heredity, Racial-Hygiene and Racial Science* therefore reminded its readers that "the incomparable achievements that the German *Volk* has obtained throughout history may be explained with reference to the high proportion, an average of 50 percent of Nordic blood that runs through the German *Volk*. This high proportion makes it also improbable that any German person would be totally free from Nordic hereditary material. Even if a person carries many physical markers of one of the other [European] races, this does not yet testify to his mental and spiritual disposition. For in most cases we are dealing with hereditary dispositions that, according to hereditary laws, mendelize independently."[43]

Finally, the non-alternating nature of the genetic material provided another opportunity for creating unity in the classroom, if not among the students themselves than at least between them and their joint forefathers. As one teacher stressed: "All the traits are [the results of] age-old hereditary material that binds all Germans directly with our ancestors back until the remotest of times. Together with the external appearance we inherited also their feelings and will and are responsible for keeping this blood and spirit pure."[44]

---

[41] "Rassenkunde in der Volksschule," 8.   [42] DIPF/BBF/Archiv Gut Ass 1400, 22.

[43] Gustav Franke, *Vererbung und Rasse: Eine Einführung in Vererbungslehre, Rassenhygiene und Rassenkunde* (Berlin: Verlag "Nationalsozialistische Erziehung," 1934), 140.

[44] DIPF/BBF/Archiv Gut Ass 1556, 42.

Mobilizing Mendelian thinking for teaching race had its opponents, too. Prominent among them was Ludwig Ferdinand Clauss, an influential racial theoretician. In 1934, he published, together with Arthur Hoffmann, a textbook designed especially for high schools, containing a teacher's manual, students' worksheets and photographs for classroom use.[45] Clauss noted that "the first thing that the keyword Race brought to students' mind was individual physical features, such as hair and eye color, for which they more or less could say, according to which rules they were inherited." For Clauss, such technical and atomistic understanding of race, which reduced it to features inherited separately and which transformed racial teaching into no more than an extension of hereditary science, was totally misguided. Students needed to learn how to see race as a holistic entity, each race endowed with its own particular soul, not as Mendelian features.[46] At least one teacher quoted Clauss' views approvingly.[47]

What Clauss seemed to have ignored was that the line between the holistic and the atomistic approaches often became blurred. The example of hair and eye color that Clauss mentioned is illustrative in this regard. Discussing the inheritance of blue eyes vs. dark eyes or of blond hair vs. dark hair was (and in many places still is) considered highly effective for teaching students about Mendelian mechanisms in humans. Successful instruction, one teacher explained, has to create vivid interest in the material studied, and this could be done more easily by using examples that would attract students' attention. The inheritance of eye or hair color – non-pathological traits that young people could identify in their own immediate and familial surrounding – was much closer to the students' world than the coat color of the guinea pig. Referring to such simple and well-known aspects of human heredity therefore allowed for "the most important conclusions to be drawn in the most impressive way."[48] Another teacher reported that, indeed, the mentioning of brown and blue eyes immediately raised great interest in the classroom, and all the students instantly turned to charting their own family trees.[49]

Yet beyond the pedagogical advantages, in the context of Nazi indoctrination these instances of hereditary patterns became loaded with racial connotations. In his 1936 booklet alternatively titled "What Must

[45] Ludwig Ferdinand Clauß and Arthur Hoffmann, *Vorschule der Rassenkunde, auf der Grundlage praktischer Menschenbeobachtung* (Erfurt: Kurt Stenger, 1934).
[46] Ludwig Ferdinand Clauß, *Rassenseelenforschung* (Erfurt: Kurt Stenger, 1934), 20–22; Ludwig Ferdinand Clauß, *Rasse und Seele* (Munich: J. F. Lehmanns, 1941), 15–19.
[47] DIPF/BBF/Archiv Gut Ass 1400, 29–30.
[48] DIPF/BBF/Archiv Gut Ass 812, "Die Vererbungslehre in der Untersekunda, 1936," 5.
[49] DIPF/BBF/Archiv Gut Ass 985, 24–26.

German Youth Know about Heredity" and "What Must the National-Socialist Know about Heredity," agricultural expert Albert Friehe explained Mendelian principles with the help of "an example that once had an ominous meaning for our *Volk*: a cross between a Jew and a Nordic woman, who has been especially exposed to Jewish lust."[50] Fortunately, by 1936 Friehe could declare that this horrible phenomenon had been prohibited in the Nuremberg Laws. Nevertheless, it provided a good opportunity for examining biological mechanisms. Considering hair color, the Nordic woman undoubtedly inherited her blond hair, as well as her blue eyes, from both of her parents (both traits being recessive, thus requiring inheritance from both parental sides). The Jew, however, was not racially pure but "a Near-Eastern-Oriental racial mixture with by-mixture of Negro and Mongol blood and hereditary material of all European races." Nevertheless, one could recognize among Jews a particular type with black hair – a deep black hair with bluish tinge that had "nothing in common" with the brown-black hair that could be found in the German population. Friehe then used a black rectangle to designate the Jewish black hair and a white rectangle to denote the blond hair color. A schematic diagram of reproduction processes demonstrated the results of crossing "a dark East European Jew and a woman of the Nordic race." The couple's children all had their father's bluish-black hair; it dominated over the blond.[51] Friehe continued to analyze the re- or self-crossing of the hybrids of the first generation; in this case, there were two *Mischlinge* that appeared black haired but carried a hidden, recessive, blond color within them. A fourth of their future progeny were expected to manifest the blond hair color; an apparent explanation thus emerged for the case of the "blond Jews." Further discussions on the inheritance of two traits and their possible combinations continued to refer back to the Jewish-German example and to its eventual racial implications.[52]

Not all educators were as explicit. One teacher assigned his students with the following "examples of dominant inheritance, that at first seem bewildering, that appear strange; precisely these [kinds of examples] give a certain charm to the solution." The examples were as follows: "1. Two dark-haired parents have a blond child. 2. Two brown-eyed [parents have] a blue-eyed [child]. 3. The father has brown, the mother blue eyes, the four children have blue eyes. How can this be explained?"

[50] Albert Friehe (1936). *Was muß die deutsche Jugend von der Vererbung wissen?* (Frankfurt a. M.: Moritz Diesterweg, 1936); Albert Friehe, *Was muß der Nationalsozialist von der Vererbung wissen?* (Frankfurt a. M.: Moritz Diesterweg, 1941).
[51] Friehe, *Was muß die deutsche Jugend*, 20–22.   [52] Ibid., 23–27.

The assignment was successful: not only did the students exhibit their proficiency in how dominant and recessive traits interacted but they also "realized by themselves that from the external appearance of a man it cannot be inferred on his inner qualities; through dominance, the heterozygosity can be covered."[53]

In a later session, the same teacher, after emphasizing to his students that "it is not races, but racial traits, that are inherited," paraphrased the Mendelian Independence Law. Using a four-on-four table (Punnett square) to analyze the possible outcomes of a cross between two organisms differing in two traits, he was able to demonstrate to his students "that only one of the 16 fields [in the table] shows the genotype of the original race P1, that this combination of traits appears with the mere probability of 1/16. With smaller numbers of descendants this pure race may not appear at all." Since among man, hundreds of different traits are considered, "the practical impossibility of regaining racial purity once a foreign hereditary material has been mixed in" became immediately evident. This, the teacher thought, laid the groundwork for racial teaching. "The German who marries a Jewess condemns all! of his descendants, mentally or physically, to bear the Jewish taint [*den jüdischen Makel zu tragen*]."[54] If this had not already been made clear, then surely by now the earlier "bewildering" examples of dominant inheritance ("father has brown, mother blue eyes") acquired their true meaning.

The fact that the seemingly neutral blond-dark or blue-brown dichotomy contained an underlying reference to two distinct racial types may also explain why in one classroom the teacher found it difficult to answer her female students' question: What about the inheritance of gray eyes? Instead of suggesting the possibility of a third hereditary factor determining gray eye color, the teacher clung to the binary alternatives and explained that "dominance does not always have to be complete; one can say only generally that the Bastard is more or less similar to the dominant parent."[55] In this classroom, as in many others, the students' homework assignment was to present a scheme for the inheritance of hair color – dark and blond; and later, so that they would internalize the independent heredity of features, their task was to draw "a scheme for the

---

[53] DIPF/BBF/Archiv Gut Ass 1556, "Methodische Behandlung der Erblehre in Untersekunda, 1936," 25.

[54] Ibid., 33–34.

[55] DIPF/ BBF/Archiv Gut Ass 1201, "Die Vererbungslehre im Unterricht der Oberprima und Untersekunda. Oberlyzeum und Lyzem, 1934," 30. The evaluator was not entirely satisfied with the teacher's reply to her students, commenting that the teacher's explanation was "factually not free from faults" (*Sachlich nicht ganz einwandfrei*).

inheritance of hair and eye color (dark haired, brown eyed x blond, blue eyed)."[56]

In addition to this specific example of Mendelian mechanisms with its overt racial connotations, German teachers routinely related the principal distinction between phenotype and genotype to Mendelian teaching in a way that had explicit antisemitic implications. Among researchers of heredity, the pair of terms genotype–phenotype, coined by the Danish researcher Wilhelm Johannsen, had a variety of meanings. For some, the two terms represented different levels of biological organization: whereas the genotype referred to something that resided inside the cells, the phenotype characterized the evolved organisms. For others, at stake were different stages of a biological process, genes being the initiators of a chain reaction that culminated with the eventual trait. Still others regarded the difference between the two as one of a degree of realization: the genotype represented merely a potential, which was materialized or checked by external impetuses, to finally yield a certain phenotype.[57]

Glossing over these subtleties, Nazi educators explained the difference between phenotype and genotype primarily, and sometimes exclusively, as an outcome of Mendelian theory. The mere fact of dominance and recessivity showed that genes and eventual traits did not coincide, because, in heterozygous organisms with dominant traits, some genes were not externally expressed. This narrow understanding of the phenotype-genotype distinction was spelled out in textbooks for teachers and conveyed repeatedly to high school students.[58] Such an equating of the phenotype-genotype distinction with dominance-recessivity enforced an extreme genetic-reductionist approach. If the reason genotype and phenotype are not the same lies only in the relations between hereditary factors, where a dominant trait conceals the recessive trait, the corollary is that in purebred or homozygous organisms there is no difference between the hereditary content and the visible traits. The traits of a homozygous organism are therefore, purportedly, identical to its genes. Implicitly, such a line of thinking presupposes that anything beyond the

---

[56] Ibid., 34–5. Similarly, see DIPF/BBF/Archiv Gut Ass 1464, 26. For a similar use of hair type and eye-color inheritance to show the results of European-African intermixture, see the following educational chart, used in Nazi classrooms and textbooks: https://collections.ushmm.org/search/catalog/irn5594.

[57] For a philosophical analysis, see Peter Taylor and Richard Lewontin, "The Genotype/Phenotype Distinction," in Edward N. Zalta (ed.), *The [Online] Stanford Encyclopedia of Philosophy* (Stanford, CA: Stanford University, 2017), https://plato.stanford.edu/archives/sum2017/entries/genotype-phenotype/.

[58] DIPF/BBF/Archiv Gut Ass 408, 33; Gut Ass 1038, 19-23; Gut Ass 985, 27; Karl Bareth and Alfred Vogel, *Erblehre und Rassenkunde für die Grund- und Hauptschule* (Bühl-Baden: Konkordia, 1937), 19, 22 and further examples below.

level of the genes is meaningless. Neither environment nor development plays any role in the translation of the genetic material into actual traits. The only thing that can cause a disruption between the genetic material (the genotype) and the manifested trait (the phenotype) is the presence of other genes.

Furthermore, this understanding also had racial implications. If the only reason for the genotype not to express itself externally is the masking of recessive traits by dominant traits, the danger of heterozygosity becomes plain. Here, Karl Bareth and Alfred Vogel's textbook on *Hereditary Theory and Racial Science for Primary and Middle-School* left very little room for interpretation:

> The *Mischlinge* deceive us through their outward appearance. Next to the visible [hereditary] dispositions they possess dispositions which are hidden from sight. These too are inherited. We distinguish genotype (*Erbbild*) from phenotype (*Erscheinungsbild*). Both are not the same. One must therefore never judge an organism by its phenotype. The most important is the **genotype**. One cannot see by the external features of an organism what hereditary dispositions are hidden in it.[59]

By definition, then, *Mischlinge*, or heterozygotes, were deceitful. Unlike pure types, whose external traits reflected their inner character, among the *Mischlinge* one could never know what sorts of genes they concealed. A few pages later, Bareth and Vogel summarize additional insights that "supply the foundation on which racial theory, hereditary health and racial fostering can be built."[60] Offspring of pure races are pure and always the same; mixing different races creates mixed-inheritance; offspring of *Mischlinge* are not similar to each other; hereditary dispositions are not all of the same strength; again, the genotype is more important than the phenotype; and never should one deduce from external appearance the inner qualities of an organism.[61] In other words, pure, homozygous organisms are constant and true to their nature; *Mischlinge*, Bastards and heterozygotes are variable and fraudulent, their outward appearance unable to testify to their essential, genetic identity, which in itself is in a state of constant imbalance.

In a similar fashion, following the discussion on dominant and recessive inheritance and with an eye on the analysis of German-Jewish intermixture, Friehe's textbook lamented the fact that "we cannot learn directly man's hereditary dispositions from the external phenotype." Indeed, it was only "through intensive ancestral research that we have the opportunity to substantiate also man's invisible genotype. The further back we follow man's ancestors, the more clearly his overall

---

[59] Bareth and Vogel, *Erblehre*, 23.   [60] Ibid., 29.   [61] Ibid., 29–30.

hereditary material will appear to us."[62] Another textbook similarly connected recessive inheritance to the inability to detect the Jewish nature of *Mischlinge*: "There are Jews that appear completely 'German.' They are children of mixed marriages." The reason for their German outlook is that the traits of the Aryan parent "are dominant. Is this *Mischling* therefore a German? No! The children or the grandchildren that this *Mischling* would grow would display absolutely Jewish racial features."[63]

Mendelian terms, mechanisms and concepts therefore became part and parcel of how young Germans learned to think about the danger lurking in mixing with Jews. These lessons were not extracurricular, but an integral part of the teaching of biology. Questions on Mendelian heredity were also common in the final biology exams for high school students. For example, in January 1934, almost all pupils who were asked, as part of their concluding exams in the Bismarck Gymnasium in Berlin, to write an essay on "The Biological Foundations of the Racial Cultivation of the *Volk*," adorned their essays with Mendelian charts, based their arguments on the analysis of these charts, and in general positioned Mendel at the center of their paper, together with Darwin and Weismann. Some students even went to the extent of attributing precedence to Mendel over Darwin by arguing that "Darwinism is largely based on Mendelian laws."[64] After the legislation of the Nuremberg Laws, in one Berlin higher secondary school for girls, the students were required to explain how the crossing of two species of *Drosophila* (fruit fly) could substantiate the scientific basis for the Nuremberg Laws, and particularly for the paragraph stating that quarter-Jews may marry full Aryans without negative hereditary implications. Many years later, the "*Mischling* second degree" Marlies Flesch-Thebesius would recall her childhood years as a student in the Anna-Schmidt high school in Frankfurt am Main, and describe her final exams in biology: "Above all, it concerned the hereditary teaching of the Augustinian priest Gregor Mendel, which laid the foundation for the entire racial teaching of the National-Socialists."[65] In the hands of Nazi educators, Mendelian

[62] Friehe, *Was muß die deutsche Jugend*, 21–22.
[63] Paul Magdeburg, *Rassenkunde und Rassenpolitik. Zahlen, Gesetze und Verordnungen*, Bildung und Nation: Schiftenreihe zur nationalpolitischen Erziehung (Leipzig: Eichblatt-Verlag (Max Zedler), 1933), 26.
[64] LAB A Rep. 020-31 Nr. 60.
[65] Marlies Flesch-Thebesius, "'Wir saßen zwischen allen Stühlen.' Als Mischling zweiten Grades in Frankfurt am Main," in Monica Kingreen (ed.), *"Nach der Kristallnacht." Jüdisches Leben und antijüdische Politik in Frankfurt am Main 1938–1945*, Schriftenreihe des Fritz Bauer Instituts, vol. 17 (Frankfurt: Campus, 1999), 415–434 (quote from p. 426).

Figure 5.6 A Hitler Youth instructor teaching the definitions of race
laid down by the Nuremberg Laws, 1937.
Source: Süddeutsche Zeitung. Photo: Scherl, Munich

teaching, racial science and racial legislation became closely intertwined;
to teachers and students alike, there was no doubt that underlying the
Nuremberg Laws were Mendel's hereditary laws.[66]

Diagnosing Race and Mendeling-Out Jewish Genes

In internal deliberations on the Nuremberg Laws, in the public domain,
and in high school education, the Nuremberg Laws were often discussed
and presented as based on Mendelian reasoning. Naturally, not everyone
was impressed by this Mendelian rhetoric. On April 13, 1938, one month
after Hitler's troops had crossed the border into Austria, officially annex-
ing the country, a Viennese middle school teacher by the name of Erwin
Meinhard sat down at his desk to write "A Memorandum on the Question of the Legal Status and Social Evaluation of Jewish *Mischlinge* First
Degree." Meinhard, himself a *Mischling* first degree, pointed out that the

---

[66] The examinations are described (in a slightly different context) in Sheila F. Weiss, *The
Nazi Symbiosis* (Chicago, IL: University of Chicago Press, 2010), 255–256.

fusion of confessional and racial-biological criteria in the wording of the Nuremberg Laws was prone to generate uncertainties and ambiguities in the handling of *Mischlinge*. He believed it was urgent to arrive at a better definition of *Mischlinge* and clarify their civil and social status; he even had several suggestions on how to do that.[67]

Meinhard began his memorandum by noting that the legal definition of Jews according to the Reich Citizenship Law shifted from a religious criterion – grandparental Jewish faith – to a racial-biological criterion – "German or related blood." He argued that if the same ideas were translated into anthropological terms, the law would read: "A Reich citizen / Reich civil servant can be only those state subjects (*Staatsange-hörige*) who do not deviate substantially in their genotype (easier would be: phenotype) from the average physical and mental racial markers present in the German and related peoples." There were many Jewish *Mischlinge*, Meinhard explained, whose traits corresponded to those of the general population; on the other hand, there were many pure Aryans who, if examined, would display significant influences of distant races – Near Eastern, Oriental and even Mongolian. Thus, instead of determining civil status according to genealogical registers, a hereditary-biological inquiry into the personality of the examinee should be given precedence. A pure Aryan could be relinquished of his civil status regardless of his origins if he displayed a-social characteristics; conversely, the status of Jewish *Mischlinge* should be determined according to their mental qualities, not their grandparental religious affiliation. Those who did not deviate substantially from the average German population should be granted full Reich citizenship; they would then regain their inner freedom, become full members of society and be granted the right to marry their fellow Germans. Such a practice, Meinhard insisted, would correspond better to the original intentions of the lawgivers and be truer to biological realities.[68]

Meinhard stressed that if his ideas were taken into consideration, they would soothe anxieties among people like him, who wholeheartedly supported the National-Socialist cause. Having completed his memorandum, Meinhard hesitated on whether he should send it and to whom. Eventually, he opted to keep it to himself. Seven months later, he changed his mind. On November 10, 1938, following *Kristallnacht* ("Night of Broken Glass"), as the fires in synagogues were still smoldering and Jews throughout Germany and Austria were being violently

---

[67] YVA M.69/10790, from Oestereichisches Staatsarchiv.   [68] Ibid.

herded into concentration camps, Meinhard sat down to write a letter that he addressed directly to Hitler's chancellery.

Reich minister Rust said in Reichenberg: "National-Socialism is applied biology!" In relation to the racial laws this saying is untrue, since: (a) a biological concept of "Aryan blood" doesn't exist. ... (b) All men are species-related ... because they are mutually fertile. But they are not racially related. This expression was intentionally avoided, because the racial relations of the peoples are not yet clear.

After referring to the lack of precise scientific knowledge on the bio-logical impact of crossing Germans and Jews, Meinhard advocated the replacement of the existing racial legislation with one that was scientific-ally grounded and biologically based. His April memorandum could explain how. "My proposal in this regard is attached."[69]

The impetus to send his letter directly to Hitler stemmed from Mein-hard's personal situation. The previous day saw the unleashing of ruth-less violence against Jews, Jewish property and Jewish religious symbols; but it also saw the twentieth anniversary of the official end of the Great War.[70] The contrast between Meindhard's emotional identification with the latter national event, and his principal exclusion from the German *Volk* as a *Mischling* first degree became unbearable. "Yesterday, the memorial for the heroes who fell for the German *Volk* was celebrated," wrote Meinhard. "On the same day, I made the decision to join the ranks of those who willingly lived, worked and felt themselves as Germans among the German people and Reich, and for whom it nevertheless became impossible to continue living here, due to a false – because of its unscientific nature – interpretation and application of the German Citizenship Law."

"Ancestral-wise, I am half-Aryan, but since my childhood I have always been nationally minded," Meinhard declared, and detailed his various services to the nation as an educator. Recently, however, and due to his official categorization as a *Mischling* first degree, he degenerated into material and mental distress. From his perspective, the situation seemed hopeless. "I voluntarily leave life, not without first sharing my thoughts on the injustice impinged on us." Folding his April memoran-dum into the envelope, Meinhard added, "Perhaps through my suicide

---

[69] Ibid.
[70] All major works on Nazi antisemitism include accounts of the *Kristallnacht*; see especially Saul Friedländer, *Nazi Germany and the Jews*, vol. 1, *The Years of Persecution 1933–1939* (New York, NY: Perennial, 1998), ch. 9; Alan E. Steinweiss, *Kristallnacht 1938* (Cambridge, MA: Belknap Press of Harvard University Press, 2009); Alon Confino, *A World without Jews: The Nazi Imagination from Persecution to Genocide* (New Haven, CT: Yale University Press, 2014).

I can contribute to making a change. Through my life I may have not
been able to reach that!" He signed the letter with a "Heil Hitler!"; and,
on top of the paper added "morituri te salutant!" – Latin for "those who
are about to die salute to you."[71]

Meinhard's (probably last) words made very little impact on Nazi
policies.[72] For the mass of people living under Nazi rule, racial affiliation
continued to be determined according to descent. Racial evaluation
through physical diagnosis remained a possibility, but it was chosen only
in exceptional cases and only as a last resort. More importantly, in the
majority of cases, anthropological examinations, when performed, were
not meant to determine directly one's own racial identity, but functioned
practically as a biological parental testing.[73] In this spirit, a letter sent on
December 1940 from the Reich Genealogical Office (*Reichsstelle für
Sippenforschung,* later *Reichssippenamt*) to Otmar Freiherr von Verschuer,
who at the time headed the University Institute for Heredity-Biology and
Racial-Hygiene in Frankfurt am Main, requested that racial evaluation
reports would summarize their findings by answering the following three
questions:

1. Can it be determined with certainty or with very high probability on
   the basis of hereditary grounds that ___ was the real father of ___, or
   can such fatherhood be seen, on the contrary, as very unlikely?
2. Is [the other disputed father] with high probability the father, or is his
   parenthood less likely?
3. Does the phenotype of the examinee have racial characteristics that
   point to a Jewish progenitor, or must it be assumed with certainty on
   the basis of the phenotype, that a German-blooded man was the
   examinee's progenitor?[74]

The order of the question could also be reversed; in a form prepared
on October 1943 by the same office, evaluators were requested to answer
the following three questions:

[71] YVA M.69/10790, from Oestereichisches Staatsarchiv.
[72] We do not know whether Meinhard indeed took his own life. What is known, however, is
that several hundred Jews committed suicide in the days surrounding the November
pogrom. See Christian Goeschel, "Suicides of German Jews in the Third Reich,"
German History 25, no. 1 (2007): 22–45, esp. 27–31.
[73] See Otto Reche, "Der Wert des erbbiologischen Abstammungsnachweises für die
richterliche Praxis," *Deutsches Recht* 9 no. 28 (1939): 1606–1612; Georg Lilienthal, "Arier
oder Jude? Die Geschichte des Erb- und Rassenkundlichen Abstammungsgutachtens," in
Peter Propping and Heinz Schott (eds.), *Wissenschaft auf Irrwegen: Biologismus –
Rassenhygiene – Eugenik* (Bonn: Bouvier, 1992), 66–84.
[74] YVA JM 11800, BArch 1509/77, fol. 72.

1. Does the examinee indicate in his phenotype, typical Jewish racial traits?
2. Can it be concluded thereon with sufficient probability that a Jew (___ name) was the progenitor of the examinee?
3. Or is it more likely to be assumed on the basis of the examination results that the German-blooded ___ (or another German-blooded man) was the progenitor of the examinee?[75]

Although the door was left open for the identification of one's racial status solely on the basis of physical markers – for instance, an anthropologist could determine, much in the spirit of Meinhard's above-mentioned proposals, that "traces of the Jewish race could not be detected in the child, whose phenotype corresponds well to the average German population" – the main goal of the anthropological inquiry was to subject the results of physical examination to the evaluation of parenthood.[76] Jewish racial traits could be used to imply that someone was Jewish, but their main target was to determine whether a certain proband indeed stemmed from a particular Jewish progenitor whose own "Jewishness" was genealogically verified and therefore uncontested. Replying to one such request for racial evaluation, Verschuer emphasized that "the phenotype of the examinee can support the assumption that a German-blooded man was Lieselotte K.'s progenitor only if a man, whose German-blooded ancestry must be certified, shows significant similarities with the examinee in the hereditary physical markers, in which the examinee displays differences from her mother."[77] Without such correlation of traits, the examinee's phenotype was principally meaningless.

This acknowledgment of the nature and limits of physical examinations was admitted at a very early stage. We have already seen it expressed in the discussions leading to the Nuremberg decree; as the examples above attest, it was also propagated to those involved in actually performing racial evaluations. The "Guidelines for the Proof of Aryan Ancestry with the Aid of Hereditary and Racial Reports" from mid-1935 made it clear that phenotype alone was not sufficient for racial assessment, and that only a racial inquiry into several generations made it possible to determine with a fair degree of certainty whether foreign

---

[75] YVA JM11797, BArch 1509/52, File 333, Doc. 6.

[76] Quote from a report sent on 6.6.1941 from Frankfurt University's (Verschuer's) Institue for Hereditary Biology and Racial Hygiene to the *Reichsippenamt* in Berlin, regarding Ida A. U., HHStAW 484/952.

[77] YVA JM 11800, BArch 1509/77, fol. 99–100. More on Verschuer's racial assessments, see Sheila F. Weiss, "The Loyal Genetic Doctor. Otmar Freiherr von Verschuer and the Institut für Erbbiologie und Rassenhygiene: Origins, Controversy, and Racial Political Practice," *Central European History* 45 no. 4 (2012): 631–668.

intrusion of blood – especially Negro or Mongolian – had taken place.[78] From the Nazi perspective, Meinhard's suggestion to determine race by measuring phenotypic similarities followed anthropological logic only superficially, but did not really make any biological sense. Official decrees would continue to stress this point throughout the years to come. "A racial evaluation of the examinee alone is out of the question, because according to experience it would not lead to any meaningful result." For the determination of descent, all the individuals in question had to be available for inquiry, and photos of all relevant family members needed to be at hand.[79]

Within these investigations of family resemblances, Mendelian reasoning was routinely and prominently applied. The first trait examined in parental valuations was blood type (A/B/O as well as M/N). Blood samples were taken from the proband, the legal father, the purported biological father and the mother, after which the possible hereditary relations of these individuals were analyzed. As with most other examined traits, such an analysis could never confirm but could sometimes refute the possibility that the purported father was the real biological progenitor. After blood type came the examination of a long series of physical markers. Here the dominance of darker pigmentation (of skin, hair or eyes) could be brought forward to evaluate the plausibility of disputed fatherhood: for example, if the proband had dark eyes and her mother "homozygous-light" eyes (*reinerbig-helle*), then the darkening hereditary factor must have stemmed from the father. Consequently, "a man with similarly homozygous-light eye color would be ruled out as the progenitor of the proband."[80] Finally, similarities in papillary ridges of the fingers were analyzed in an ostensibly Mendelian manner, the purported underlying Mendelian alleles ("genotype: VvRRUu") of every examined person noted and their relations analyzed, once again, to verify that the purported father could indeed be the real father.[81]

---

[78] "Richtlinien über den Nachweis der arischen Abstammung mittels erb- und rassenkundlicher Gutachten," YVA JM11797, BArch 1509/42, fol. 125–126.

[79] YVA JM11797, BArch 1509/43, fol. 53, 58.

[80] HHStAW 484/786, Evaluation of Irma L. F., 15.09.43. As noted above (p. 91), "*reinerbig*" (purely heritable) was the German term for homozygosity. For the use of similar reasoning to discuss hair color, see HHStAW 484/782, report from 31.10.36 and 484/775, report from 29.09.41: "The child's mother had light-blond hair. Since the child had dark-brown hair, a man with dark hair is to be presumed as a father."

[81] HHStAW 484/776, 777. On the theory underlying the Mendelian analysis of fingerprints, see Kristine Bonnevie, "Was lehrt uns die Embryologie der Papillarmuster über ihre Bedeutung als Rassen- und Familiencharakter? Teil III. Zur Genetik des quantitativen Wertes der Papillarmuster," *Zeitschrift für induktive Abstammungs- und Vererbungslehre* 59 (1931): 1–60; Wolfgang Abel, "Über Störungen der Papillarmuster I. Gestörte Papillarmuster in Verbindung mit einigen körperlichen und geistigen Anomalien," *Zeitschrift für Morphologie und Anthropologie* 36, no. 1 (1936): 1–38.

In addition to these direct applications of Mendelian thinking to racially oriented paternal evaluations, two Mendelian mechanisms became a particular cause of Nazi concerns. Both were related to the persistent existence of evasive, recessive Jewish genes. The first concern particularized the principle that the phenotype was a poor indication of racial identity, by arguing that it was Jewish traits in *Mischlinge*, not simply racial traits in general, that evaded direct inspection. On November 9, 1942, Friedrich Knost, Director of the Reich Genealogical Office, issued a circular stressing that lack of visible Jewish markers could not be used to challenge one's racial status: "as a rule, no definitive judgment on an examinee's Jewish or non-Jewish descent can be given solely on the basis of racial phenotype."[82] A ready-made template for a formal response letter clarified to potential applicants that the authorities were "unable to meet your application to go through a hereditary and racial examination on the basis of your and your relatives' external appearance. ... Even if an anthropological inquiry would substantiate your statement that you display no Jewish racial markers, this statement alone cannot invalidate your legal ancestry, because Jewish traits often do not become visible among *Mischlinge*."[83]

Therefore, in addition to high school students, applicants who tried to dispute their (or their relatives') Jewish status also became a target for racial indoctrination, their correspondences with Nazi authorities turning into another venue through which views on Jewish recessive undetectability could be communicated. For example, in 1942 Max S. from Breslau attempted to challenge the racial status of his daughter, claiming that his daughter's mother, his late first wife, was in fact not the child of a certain Jewess. The reply that he received denied the possibility of conducting a racial examination of the daughter, since both the mother and the Jewish grandmother were already dead, and comparison of their traits with those of the daughter was no longer possible. The daughter's status was therefore to remain a *Mischling* second degree. "A racial examination of your daughter alone is also out of the question, because according to experience, Jewish traits often do not become visible among *Mischlinge*," the letter explained. Other applicants received similar replies.[84]

Politically speaking, the *Mischlinge* problem was a difficult one to solve because many of the individuals in question were fully integrated into German society, including the armed forces. Harsh treatment of the

[82] YVA HM11797, BArch 1509/42, fol. 56–57.
[83] YVA JM11797, BArch 1509/43, fol. 48.
[84] YVA JM11797, BArch 1509/43, fol. 52–56.

Mischlinge could potentially arouse public discontent, as many of them had relatives, spouses or children who were themselves considered German-blooded. Furthermore, any official alteration of the status of *Mischlinge* was bound to create endless bureaucratic complications.[85] But from the biological point of view, the real problem with the *Mischlinge* was not the invisibility of their recessive traits, which made direct racial evaluation impossible, but the reverse: the inevitable future reemergence of those same traits if *Mischlinge* married each other. A short commentary on the Nuremberg Laws that appeared in *Volk und Rasse* in 1936 expressed these fears publicly. The author of the commentary supplied the alleged grounding for the legal prohibition of second-degree *Mischlinge* to marry among themselves. "Marriage between quarter-Jews is forbidden," he reasoned, "because the offspring of such matrimony have a higher probability of exhibiting the Jewish hereditary mass."[86]

Why was this purportedly so? Since the author did not explain, we may assume that he considered the answer self-evident, although, for the uninitiated, this statement may seem counterintuitive. If, for example, the proportion of "Jewish blood" in an individual had been considered to be the decisive factor for the manifestation of Jewish traits, the offspring of marriages between two quarter-Jews would have been two-eighth Jews, and should have been evaluated just like the quarter-Jews themselves. Conversely, if the generational distance of an individual from his/her Jewish ancestor had been of any significance, the children of two quarter-Jews would actually be in a better state than their parents, because their closest Jewish ancestor would have been their great-grandparent, not their grandparent. The "one drop of blood" rule also cannot help us explain why two-eighth Jews were more likely to express their Jewish character than quarter Jews. Evidently, neither generational distance nor blood proportion nor blood contamination was at stake. To the contrary: for those immersed in Mendelian reasoning, it was no wonder that generational distance was of little importance; Mendelian theory outwardly rejected the idea that the influence of genes waned from one generation to the next. On the other hand, one of the core principles of Mendelian thought was that recessive traits may only manifest themselves when the respective genes are inherited from both parental sides. This can explain why the offspring of two quarter-Jews "have a higher probability of exhibiting the Jewish hereditary mass." It was among

[85] Jeremy Noakes, "The Development of Nazi Policy towards the German-Jewish 'Mischlinge' 1933–1945," *Leo Baeck Institute Yearbook 34* (1989): 291–354.
[86] "Aus Rassenhygiene und Bevölkerungspolitik. Die Durchführungsbestimmungen zu den Nürnberger Gesetzen," *Volk und Rasse* 11, no. 1 (1936), 23.

them – but not among the quarter-Jews themselves – that Jewish dispositions could be inherited from both parental sides, re-combine, and express outwardly one's Jewishness.

The fear of recombining Jewish genes continued to haunt the Third Reich's leadership throughout the following years. On June 22, 1941, Germany attacked Soviet Russia, and the mass killing of Jews began. From the summer of 1941, the SS had forcefully advocated the inclusion of first-degree *Mischlinge* in the Final Solution, and high-ranking Nazi officials debated on whether "evacuation" (that is, murder), ghettoization or sterilization was the preferred way to deal with those *Mischlinge*.[87] In early October 1941, Hans Lammers, Head of the Reich Chancellery, discussed the fate of the *Mischlinge* with Walter Gross, who was the Head of the Nazi Party's Racial Policy Office. Lammers (according to Gross' account) thought that first-degree *Mischlinge* should be sterilized and that the marriages of second-degree *Mischlinge* should be restricted so as to "prevent, under all circumstances, the marriage of mixed-breeds of the second degree among themselves, because of the danger of transmitting Jewish characteristics in accordance with Mendel's laws." Gross had an ingenious solution to the problem; he replied that "one should thoroughly examine whether it was more advantageous to spread Jewish traits throughout the entire *Volk* [by allowing *Mischlige* second degree to marry only the German-blooded] rather than isolating them among a limited section of the community from which persons possessing an accumulation of Jewish characteristics could occasionally appear, who in turn could be eradicated in some way."[88] Noticeably, both sides of the conversation based their reasoning on Mendelian notions: Lammers by reiterating the idea that *Mischlinge* second degree harbor Jewish genes in a recessive form and that allowing them to marry each other would raise the danger that those genes would regain their homozygous nature "in accordance with Mendel's laws"; Gross by following the same logic but reaching the opposite conclusion. For him, the Jewish homozygosity that Lammers tried to avoid could be transformed from a risk into an opportunity, since through the counterintuitive measure of accumulating Jewish genes it allowed for the phenotypic manifestation, identification and extermination of these genes – along with their carriers.

---

[87] Noakes, "The Development of Nazi Policy," 338–340.
[88] Quoted in Saul Friedländer, *Nazi Germany and the Jews*, vol. 2, *The Years of Extermination* (New York, NY: HarperCollins, 2007), 292, originally from John Mendelsohn (ed.), *The Holocaust: Selected Documents in Eighteen Volumes*, vol. 2 (New York, NY: Garland, 1982), 284–285.

Admittedly, Gross' proposal was quite exceptional (even if not entirely unprecedented).[89] Anthropologists were eager to find methods that would help identify the presence of recessive genes among heterozygous carriers, but until they did, the overall policies of Nazi authorities were to avoid the recombining of pathological genes – and Jewish genes were naturally regarded as such – at all cost.[90] The *Mischlinge* problem was also an important part of the agenda of the Wannsee conference, convened on January 20, 1942, to coordinate the Final Solution of the Jewish Question. As its protocol attests, the second half of the meeting was devoted almost in its entirety to discussing the fate of the *Mischlinge*. In the spirit of the Nuremberg Laws, the status of second-degree *Mischlinge* was to be equated with that of German blood – unless a *Mischling* gave reasons to suspect that he was especially imprinted by Jewish influence. This was true in cases when the *Mischling* in question had a "racially especially unfavorable phenotype, that already externally sorts him with the Jews"; if he had an "especially lousy police and political record, that makes it evident that he feels and behaves like a Jew"; or (and prior to these) a *Mischling* was to be equated with a Jew if he "stemmed from bastard marriage (both sides *Mischlinge*)." Furthermore, with regard to children born from marriages between first-degree and second-degree *Mischlinge*, the Wannsee protocol stated explicitly that "racially, such children display as a rule a stronger Jewish blood-impact (*Bluteinschlag*) than the Jewish *Mischlinge* of the second degree." Thus, the Mendelian idea that two-sided inheritance of (Jewish) traits was particularly harmful

[89] For example, when discussing the problem of recessively determined feeble-mindedness, geneticist Reginald Ruggles Gates wrote in 1923 that "any tendency for those transmitting feeblemindedness to intermarry will have the desirable effect of bringing it to the surface where the individual can be segregated, rather than spreading the condition subterraneously by marriage with sound stocks." Reginald Ruggles Gates, *Heredity and Eugenics* (London: Constable, 1923), 159–160.

[90] See the excerpts from Hans Grebe, "Der Nachweis der Heterozygoten bei rezessiven Erbleiden," *Der Erbarzt* 11, (1943): 1–9, in Hans-Walter Schmuhl, *The Kaiser Wilhelm Institute for Anthropology, Human Heredity, and Eugenics, 1927–1945: Crossing Boundaries*, Boston Studies in the Philosophy of Science, vol. 259 (Dordrecht: Springer, 2008), 291: "But should it be possible to find a way to recognize the heterozygotes in the future, for other recessive genetic conditions as well?" The importance of this question for the "practical care of genes and race" "could not be underestimated." The 1950s would see the development of biochemical methods enabling the identification of heterozygotes, accompanied by renewed eugenic discussions much in the spirit of those mentioned above. See Diane B. Paul, "Genes and Contagious Disease: The Rise and Fall of a Metaphor," in Diane B. Paul, *The Politics of Heredity: Essays on Eugenics, Biomedicine, and the Nature-Nurture Debate* (Albany, NY: SUNY Press, 1988), 157–171; Keith Wailoo and Stephen Pemberton, *The Troubled Dream of Genetic Medicine: Ethnicity and Innovation in Tay-Sachs, Cystic Fibrosis, and Sickle Cell Disease* (Baltimore, MD: Johns Hopkins University Press, 2006).

also found its way into the formal coordination of the Final Solution, sealing the fate of certain categories of *Mischlinge*.

The same perception continued to inform Nazi policy in the year that followed. In the summer of 1943, Martin Bormann, head of Hitler's party chancellery, compiled a circular on the "Assessment of the Hereditary Factors of Jewish Mischlinge 2nd Degree in their Political Evaluation by the [Nazi] Party."[91] Bormann has already proved to be a hardliner in all matters related to *Mischlinge*; he also alluded to the tendency of *Mischlinge* descendants to mendel-out in subsequent generations.[92] As of June 1943, proposals had been made to draft *Mischlinge* first degree who had been exempted from military service to aid with the war effort. Special labor battalions were to be formed from such *Mischlinge* and were to be assigned the task of repairing damage caused by air raids.[93] As the circumstances of the war became less and less favorable, the need to recruit every available resource to support the military effort seemed ever more pressing. The question that Bormann addressed concerned the consequential need to determine the political reliability of second-degree *Mischlinge*. Could they be placed in positions of trust?

Acknowledging that second-degree *Mischlinge* were commonly treated as if they were German-blooded, Bormann explained that it was nevertheless a mistake to determine categorically that such *Mischlinge* only had one-fourth Jewish blood. The part of the Jewish blood they might have inherited could in reality fluctuate, from one-half to less than a fourth. Experience showed, Bormann argued, that the actual degree of penetration of Jewish blood could be readily assessed by the phenotype; evaluation according to the apparent features was therefore recommended for determining whether a second-degree *Mischling* could or could not be entrusted with a sensitive role. Bormann emphasized, however, that "in the case of a Jewish *Mischling* 2nd degree that stems from a so-called bastard-marriage, that is, a marriage where both partners are Jewish *Mischlinge* (e.g. the father is a *Mischling* 2nd degree, the mother *Mischling* 1st degree), it can be readily assumed even without closer inspection of the phenotype that the Jewish blood part prevails. In the case of such *Mischlinge* it is always to be decided against exceptional treatment [in their favor]."[94] Like the other statements analyzed above, these assertions cannot be understood unless we acknowledge their underlying Mendelian suppositions.

---

[91] YVA R.2/30, identifier 3667626, p. 151–153, "Bewertung der Erbanlagen von jüdischen Mischlingen 2. Grades bei ihrer politischen Beurteilung durch die Partei," 22.8.43.
[92] Noakes, "The Development Nazi Policy," 334.   [93] Ibid., 351.
[94] This passage was emphasized in its entirety in Bormann's circular.

The same Mendelian principles also set no principal limit to the number of generations that recessive traits could be inherited without being detected. Hitler himself referred on several different occasions to the resulting persistent, almost unstoppable proliferation of Jewish genes. On December 1, 1941, he addressed the tendency of Jewish *Mischlinge* to copulate again with Jews. Fortunately, he believed, nature itself took care of eradicating the damaging elements, and the Jewish parts were usually out-mendeled in the seventh, eighth or ninth generation, so that blood purity was regained.[95] Half a year later, Hitler sounded less optimistic, claiming this time that "experience shows that descendants of Jewish [*Mischlinge*] even after four, five, six generations continue to out-mendel pure Jews. These out-mendeled Jews constitute a great danger!"[96] Returning to the same topic two days later, Hitler referred to the "many blond, blue-eyed Jews, among whom will always be found the spokespersons for Germanisation. But precisely among Jews it has been proven without doubt that, even if the appearance and the character-tendencies have been totally dispersed for one or two generations, in the next generation they nevertheless uncover themselves again most clearly and reappear."[97] Two months later, in July 1942, Hitler insisted again that "among *Mischlinge* – even if the amount of Jewish blood is very small – in the course of generations, over and over again, a racially pure Jew out-mendels. The Jewish nation (*Volkstum*) is simply tougher."[98]

These remarks by Hitler indicate that he, too, accepted the idea that Jewish traits were inherited independently, that they could be undetected or unmanifested for several generations and that they could then recombine and reemerge; and that he related these ideas explicitly to Mendel's teaching. For Hitler, whose antisemitic anxieties drew on multiple sources and acquired changing emphases under different circumstances, Mendelism provided another foundation for fear of the biological powers, or racial tenacity, of Jews.

In the spring of 1943, when the extermination of Jews was at its peak, Bormann received a letter from Heinrich Himmler. Referring to a previous discussion the two had held on Jewish *Mischlinge*, Himmler mentioned a proposal by Bruno K. Schulz, a racial-anthropologist and high-ranking officer in the SS Race and Settlement Main Office (RuSHA). Contrary to the spirit of Nuremberg, yet echoing the spirit

---

[95] Henry Picker, *Hitler's Tischgespräche im Führerhauptquartier, 1941–1942* (Munich: Deutscher Taschenbuch Verlag, 1968), 36; Friedländer, *Years of Extermination*, 278.
[96] Picker, *Tischgespräche*, 125, conversation from 10.05.1942.
[97] Ibid., 135–136, conversation from 12.05.1942.
[98] Ibid., 204, conversation from 01.07.1942.

of Wannsee, Schulz suggested that *Mischlinge* second degree be racially examined so that the treatment of those whose phenotype betrayed Jewish traits could be equated with either that of the *Mischlinge* first degree or with Jews.[99] Himmler asserted the following:

I consider such [racial] examinations absolutely essential. Perhaps not only for *Mischlinge* second degree, but also for *Mischlinge* of higher degrees. Only between the two of us – we must apply the same measures here as one does in plant or animal breeding. The offspring of such *Mischling* families must be thoroughly racially examined by independent institutions for several generations at least (3 or 4 generations) and, in case of racial inferiority, sterilized, and thereby prevented from further reproduction.[100]

The war ended before the Nazi leadership had the chance to fulfill all of its racial-hygienic fantasies. What should be clear by now is that these fantasies themselves incorporated Mendelian notions, ideas, perceptions, mechanisms and terminologies. Evidently, Mendelian thinking would not have met such an open embrace from Nazi thinkers if it did not merge so easily with other antisemitic concerns. Central among these was the Nazi preoccupation with "the Jew inside us."[101] "When will a new Luther appear ... and free us from the Jew inside us and outside us [...] ?" pleaded the protagonist of Artur Dinter's 1917 best seller, *Sin against the Blood*.[102] Point 24 of the Nazi Party program similarly referred to the need to combat the Jewish-materialistic spirit within and around us. "What would we gain if we killed all the Jews around us, but left the Jew in ourselves still alive?" asked a *völkisch* leader in 1924.[103] "The Jew inside" was an eponym for the fight against materialistic or international spirit, or any other tendency that the Nazis deemed un-German. Mendelian thinking turned this language from metaphor into a concrete description of biological reality; within the national body, it argued, there

---

[99] On Schulz and his proposal, see Isabel Heinemann, "Ambivalente Sozialingenieure? Die Rasseexperten der SS," in Gerhard Hirschfeld and Tobias Jersak (eds.), *Karrieren im Nationalsozialismus. Funktionseliten zwischen Mitwirkung und Distanz* (Frankfurt a. M.: Campus, 2004), 73–95 (here 81).

[100] See Helmut Heiber (ed.), *Reichsführer! ... Briefe an und von Himmler* (Stuttgart: Deutsche Verlags-Anstalt, 1968), 213. The letter, dated May 22, 1943, appears in a slightly different translation in Noakes, "The Development of Nazi Policy," 347–348.

[101] Cornelia Essner, "Nazi Anti-Semitism and the Question of 'Jewish Blood,'" in Christopher H. Johnson, Bernhard Jussen, David Warren Sabean and Simon Teuscher (eds.), *Blood and Kinship: Matter for Metaphor from Ancient Rome to the Present* (New York, NY: Berghahn, 2013), 227–243.

[102] Artur Dinter, *Die Sünde wider das Blut* (Leipzig: Matthes und Thost, 1920), 137; translation according to Essner, "Nazi Anti-Semitism," 228.

[103] Quotes in ibid, 230. An insightful analysis of the later political function of the cultural processes of *Entjudung* is provided by Peter Longerich, *Holocaust: The Nazi Persecution and Murder of the Jews* (Oxford: Oxford University Press, 2010), 77–85.

were indeed Jewish traits that were still viable and that needed to be eradicated – or otherwise, in later generations, they would resurface.

The roots of these cultural anxieties ran even deeper. With the progress of Jewish emancipation in the nineteenth century, the fact that Jews became indistinguishable from their Christian "hosts" was perceived by antisemites as a genuine societal threat.[104] Secularization and assimilation aggravated the precariousness of Jewish invisibility. Linking Jews and Jewish *Mischlinge* to the threat of recessive genes fitted perfectly into this framework of thought. As noted in Chapter 3, the affinity between the meanings attributed to recessive traits and popular antisemitic fantasies helped to substantiate both antisemitism and biology as related and mutually enforcing. As a final example, we may consider the popular 1940 Nazi propaganda film *Jud Süß*. The movie tells the story of the city of Württemberg, which functions as a symbol for the German *Volkskörper*. The city remained secure so long as it kept Jews at a safe distance. The opening scene depicts an act of inheritance – a legal, not a biological one: a new Duke is crowned, receiving the right and obligation to rule from his deceased father. But the infallible character of this new Duke, his lust for niceties and desire to seduce and spoil women, lead him to open the gates of the city to a Jewish merchant, Joseph Süß Oppenheimer, who could provide him with jewelry and financial backing. The Jew enters the city of Württemberg in disguise, having shaved his beard, changed his clothes and taken a ride in the carriage of a young Christian woman, whom he would later attempt to seduce and finally rape. His entrance into Württemberg results in the complete contamination of the city by Jews. When at the end of the movie Süß is put on trial, he is charged with bringing into the city "lies, lies, and ever more lies." The cunning nature of Jews (the "masters of the Lie"), their tendency to delude and deceive, their habit of penetrating into the national body by exploiting the innocence of female carriers and by adopting German-Christian customs, behaviors or names – all these antisemitic images corresponded well with the idea that at least some of the Jewish traits were recessive, hidden, unavailable for direct inspection or simply

---

[104] Cf. Steven M. Lowenstein, Paul Mendes-Flohr, Peter Pulzer and Monika Richarz (eds.), *Deutsch-jüdische Geschichte*, vol. III, *Umstrittene Integration 1871–1918* (Munich: C. H. Beck, 2000), 151ff. See also the discussion in Zygmunt Bauman, *Modernity and the Holocaust* (Ithaca, NY: Cornell University Press, 1989), ch. 2–3; Sander L. Gilman, "The Jewish Nose," in Gilman, *The Jew's Body* (New York, NY/London: Routledge, 1991), 169–193.

"covered" (*überdeckt*), the German word for "recessive."[105] Identifying and eradicating recessive Jewish genes therefore became one of the central goals of biologically minded Nazis. And it was Mendel's laws that both defined the problem as well as outlined its possible solutions on the path toward national purification and racial regeneration.

---

[105] See the antisemitic Ernst Hiemer, *Der Jude im Sprichwort der Völker* (Nuremberg: Der Stürmer Buchverlag, 1942), ch. 4: "Meister der Lüge."

Epilogue: Social Mendelism beyond the Nazis

Postwar Mendelian Racial-Hygiene

At the close of the war, foreign military powers formally abolished Nazi racial discrimination laws in all the territories they possessed, and criminalized Nazi policies that had been based on racial or religious hatred. Some of the perpetrators were prosecuted, and survivors gradually became entitled to certain forms of material compensation. Doctors who took part in the sterilization of Jews, Sinti or Roma could therefore find themselves facing the bench, their actions constituting what the allies now defined as "crimes against humanity."[1]

Not so with those who sterilized Germans. Population policy measures, eugenics and sterilization, as long as they were not motivated by racial or religious animosity, were not regarded by the allies as reprehensible or unlawful. Practically speaking, sterilization operations were discontinued, and in the American occupation zone the Higher Hereditary Court was banned from meeting. Hereditary Courts also remained dormant in other occupied territories in western Germany; in the Soviet Occupation Zone and in Bavaria, the Sterilization Law itself was revoked, and in Baden-Württemberg and in Hessen it was suspended. But in most German lands the law remained in force, at least *de jure*.[2] Moreover, the influx of medical reports with recommendations for sterilization did not cease: in Bremen alone, for example, between October 1945 and

---

[1] George J. Annas and Michael A. Gordin (eds.), *The Nazi Doctors and the Nuremberg Code* (New York, NY: Oxford University Press, 1992); Arieh J. Kochavi, *Prelude to Nuremberg: Allied War Crimes Policy and the Question of Punishment* (Chapel Hill: University of North Carolina Press, 1998); Paul J. Weindling, *Nazi Medicine and the Nuremberg Trials: From Medical War Crimes to Informed Consent* (New York, NY: Palgrave Macmillan, 2005).

[2] On the legal entanglement and the state of the Sterilization Law after the war, see Roland Zielke, *Sterilisation per Gesetz. Die Gesetzesinitiativen zur Unfruchtbarmachung in den Akten der Bundesministerialverwaltung (1949–1976)* (Berlin: Buchmacherei, 2006), 37–44; Daphne Hahn, *Modernisierung und Biopolitik. Sterilisation und Schwangerschaftsabbruch in Deutschland nach 1945* (Frankfurt a. M.: Campus, 2000), 52–53; Stefanie Westermann, *Verschwiegenes Leid. Der Umgang mit den NS-Zwangssterilisationen in der Bundesrepublik Deutschland* (Cologne: Böhlau, 2010), 60–74.

May 1949, appeals for no less than 329 sterilizations arrived at the city medical officer's desk.[3]

Under the active encouragement of military authorities, German doctors and jurists in the American Occupation Zone attempted to re-establish a workable legal framework that would enable them to continue combating what they regarded as looming genetic dangers. To resolve the legal ambiguities, they worked together to formulate a new sterilization law. Psychiatrist Werner Villinger coordinated these efforts from Stuttgart. In addition to his impressive record of sterilization operations as chief physician of the mental hospital at Bethel, Villinger's résumé included a term as Higher Hereditary Court judge at Hamm, performing medical experiments on psychiatric patients, and providing expert advice to the Nazi Euthanasia program. He would go on to a successful career in the Federal Republic of Germany, which was only brought to an abrupt end when he fell to his death while hiking in 1961, just after news of his involvement in Nazi crimes began to surface.[4] Two years after the war, however, he was preoccupied with putting sterilization policies back on track. He was not alone. Other eugenicists and many official bodies, such as the Wiesbaden Health Committee, the Hessian Ministry of Justice, the Hessian Department of Health, the Bremen state health office and the Stuttgart Ministry of Interior, took active roles in commenting on a "Draft for a Sterilization and Refertilization Law," which began to take form in the summer of 1947.[5]

The wording of the draft drew heavily on the existing Sterilization Law, while making several notable changes. Voluntarism replaced coercion, and the list of diseases that had featured in the Nazi law was removed. Eugenicists nevertheless feared that these changes alone would not be sufficient to convince the public, both at home and abroad, of the necessity and legitimacy of the reformulated law. They therefore set out to delineate a justification for the planned legislation. A first draft was drawn up by the Munich psychiatrist Bruno Schulz, who, "like another

---

[3] Asmus Nitschke, *Die "Erbpolizei" im Nationalsozialismus: Zur Alltagsgeschichte der Gesundheitsämter im Dritten Reich* (Opladen/Wiesbaden: Westdeutscher Verlag, 1999), 267.

[4] Hans-Walter Schmuhl, "Zwischen vorauseilendem Gehorsam und halbherziger Verweigerung. Werner Villinger und die nationalsozialistischen Medizinverbrechen," *Nervenarzt* 73, no. 11 (2002): 1058; Martin Holtkamp, *Werner Villinger (1887–1961). Die Kontinuität des Minderwertigkeitsgedankens in der Jugend- und Sozialpsychiatrie* (Husum: Matthiesen, 2002); in English, see Thomas Röder, Volker Kubillus and Anthony Burwell, *Psychiatrists: The Men behind Hitler* (Los Angeles, CA: Freedom Publishing, 1995), 98–103.

[5] See the working paper addressed to the members of the Wiesbaden Health Committee, prepared in August 1947, in MPIP-GDA 79.

pioneer of modern genetics, Gregor Mendel, was the son of a gardener."[6] Schulz took Rüdin's place as head of the Genealogical-Demographic Department when Rüdin was on leave in Basel from 1925–1928. He continued to work with Rüdin, published influential and internationally renowned statistical studies of genetic psychiatry, and replaced Rüdin again in 1945, when Rüdin was temporarily interned. In October 1947, Schulz composed a "Proposal for Grounds for the Sterilization Law Draft."[7] It embodied, and anticipated, the rhetorical strategy that German eugenicists were to adopt from that point onward. It began as follows:

Germany received its first law on the sterilization of the hereditarily diseased in 1933, under the National Socialist government. It might seem surprising that after the overthrow of National Socialism, the old sterilization law was suspended and immediately a new one was proposed. However, the idea of sterilizing the hereditarily ill is by no means something peculiar to National Socialism alone. Thus, the first sterilization law was issued in 1907 in the North American state of Indiana, and today such laws exist, some of them forced sterilization laws, in 27 states of the USA. The first legal regulation of eugenic sterilization in Europe took place in 1928 in the Swiss canton of Vaud, followed by Denmark in 1929. A corresponding law was in preparation in Germany in 1932. It was supported with particular urgency by the Social Democrats.[8]

Having established that there was nothing particularly Nazi about a sterilization law, Schulz moved on to discuss the pressing scientific and medical realities.[9] Experience, he argued, showed that certain diseases that caused misery to individuals and to their families were hereditary; science had managed to discover the mode of heredity underlying these

---

[6] This was how Schulz was described by his colleague, Franz Kallmann following Schulz's death. Franz J. Kallmann, "In Memoriam Bruno Schulz, 1890–1958," *Archiv für Psychiatrie und Nervenkrankheiten* 197, no. 2 (1958): 121–123. The psychiatrist Bruno Schulz should not be confused with the SS officer and anthropologist Bruno K. Schulz, mentioned in the previous chapter; and neither of them has anything to do with the Polish-Jewish prose writer who carried the same name.

[7] MPIP-GDA 79, "Vorschlag einer Begründung zu dem Gesetzentwurf über Sterilisierung."

[8] Ibid.

[9] The claim that the sterilization policy need not be regarded as "Nazi" would continue to appear in medical, judicial and public spheres in the following decades, with negative implications for those who would seek recognition or compensation for their suffering. See, for example, Hans Nachtsheim, "Die Frage der Sterilisation vom Standpunkt der Erbbiologen," *Berliner Gesundheitsblatt* 1, no. 24 (1950): 603–604; Nachtsheim, *Für und wider die Sterilisierung aus eugenischer Indikation* (Stuttgart: Georg Thieme, 1952); Westermann, *Verschwiegenes Leid*, 60–80. Interestingly, in comparing the Sterilization Law with the parallel U.S. policy, the Germans were taking their cue from the Americans themselves, whose Military Tribunal openly declared on July 2, 1947, that there was nothing illegal about sterilization. See Zielke, *Sterilisation per Gesetz*, 39.

diseases; and sterilization was the only certain way to prevent carriers of such diseases from propagating. Schulz noted that "even when the respective persons, in times of quiet reflection, have the firm intention to avoid procreating, they are nevertheless not always able to control themselves as is necessary. This applies in particular when certain mental disturbances exist, especially when it comes to feeble-mindedness."[10] Sterilization was therefore unavoidable.

Schulz further argued that when considered from a purely scientific perspective, the 1933 law seemed entirely appropriate. Nevertheless, the new draft stressed that only the sterilization candidate or his legal guardians could authorize sterilization. This modification was not necessarily a hindrance, but an opportunity: "The principle of voluntarism being thereby safeguarded, it will allow for the substantial widening of the circle of persons that can be considered eligible for eugenic sterilization."[11] Sterilizing all cases of epilepsy – even against the will of those afflicted – could justly be disputed; but neither would it make much sense to prevent epileptics who wish to be sterilized from doing so, even if the exact hereditary course of their disease was not entirely clear.

Yes, one asks oneself, whether in certain cases, individuals who are phenotypically completely healthy should not also have the right to be sterilized, that is, if they are to be safely regarded as carriers of the hereditary factor for a severe hereditary disease.

Think of the parents of a child who suffers from amaurotic idiocy. Here it can be determined with certainty, that any child of such parents has the average prospect of one fourth to become ill with this severe hereditary disease. His prospects to be hereditarily and externally healthy are one fourth. His prospect of being externally healthy but, again, a carrier of the diseased hereditary disposition, just like his parents, are two-fourths. But precisely such – hidden – carriers present a danger to the general public, for they unknowingly transmit further the disposition to the disease.[12]

At the bottom of the page, Schulz scribbled a small diagram: a basic scheme of recessive Mendelian inheritance (Figure 6.1). Omitted from the copy Schulz would later forward to Villinger for consideration, this diagram was not destined to communicate his ideas further, but only to help Schulz himself reason about the issue at hand, properly compute the ratios mentioned in his literal description, and explain why carriers of recessive diseases, such as amaurotic idiocy, should be allowed the privilege of being sterilized. On October 30, 1947, Schulz sent his justification draft to Villinger.[13]

---

[10] MPIP-GDA 79, "Vorschlag."   [11] Ibid.   [12] Ibid.
[13] MPIP-GDA 79, Schulz to Villinger, 30.10.1947.

- 3 -

derum Träger der krankhaften Erbanlage zu sein,wie die Eltern selbst,
beträgt zwei Viertel.Grade derartige - verdeckte - Anlageträger aber
stellen eine Gefahr für die Allgemeinheit dar,da sie unerkannt die
Anlage zu dem Leiden immer weiter vererben können.

Beschränkt man sich hier strikt auf die Fälle,in denen die mit
Sicherheit als Träger der Krankheitsanlage festgestellten Personen
selbst den Wunsch zur Sterilisierung haben,so ist nicht einzusehen,
weshalb diesem Wunsche,der dem Wohl der Allgemeinheit entgegenkommt,
nicht willfahrt werden sollte.

Man könnte bei dieser Sachlage fragen,warum man nicht über=
haupt die Sterilisation ganz allein von dem Wunsche des einzelnen
abhängig sein lassen und auf eine gesetzliche Regelung verzichten
sollte.Das erscheint jedoch deshalb bedenklich,weil es sich bei der
Sterilisation um eine verhältnismässig einfach durchzuführende und
gefahrlose Operation handelt,zu der sich jemand in einer augenblick=
lichen Stimmung oder Verstimmung leicht entschliessen könnte,während
er nach der Durchführung der Operation bedauert,nun der Fortpflan=
zungsfähigkeit beraubt zu sein,und zwar mit Wahrscheinlichkeit un=
wiederbringlich,da sich der Erfolg der Operation nur in einem Bruch=
teil der Fälle wieder rückgängig machen lässt.Um nach Möglichkeit
jeden Missbrauch,jeden Irrtum und jede Übereilung zu verhindern,
scheinen grade bei einer derartigen Operation gewisse Sicherungen er=
forderlich,wie sie das hier neu in Vorschlag gebrachte Gesetz bieten
soll.Und schon gar nicht wird man dort,wo der für die Sterilisation
in Betracht Kommende an einer geistigen Störung leidet,ihm und einem
einzelnen Arzt allein die Entscheidung überlassen können.Anderer=
seits erschien es angezeigt,in den Fällen,in denen der Betroffene ,
da nicht oder nicht voll geschäftsfähig,sondern sein gesetzlicher Ver
treter die Einwilligung zur Sterilisation zu erteilen haben wird,
auch noch die Zustimmung des Vormundschaftsgerichtes zur Bedingung
zu machen,wobei wohl zu vermuten ist,daß das Vormundschaftsgericht
mit der Erteilung der Zustimmung dann besonders leicht zurückhaltend
sein wird,wenn damit zu rechnen ist,daß der Betroffene in absehbarer
Zeit als voll geschäftsfähig anzusehen sein wird.

Figure 6.1 A 1947 draft for justification of a postwar sterilization law.
Note the Mendelian diagram at the bottom.
Source: MPIP-GDA 79 ("Vorschlag einer Begründung zu dem Gesetzentwurf
über Sterilisierung")

In addition to Schulz, another scholar who became involved in commenting on the upcoming law was Fritz Lenz. Lenz was clearly unhappy with the emerging draft. On November 10, he detailed his reservations in a letter to Georg Stertz, a psychiatrist whose career had been put on hold in 1937 because of his marriage to a Jewish woman – the daughter of the famous neuropathologist, Alois Alzheimer. In 1946, Stertz was appointed director of the psychiatric clinic of the University of Munich. Lenz wrote to Stertz, "Above all I find it not good, that voluntary sterilization should be restricted to cases of '**severe**' hereditary diseases. Carriers of severe hereditary disorders anyway breed little or not at all." Therefore, "from a eugenic perspective, sterilizing the severely hereditarily diseased is not urgent. A hundred times more important would be the sterilization of slightly feeble-minded and a-social psychopaths." Moreover, "the female carriers of the gene for hemophilia and the gene for Leber's congenital amaurosis [a rare genetic eye disorder] would not be candidates for sterilization according to the Stuttgart draft, even though on average half of their sons would become sick." Lenz thought that the law should explicitly target not the sick themselves, but those whose children were expected to become "feeble-minded, deformed, a-social or otherwise disposed to disease." It was obvious for Lenz that neither the state nor any individual had any interest in rearing children with such diseases.[14]

Adjusting himself to the spirit of the time, Lenz did not fail to emphasize that sterilization should only be implemented if it was according to the will of the sterilized persons or their legal guardians. The fact that the 1933 law was compulsory was "a grave psychological mistake." When it came to the feeble-minded, it was the consent of parents that was to be sought; one simply could not wait until the feeble-minded were adults because "feeble-minded young girls often beget children already when they are 14 or 15." Be that as it may, Lenz thought that it would be wiser to leave the entire matter of sterilization unresolved for the time being, until the general atmosphere toward such matters improved.[15]

Disregarding the last point, Stertz found Lenz' proposals "not bad."[16] Schulz, who took part in the exchange of views, was also in agreement with Lenz' position. On December 3, 1947, Schulz reported to Lenz that he had appeared in person at the meeting of the Stuttgart committee, and that he had also written to Villinger to express his reservations on the proposed law. "As you can see, in particular I speak about the fact that phenotypically completely healthy persons, under certain conditions,

[14] MPIP-GDA 79, Lenz to Stertz, 10.11.1947.   [15] Ibid.
[16] Ibid., Stertz to Schulz, 26.11.1947.

should nevertheless be candidates for sterilization." When speaking at the meeting, he gave the example of hemophilia to explain the latter point.[17] Lenz' views were also presented to Villinger by Stertz during a meeting held in January 1948.[18] These combined efforts seemed to have succeeded: contrary to the original sterilization law draft, that targeted any "person, who suffers from a severe hereditary disease," the revised version from February 9, 1948 explicitly included as sterilization candidates also those who were not themselves ill, but were "carriers of a severe hereditary disease."[19]

Schulz' justification draft was also reworked by Villinger. The revised version retained in its opening paragraph the reference to sterilization laws outside of Germany, "most of which [were] in the United States of America," and proceeded to emphasize the scientifically based need to prevent the suffering of individuals and of their carers. "Accordingly, it would be best from the standpoint of humanitarianism, as well as from that of the cultivation of health, if carriers of such diseased hereditary dispositions abstain from procreating. To make it easier for them to safely prevent their own procreation and to rule out abusive sterilizations, voluntary sterilization shall be regulated and allowed with the help of the proposed law."[20] Eighteen years earlier, Luxenburger referred to the "jabbering of humanitarianism" (Humanitätsdudelei) as "a degenerate child" that stood in the way of implementing radical eugenic measures to check the procreation of the mentally ill.[21] Now, it was humanitarianism itself, no longer negatively characterized, that provided the justification for a wide-ranging policy, one that would target even those who were healthy, but who carried within them malignant genes. To compensate for past injustices, Villinger further explained, the law was to be bound up with the regulation of refertilization for those who were unjustly sterilized (tubal ligation in women, for example, was reversible, at least in principle).[22] Several years later, Hans Nachtsheim, who fully supported reenacting sterilization despite its bad reputation and "misuse" by the Nazis, openly admitted that "practically, refertilization is

---

[17] Ibid., Schulz to Lenz, 03.12.1947.   [18] Ibid., Anonymous to Lenz, 30.01.1948.
[19] Ibid., Entwurf, Gesetz über Sterilisirung und Refertilisierung, Stuttgart, 09.02.48.
[20] Ibid., "Entwurf zu einer Begründung des Sterilisierung- und Refertilisierungsgesetzes," v. Prof. Villinger.
[21] Hans Luxenburger, "Die wichtigsten Ergebnisse der psychiatrischen Erbforschung und ihre Bedeutung für die eugenische Praxis (Vortrag, gehalten in der Sitzung der Münchner gynäkologischen Gesellschaft, am 23. 1. 1930)," Archiv für Gynäkologie 141 (1930): 237–254, esp. 242, 254.
[22] MPIP-GDA 79, "Entwurf zu einer Begründung," v. Prof. Villinger.

meaningless. … Such an important eugenic law for combatting hereditary diseases should not be bound up with such an insignificant method."[23]

The sterilization and refertilization law never materialized; but neither was the Nazi law revoked. The depiction of the sterilization campaign as non-Nazi, part of international and in particular U.S. eugenics, served multiple purposes. It exculpated those involved in implementing the Nazi sterilization policy and made it easier for them to try to further pursue similar policies. It also undermined the victims' entitlement to recognition and compensation. One doctor questioned the motives of an association that represented the victims of sterilization: "What other purposes does this association have, who are its founders and leaders?" Sterilization, for that doctor, was fully justified on "humane, moral, economic, societal and religious grounds." Complaints from the sterilized were unjustified: "Was the [sterilization] procedure negligent?"[24] In a cruel twist of judicial logic, only those whose sterilization procedure violated the wording of the 1933 law could be regarded as victims of Nazism. In the decades that would follow, sterilization victims would need to fight for public recognition of their suffering, to very limited avail. A 1961 Parliamentary Committee for Compensation would summon Villinger and Nachtsheim to testify as specialists; they would argue in favor of re-allowing forced sterilizations, and Villinger would even claim that compensating the sterilized would only increase their suffering, destabilizing their psychiatric state by creating "compensation neuroses."[25]

---

[23] Hans Nachtsheim, "Sterilisation aus eugenischer Indikation," *Rheinisches Ärzteblatt* 10, no. 10 (1956), 215–219 (quote from p. 217).

[24] A. W. Zehlendorf, "Sterilisation Erbkranker," *Berliner Gesundheitsblatt* 1, no. 19 (1950), 486.

[25] Rolf Surmann, "Rehabilitation and Indemnification for the Victims of Forced Sterilization and 'Euthanasie.' The West German Policies of 'Compensation' ('Wiedergutmachung')," in Volker Roelcke, Sascha Topp and Etienne Lepicard (eds.), *Silence, Scapegoats, Self-Reflection: The Shadow of Nazi Medical Crimes on Medicine and Bioethics* (Göttingen: V & R Unipress, 2014), 113–127; Dorothea S. Buck-Zerchin, "Seventy Years of Coercion in Psychiatric Institutions, Experienced and Witnessed," in Thomas W. Kallert, Juan E. Mezzich and John Monahan (eds.), *Coercive Treatment in Psychiatry: Clinical, Legal and Ethical Aspects* (Chichester: John Wiley & Sons, 2011), 235–243 (esp. p. 238). More than 50 years after the fall of the Nazi regime, decisions made by Hereditary Courts remained in force; only in 1998 were they legally revoked by a decision of the German Bundestag. See also Klaus Dörner, *Gestern minderwertig – heute gleichwertig?* (Gütersloh: Jakob van Hoddis, 1986); Christian Pross, *Paying for the Past: The Struggle over Reparations for Surviving Victims of the Nazi Terror* (Baltimore, MD: Johns Hopkins University Press, 1998); Stefanie Michaela Baumann and Andreas Scheulen, "Zur Rechtslage und Rechtsentwicklung des Erbgesundheitsgesetzes 1934," in Margret Hamm (ed.),

The ease with which Nazi psychiatrists, deeply involved in the sterilization policy, revived their eugenic efforts, is difficult to grasp. The smooth transition from National-Socialist moral justifications to Western-Liberal humanitarianism is equally startling.[26] No less remarkable, however, is the nature of what survived the change in rhetoric, what was considered still pertinent and legitimate under the refurnished political discourse that postwar eugenicists were quick to master. Beyond the paternalistic disdain toward the feeble-minded and the relentless efforts to contain female (promiscuous) sexuality, these discussions attest to the continuing fear of recessive diseases and the desperate attempt to exploit the opportunity – missed by the Nazi law – to sterilize also healthy carriers of hidden genetic pathologies. As Villinger stressed repeatedly, one's actual phenotypic state, and one's belonging to a certain group of disease categories, need never be taken into account in its own right. Without exception, what mattered was the inheritability of the diseases, not an individual's state of health. It was the health of genes, not of people, that was at stake.

Finally, these discussions point to the perpetual and uninterrupted centrality of the use of Mendelian models to characterize, and reason about, the nature of the eugenic challenge. Most striking of all, these postwar discussions among Nazi eugenicists testify to the seamless retention of the network of cultural associations that came along with Mendelian teaching, when applied to the eugenic domain. Schulz' choice of amaurotic idiocy as the explanatory model for the need for sterilization is of particular importance in this regard. Amaurotic idiocy, or, Tay-Sachs, was not only a severe disease that followed simple recessive Mendelian patterns; it was also (and still is) a genetic disease identified with a particular "racial" group: Ashkenazi Jews. Far from being a socially neutral choice, Schulz' reference to amaurotic idiocy to explain the need

Lebensunwert – zerstörtes Leben. Zwangssterilisation und "Euthanasie" (Frankfurt a. M.: VAS, 2005), 212–219

[26] See in this respect also Stefanie Westermann, "'Die Gemeinschaft hat ein Interesse daran, dass sie nicht mit Erbkranken verseucht wird' – Der Umgang mit den nationalsozialistischen Zwangssterilisationen und die Diskussion über eugenische (Zwangs-)Maßnahmen in der Bundesrepublik," in Westermann, Richard Kühl and Dominik Groß (eds.), *Medizin im Dienst der "Erbgesundheit." Beitrag zur Geschichte der Eugenik und "Rassenhygiene"* (Berlin: LIT, 2009), 169–199; Anne Cottebrune, "Eugenische Konzepte in der westdeutschen Humangenetik, 1945–1980," *Journal of Modern European History* 10, no. 4, special issue: Eugenics after 1945 (2012): 500–518; Cottebrune, "The Emergence of Genetic Counselling in the Federal Republic of Germany: Continuity and Change in the Narratives of Human Geneticists, c. 1968–80," in Bernd Gausemeier, Staffan Müller-Wille and Edmund Ramsden (eds.), *Human Heredity in the Twentieth Century* (London: Pickering and Chatto, 2013), 193–204.

for sterilization therefore fortified the links between Mendelian genetics, racial characterization, medical danger and the need for eugenic action. By choosing a "Jewish disease" to justify the need for sterilization, Schulz outlined the path that would come to dominate racial thinking in medical genetics in years to come: explicit racial categorization was out; implicit racially associated genetics was in.

Meanwhile, beyond the German Frontier

As disturbing as these postwar developments may be, Nazi eugenicists were accurate in their representation of the sterilization campaign as neither peculiarly Nazi nor even a German invention. Indeed, it had precedents that began well before the Nazis even came into being and far from Germany's borders. Eugenics in general was a multinational and international movement, as well as a loose framework for many different and often conflicting political and social organizations seeking to reform or heal the national and social body.[27] Was Social Mendelism also an international or cross-national phenomenon? What cultural or social meanings did Mendelism acquire in other national settings and how, if at all, was it integrated into national improvement plans beyond Germany?

A comprehensive answer to these questions would require a long series of specialized studies. The existing literature offers only preliminary indications, or working hypotheses, for future research endeavors in this field. Waiving in advance any pretense to exhaustiveness and building primarily upon the already available works, what follows is limited to the examination of the Mendelian-eugenic complex in certain key states.

Historians have already identified several different trends in the relationship between eugenics and Mendelian teaching. One variant of eugenics that was generally antagonistic to Mendelism was the French-Italian "Latin eugenics." In France, most researchers of heredity did not look favorably on Mendel's theory; in French eugenic circles, it was neo-Lamarckian ideas that prevailed. The French preoccupation with degeneration and the declining birth rate found its remedy in the science

---

[27] On eugenics as an international movement, see Alison Bashford, "Internationalism, Cosmopolitanism, and Eugenics," in Alison Bashford and Philippa Levine (eds.), *The Oxford Handbook of the History of Eugenics* (New York, NY: Oxford University Press, 2010) [hereafter: *Oxford Handbook*], 154–172; Stefan Kühl, *For the Betterment of the Race: The Rise and Fall of the International Movement for Eugenics and Racial Hygiene* (New York, NY: Palgrave Macmillan, 2013); Garland E. Allen, "Eugenics as an International Movement," in James D. Wright (ed.), *International Encyclopedia of the Social & Behavioral Sciences*, 2nd ed., vol. 8 (Oxford: Elsevier, 2015), 224–232.

of *puériculture*, which focused on improving the environmental conditions needed for babies to develop safely and properly, not on the selection of genes or the racial management of reproduction choices. To ensure successful pregnancies, the health, welfare and emotional well-being of mothers had to be improved. In addition, neo-Lamarckian mechanisms supported an emphasis on the sound bodies and minds of the parents, since parental qualities were believed to be transmitted to their future children.[28]

A similarly environmentalist, social-hygienic tendency characterized Italian discourse on eugenics. Italian eugenicists openly opposed the rigidness of Mendelian ideas, and it was only in the 1940s and under the influence of Nazi Germany that Mendelian thinking began to receive wider acceptance in Italy.[29] The picture was largely similar in Latin America – in Cuba, Puerto Rico, Mexico, Argentina and Brazil – although questions about the implications of racial mixture were much more acute there. With respect to racial-anthropology in the aforementioned nations, the demarcation between Lamarckism and Mendelism was not always very clear. For example, in Brazil, influential eugenicist Renato Kehl established close contacts with Fischer in Germany and with Davenport in the United States, and used Mendelian reasoning in his discussion on the impact of racial-crossing; anthropologist Edgar Roquette-Pinto and biologist Octavio Dominguez similarly adopted Mendelian research programs and related them to the larger eugenic discourse. But in general, the scientific communities in those nations adopted Mendelian ideas relatively peripherally and belatedly, or else subjected them to a larger environmentalist framework.[30] Thus, as

---

[28] On the French antagonism toward Mendelism, see Jean Gayon and Richard M. Burian, "France in the Era of Mendelism," *Comptes Rendus de l'Académie des Sciences – Series III – Sciences de la Vie* 324 (2000): 1097–1106; Richard M. Burian, Jean Gayon and Doris Zallen, "The Singular Fate of Genetics in the History of French Biology, 1900–1940," *Journal of the History of Biology* 21, no. 3 (1988): 357–402; On French eugenics, see William H. Schneider, *Quality and Quantity: The Quest for Biological Regeneration in Twentieth-Century France* (Cambridge: Cambridge University Press, 1990), 55–83; Richard S. Fogarty and Michael A. Osborne, "Eugenics in France and the Colonies," in *Oxford Handbook*, 332–346. For another view, which sees the medical profession in France as more receptive to Mendelian notions, see Jean-Paul Gaudillière, "Mendelism and Medicine: Controlling Human Inheritance in Local Contexts, 1920–1960," *Comptes Rendus de l'Académie des Sciences – Series III – Sciences de la Vie* 323, no. 12 (2000): 1117–1126.

[29] Maria Sophia Quine, "The First-Wave Eugenic Revolution in Southern Europe: Science sans Frontières," in *Oxford Handbook*, 377–397; Francesco Cassata, *Building the New Man: Eugenics, Racial Science and Genetics in Twentieth-Century Italy* (Budapest/New York, NY: CEU Press, 2011).

[30] Nancy L. Stepan, *The Hour of Eugenics: Race, Gender, and Nation in Latin America* (Ithaca, NY/London: Cornell University Press, 1991); Patience A. Schell, "Eugenics Policy and Practice in Cuba, Puerto Rico, and Mexico," in *Oxford Handbook*, 477–492;

several historians have already observed, France, Italy and Latin America saw the development not of Social-Mendelian eugenics, but of a Social-Lamarckian one.[31]

What role did Mendelian theory acquire in Russia and in the Soviet Union? The development, institutionalization and reception of eugenics, on the one hand, and of Mendelism, on the other hand, in the Soviet Union were both in a constant state of flux, partly reflecting the great social and political transformations that Russian society itself was experiencing at the time. Mendelian theory was initially endorsed with great enthusiasm in Imperial Russia, and its major protagonists (Nikolai Klo't-sov and Iurii Filipchenko) also became the leaders of eugenics in Russia. In the early 1920s, after the stormy years of war(s) and revolution(s), eugenics continued to extend its influence, attaining its own Russian, proletarian and Bolshevik flavors (focusing, for example, on the genea-logical investigation of creative geniuses much more than on promoting negative eugenic measures). In the mid-1920s, eugenics began to be criticized for its ostensible capitalist, Western and bourgeois character; into the debate was also thrown the question on the materialist or idealist, Marxist or anti-Marxist nature of Lamarckian and Mendelian theories of inheritance. This trend became much more severe in the late 1920s with the consolidation of Stalinism (the "Great Break"). Eugenics was branded as an illegitimate pursuit; for a while, however, genetics, and even medical genetics, remained viable. But attacks on Mendelian genetic methods continued and later accelerated during the 1930s, as the neo-Lamarckian agrobiologist Trofim Lysenko grew in influence.[32] Throughout these turbulent developments, did Mendelian terms, notions, metaphors or mechanisms manage to take hold in society and acquire cultural overtones, beyond the relatively narrow confines of genetic laboratories? Only further research will tell.

Gilberto Hochman, Nísia Trindade Lima and Marcos Chor Maio, "The Path of Eugenics in Brazil: Dilemmas of Miscegenation," in *Oxford Handbook*, 463–510.

[31] Nevertheless, in my view it is doubtful that Lamarckian ideas brought with them a set of terms, concepts, examples and mechanisms comparable in their richness, complexity and specificity to those provided by Mendelism - or, for that matter, Darwinism. In this sense, adding new "Social-isms" needs to be justified in each and every case. This book argues that the term is justified with respect to Mendelism, but from this it does not follow automatically that the same applies for other biological theories, or thinkers (e.g., Social Weismannism, Social Dobzhanskyism, Social Emelonism, etc.).

[32] Mark B. Adams, "Eugenics in Russia 1900–1940," in Adams (ed.), *The Wellborn Science: Eugenics in Germany, France, Brazil, and Russia* (New York, NY: Oxford University Press, 1900), 153–216; Nikolai Krementsov, "From 'Beastly Philosophy' to Medical Genetics: Eugenics in Russia and the Soviet Union," *Annals of Science* 68, no. 1 (2011): 61–92; Krementsov, "Eugenics in Russia and the Soviet Union," in *Oxford Handbook*, 413–429.

Eugenics in Denmark – the first European country to introduce, in 1929, eugenic sterilization – provides another test case for the possible combinations of Mendelian thinking and social reform.[33] The introduction of formal requirements for marriage certificates for the mentally infirm and later their sterilization were seen in Denmark as part of a social package, and were promoted by the Social Democratic Party. Mendelism found little expression in Danish racial discourse, primarily because physical anthropology and racial thinking were generally marginal there. In the field of psychiatry, psychiatrist August Wimmer "attempted Mendelian analysis of mental illnesses as early as 1920"; Wimmer was also an early advocate of eugenics. Denmark was also home to the internationally influential botanist and geneticist, Wilhelm Johannsen. Johannsen held complex views on the holistic nature of the genotype, which in turn led him to develop a critical stance toward eugenic programs based on crude Mendelian reductionism. Both Wimmer and Johannsen served as genetic experts in a governmental commission on sterilization and castration; in its 1926 official report, Johannsen relied on Mendelian logic (the Hardy-Weinberg equilibrium) to point at the futility of eugenic sterilization; whereas Wimmer used Mendelian reasoning to argue for the opposite, that is, for the need to implement eugenic measures.[34]

In Britain, the birthplace of eugenics, Mendelism was initially rejected by most leading eugenicists, who, as devoted biometricians, openly criticized the oversimplistic and anti-statistical style of reasoning propagated by their U.S. counterparts, most notably Davenport.[35] British biometricians saw Mendelism as incompatible with Darwinian gradual evolution, and as contradictory to the dynamics of heredity transmission, whose patterns they studied with great statistical care – and complexity. At the other end of the spectrum, Bateson, who was the most vocal proponent of Mendelism in Britain, was generally reticent with regard

---

[33] Norway and Sweden are also interesting in this regard; see below, p. 223.

[34] Bent Sigurd Hansen, "Something Rotten in the State of Denmark: Eugenics and the Ascent of the Welfare State," in Gunnar Broberg and Nils Roll-Hansen (eds.) *Eugenics and the Welfare State. Sterilization Policy in Denmark, Sweden, Norway and Finland* (East Lansing: Michigan State University Press, 1996), 9–76 (quote from p. 26); Staffan Müller-Wille, "Leaving inheritance Behind: Wilhelm Johannsen and the Politics of Mendelism," *Conference: A Cultural History of Heredity IV: Heredity in the Century of the Gene*, Max-Planck Institute Preprint Series 343 (2008): 7–18; Raphael Falk, "Commentary: A Century of Mendelism: On Johannsen's Genotype Conception," *International Journal of Epidemiology* 43, no. 4 (2014): 1002–1007; Nils Roll-Hansen, "Commentary: Wilhelm Johannsen and the Problem of Heredity at the Turn of the 19th Century," *International Journal of Epidemiology* 43, no. 4 (2014): 1007–1013.

[35] Exemplary in this regard is David Heron, *Mendelism and the Problem of Mental Defect*, vol. I, *A Criticism of Recent American Work* (London: Dulau, 1913).

to eugenic policies. He saw mutations as favorable to human progress and argued, "If populations were homogenous civilisation would stop." He further warned against trying to directly and simplistically apply genetic knowledge to reshape human society.[36] The gradual amalgamation of Mendelian and Darwinian principles, in Roland A. Fisher's 1918 paper and in other works, paved the way toward population genetics and to the later development of Modern Evolutionary Synthesis. Grounded in eugenic concerns and informed by Mendelian suppositions, Fisher's population genetics may be regarded as an alternative version of Social-Mendelian thinking, one that built on, yet simultaneously surpassed and diverted from, the set of concepts surveyed here.[37]

At the same time, Social-Mendelian thinking of the kinds analyzed in this book also existed in certain wings of British eugenics. However, in Britain it was class more than racial-antagonism that provided the motivating force for the eugenic movement.[38] As it turned out, Mendelism was flexible enough to accommodate the "working class menace." For example, writing to *The Eugenics Review* in 1940, Leonard Darwin, Charles Darwin's son and a eugenicist in his own right (among other

---

[36] Rosemary D. Harvey, "Pioneers of Genetics: A Comparison of the Attitudes of William Bateson and Erwin Baur to Eugenics," *Notes and Records of the Royal Society of London* 49, no. 1 (1995): 105–117. Bateson's quote is taken from his 1914 Presidential Address to the British Association in Melbourne, Australia; see Beatrice Durham Bateson, *William Bateson, F. R. S., Naturalist: His Essays & Addresses* (Cambridge: Cambridge University Press, 1928), 308.

[37] Ronald A. Fisher, "XV.—The Correlation between Relatives on the Supposition of Mendelian Inheritance," *Transactions of the Royal Society of Edinburgh* 52, no. 2 (1918): 399–433; Bernard Norton, "Fisher's Entrance into Evolutionary Science: The Role of Eugenics," in Marjorie Grene (ed.), *Dimensions of Darwinism. Themes and Counterthemes in Twentieth-Century Evolutionary Theory* (Cambridge/Paris: Cambridge University Press, 1983), 19–30. On the Modern Synthesis, see Julian Huxley, *Evolution: The Modern Synthesis* (London: George Allen & Unwin, 1942); Ernst Mayr and William B. Provine (eds.), *The Evolutionary Synthesis: Perspectives on the Unification of Biology* (Cambridge, MA: Harvard University Press, 1998).

[38] As the social biologist Lancelot Hogben stated in 1939: "In Germany the Jew is the scapegoat. In Britain the entire working class is the menace." Although this statement was intended to defame eugenics in Britain, it captured correctly the overall orientation of British eugenics. See Pauline Mazumdar, *Eugenics, Human Genetics and Human Failings: The Eugenics Society, Its Sources and Its Critics in Britain* (London: Routledge, 1992). That said, British eugenicists were certainly not free from racial prejudice, which informed their attitudes toward Jews, Blacks and other "others." Some of them also expressed outspokenly their fears from the results of racial miscegenation. See Dan Stone, "Race in British Eugenics," *European History Quarterly* 31 (2000), 397–425; Stone, *Breeding Superman: Nietzsche, Race and Eugenics in Edwardian and Interwar Britain* (Liverpool: Liverpool University Press, 2002); Gavin Schaffer, *Racial Science and British Society, 1930–1962* (Basingstoke: Palgrave Macmillan, 2008). On the British preoccupation with race mixture, see also Lucy Bland and Lesley A. Hall, "Eugenics in Britain: The View from the Metropole," in *Oxford Handbook*, 213–227.

accolades, he was president of the British Eugenics Education Society between 1911 and 1928), insisted on the need to put eugenic measures into action immediately. He argued that, if implemented correctly, such eugenic measures would promote social equality. But he warned that social or economic reforms that would give better opportunities to individuals based on their manifested traits would neglect "the difference between them in regard to their recessive qualities." It was only at the level of whole groups that "the effect of the presence of recessive genes becomes more apparent." Therefore, "to equalize the care given to the poorer strata with that obtainable by the well-to-do would be a most desirable reform; but if the poverty of the poor as a class is to any extent dependent on the presence in excess of inferior recessive genes, such a reform would produce dysgenic results by keeping more of them alive."[39] As these pronouncements illustrate, not only did Mendelian concepts find their way into the discussion on solving social problems in Britain; they were also adapted to address local concerns. Admittedly, however, they never acquired the kind of prominence that they received in Germany.

In the United States, in contrast, Mendelian reasoning had a profound influence on eugenic thinking. There, as in Germany, adherence to Mendelian principles and the promotion of racial/eugenic visions were closely intertwined. Charles B. Davenport – unlike Bateson – was both the most vocal U.S. advocate of Mendelism and, at the same time, the recognized leader of eugenics. His counterpart in the field of mental diseases – the U.S. equivalent of Rüdin – was Aaron Rosanoff. To these we may add other scholars, such as Harry H. Laughlin or the ophthalmologist and eugenicist Lucien Howe, who lived on an estate he called "Mendel's Farm" and referred to Mendel as "our saint."[40] These scholars, who were pivotal in shaping and developing U.S. eugenics, were criticized early on, both at home and abroad, for their tendency toward crude Mendelian simplifications. Later geneticists would take comfort in the fanaticism of Davenport and his peers, which would allegedly prove that true, self-critical science was much less prone to dogmatism and radicalism. Nevertheless, Mendelian notions also influenced the work of Davenport's critics, some of whom had undisputed academic credentials. Anxiety related to the nature and impact of recessive traits, for example, or stemming from purportedly disharmonious racial crossings, found expression in the works of the most esteemed

---

[39] Leonard Darwin, "The Geneticists' Manifesto," *The Eugenics Review*, 31, no. 4 (1940): 229–230.
[40] See Howe's letter to Laughlin at www.dnalc.org/view/10318-Lucien-Howe-letter-to-Harry-Laughlin-about-Mendel-s.

geneticists. One of them was Edward M. East, who warned in 1923 that the white race should avoid crossing with the negro race, who "as a whole is possessed of undesirable transmissible qualities both physical and mental." In his 1927 book *Heredity and Human Affairs* he explained that "inbreeding may weaken a house by uncovering a whole closet full of family skeletons – recessive abnormalities."[41] The same year, addressing the annual meeting of the American National Tuberculosis Association, geneticist Herbert S. Jennings, after having established that, "most gene defects are recessive," said: "A defective gene – such a thing as produces diabetes, cretinism, feeble-mindedness – is a frightful thing; it is the embodiment, the material realization of a demon of evil; a living self-perpetuating creature, invisible, impalpable, that blasts the human being in bud or in leaf. Such a thing must be stopped wherever it is recognized." Jennings therefore thought that "so fast as we can discover individuals that bear seriously defective genes – whether themselves personally defective or not – so rapidly must those individuals be brought to cease reproduction."[42] These views are virtually indistinguishable from those analyzed in the previous chapters.

There are also multiple parallels between Germany and the United States with regard to the societal dispersal of Mendelian thinking, only that in the United States it was the "Negroes" who, in the mind of racial thinkers, occupied a social position similar in many senses to that of Jews in German antisemitism. Despite the obvious differences, or maybe because of them, the parallels between the cultural depiction of the Negro in America and that of Jews in Germany are striking.[43] Of whom was it told that he "exhibits by nature a pliability, a readiness to accommodate himself to circumstance, a proneness to imitate those among whom he lives"? These words were written in August 1863 by Louis Agassiz to describe Black slaves; but they might have been similarly pronounced in Weimar Germany by antisemites describing the nature of the Jew.[44]

---

[41] Edward M. East, *Mankind at the Crossroads* (New York, NY: Charles Scribner's Sons, 1923), 133; East, *Heredity and Human Affairs* (New York, NY: Charles Scribner's Sons, 1927), 68; Bentley Glass, "Geneticists Embattled: Their Stand against Rampant Eugenics and Racism in America during the 1920s and 1930s," *Proceedings of the American Philosophical Society* 130, no. 1 (1986): 130–154.

[42] Herbert S. Jennings, "Public Health Progress and Race Progress – Are They Incompatible?" *Science* 66, no. 1698 (1927): 45–50 (quote from pp. 47–48).

[43] Many of the similarities between the social and cultural depictions of Jews in Germany and of Negroes in America have been highlighted in the works of Sander Gilman. See Sander L. Gilman, *Difference and Pathology: Stereotypes of Sexuality, Race, and Madness* (Ithaca, NY: Cornell University Press, 1985); Gilman, *The Jew's Body* (New York, NY/London: Routledge, 1991).

[44] See the letter of August 9, 1983, in Elizabeth Cary Agassiz, *Louis Agassiz – His Life and Correspondence* (New York, NY: Houghton, Mifflin and Company, 1885), 597.

At times, antisemitism and "negrophobia" were not only structurally similar but also practically combined.[45] Was there therefore a Mendelization of U.S. thought on "Negroes," similar in kind to the one analyzed above with regards to the Jews? In his 2008 book on eugenics in Virginia, Gregory Michael Dorr provides some evidence that there was; he even implies – but stops short of actually showing – that the Mendelian Law of Segregation was associated with the U.S. racial "laws of segregation."[46] It would indeed be of great interest to know whether such an association was a functional one for contemporaries, one that legitimized Mendel's teaching for racial thinkers and imbued Mendelian genetics with concrete racial meanings. This, however, would require further research.

While the extent and exact nature of the Mendelization of racial thought in the United States still needs to be determined, there is no doubt that Mendelian reasoning was central to the discussions on the physically and mentally impaired. Examples abound, but for the purpose of illustration, one should suffice. Under the title "Mendel's Law. A Plea for a Better Race of Men," the following poem was published three times in the *Virginia Medical Monthly* in the 1920s and 1930s, and quoted in full by Superintendent John N. Thomas of the Louisiana Hospital for the Insane in his 1922 annual report:

> Oh, why are you men so foolish —
>   You breeders who breed our men
> Let the fools, the weaklings and crazy
>   Keep breeding and breeding again?
> The criminal, deformed, and the misfit,
>   Dependent, diseased and the rest —
> As we breed the human family
>   The worst is as good as the best.
>
> Go to the house of some farmer,
>   Look through his barns and sheds,
> Look at his horses and cattle,
>   Even his hogs are thoroughbreds;
> Then look at his stamp on his children,
>   Lowbrowed with the monkey jaw,
> Ape handed, and silly, and foolish—
>   Bred true to Mendel's law.

---

[45] For one example, see Gregory Michael Dorr, *Segregation's Science: Eugenics and Society in Virginia* (Charlottesville: University of Virginia Press, 2008), 174–176.

[46] Dorr, *Segregation's Science*, Introduction.

Go to some homes in the village,
    Look at the garden beds,
The cabbage, the lettuce and turnips,
    Even the beets are thoroughbreds;
Then look at the many children
    With hands like the monkey's paw
Bowlegged, flat headed, and foolish —
    Bred true to Mendel's law.

This is the law of Mendel,
    And often he makes it plain,
Defectives will breed defectives
    And the insane breed insane.
Oh, why do we allow these people
    to breed back to the monkey's nest,
To increase our country's burdens
    When we should only breed the best?

Oh, you wise men take up the burden,
    And make this your loudest creed,
Sterilize the misfits promptly—
    All not fit to breed.
Then our race will be strengthened and bettered,
    And our men and our women be blest,
Not apish, repulsive and foolish,
    For the best will breed the best.[47]

This poem, and recent works such as Tamsen Wolff's *Mendel's Theatre*, indicate that Mendelian notions far transgressed the walls of Davenport's laboratories in Cold Spring Harbor; and that, accordingly, there is still much to reveal about how Mendelian thinking became embedded in U.S. cultural and social discourse on race, disease and national prophylaxis.[48]

Thus, from a wider international perspective it is evident that Mendelian thinking became crucial for the development of eugenic ideas in the Anglo-Saxon and German-speaking world. This seemingly trivial assertion may look contradictory to a statement made in 1990 by Mark B. Adams, who dismissed the idea that "eugenics was somehow intrinsically bound up with Mendelian genetics" as a "myth."[49] This myth did

---

[47] The poem is available at www.dnalc.org/view/11212-Mendel-s-Law-Poem-by-Joseph-DeJarnette-MD-witness-in-Buck-vs-Bell-case.html; for context, see Edward J. Larson, *Sex, Race and Science. Eugenics in the Deep South* (Baltimore, MD: Johns Hopkins University Press, 1995), 44–45; Dorr, *Segregation's Science*, 123–124.

[48] Tamsen Wolff, *Mendel's Theatre: Heredity, Eugenics, and Early Twentieth-Century American Drama* (Basingstoke: Palgrave Macmillan, 2009).

[49] Mark B. Adams, "Toward a Comparative History of Eugenics," in Adams (ed.), *The Wellborn Science*, 217–231 (quote from p. 218).

not develop without reason. Precisely in the nations that attracted the attention of an earlier generation of historians, the rise of Mendelian heredity had clear and palpable implications for the development of eugenic thought. In those nations, which were also world leaders in scientific matters, it is far from coincidental that eugenic societies, committees, laboratories, conferences and exhibitions proliferated precisely at the time when Mendelian research was booming – 1905–1913 – and that influential racial-eugenicists achieved their initial fame through their research on Mendelian transmission patterns in humans.[50] Conceptually, at the core of both eugenic impetus and Mendelian teaching lay concerns over purity at both the individual and supra-individual levels, purity that could only be achieved through the identification of hidden genetic components, and that simultaneously necessitated and enabled the management and control of reproduction at the level of the population. Eugenics and Mendelism were therefore, quite clearly, mutually reinforcing.

As the case of Latin eugenics (among others) shows, this by no way implies that eugenics could not have developed without Mendelian theory, or that there was anything inherently racist, reductionist, deterministic or fatalistic in Mendel's theory itself. To take two examples that have not yet been mentioned: the leader of Swedish eugenics in its earliest days was the racist and antisemite Hermann Lundborg, a scholar who acquired world fame with his 1913 study that showed that myoclonic epilepsy followed recessive Mendelian patterns. His contemporary, the Swedish geneticist Nils von Hofsten, who was central in promoting the sterilization program in Sweden, referred to Mendelism to argue that some of the fantasies regarding the utility of sterilization were exaggerated. He warned in his 1919 standard book on genetics that "in many cases even the most radical program of racial-hygiene cannot wipe out an unfavorable quality in man, and it has less potential to check its spreading than most people, ignorant of the Mendelian laws, believe."[51] In neighboring Norway, psychiatrist Ragnar Vogt similarly relied on the recessive nature of mental illnesses to argue **against** the benefit of sterilization; he and other geneticists used Mendelian logic to attack Jon Alfred Mjøen's racial-hygienic programs.[52] These examples are sufficient to demonstrate that, like other scientific theories, Mendelism could be taken in various

---

[50] For a concise timeline of the international institutionalization of eugenics, see Alison Bashford and Philippa Levine, *Oxford Handbook*, 559ff.

[51] Mattias Tydén, "The Scandinavian States: Reformed Eugenics Applied," in *Oxford Handbook*, 363–376 (quotes from pp. 367–368).

[52] Nils Roll-Hansen, "Norwegian Eugenics: Sterilization as Social Reform," in Broberg and Roll-Hansen, *Eugenics and the Welfare State*, 151–194.

directions. As a result, the scrutiny of each particular national-cultural context is indispensable to a full appreciation of the true and varied historical meaning of Social Mendelism as an international and multi-national framework of thought.[53]

Depoliticizing Mendel

Today, the common view of Mendelism is that of a socially neutral, ideology-free and essentially technical theory on the regularities of heredity. Given all we have learned in the previous chapters, this view may seem outwardly paradoxical. In reality, this view in itself is a Western legacy of the postwar years. In what follows, we will track how this view became established, and what preceded it.

Throughout the first half of the twentieth century, Mendelism was not only laden with implicit cultural and social meanings; it was also recognized as such and was regarded as a theory with clear ideological and even political ramifications. This was true, first of all, in Germany itself, where biological theories were quick to acquire political overtones. As Robert Proctor and others have shown, from the 1920s onward, German eugenicists associated Lamarckism with liberal and Marxist dogmas that insisted – in spite of contradictory evidence – that human nature was malleable and that by amending education, or cultivating class consciousness, or eradicating poverty, or reducing inequality, society's problems could be solved, social maladies could be healed, and the challenges of modernity could be met. Weismann's and Mendel's theories, eugenicists countered, supported the opposite view, one that explained social differences as natural and innate, and that ascribed to the environment limited powers over human capacities. For German right-wing thinkers, the difference was not simply a divergence of opinions. Since Mendelism was right and Lamarckism wrong, Mendelism could help refute Marxist-socialist-communist or liberal theories altogether and, moreover, expose that they were essentially ideologically driven, scientifically unsound, and, in short, no more than wishful thinking.[54]

Against this background, when the Austrian Lamarckian biologist and advocate of socialism and pacifism, Paul Kammerer, was shown to have falsified his experimental results, German racial-hygienists celebrated.

---

[53] A useful examination of various possible combinations of eugenics, politics and theories of heredity is offered in Maurizio Meloni, *Political Biology* (Basingstoke: Palgrave Macmillan, 2016), 93–135; for a rich historical analysis, see Marius Turda, *Modernism and Eugenics* (New York, NY: Palgrave Macmillan, 2010).

[54] Robert Proctor, *Racial Hygiene: Medicine under the Nazis* (Cambridge, MA: Harvard University Press, 1988), 30–38.

When Kammerer subsequently took his own life, Lamarckism received a fatal blow, at least in the public's mind. To make things even worse (for Lamarckists), Kammerer was also Jewish. In the eyes of Lenz, it was only natural that Lamarckian ideas would appeal to Jews: only by believing that acquired characteristics can be inherited could Jews hope to assimilate in their Gentile surrounding and escape their racial fate. This, of course, was an illusion, acknowledged also by the German-Jewish physician Fritz Kahn: "Jews do not become Germans by writing books on Goethe."[55]

In many senses, these perceptions of German eugenicists mirrored similar (yet opposite) views that developed in the Soviet Union. Kammerer himself had claimed that adherence to concepts of hard heredity would make humans, "slaves of the past instead of masters of the future;" and factions within the Soviet genetic community endorsed this position and forcibly argued for abandoning Mendelism (and eugenics) and advancing neo-Lamarckian works. A potential remedy for the dichotomy between the competing theories seemed to have been suggested by Hermann J. Muller's 1927 discovery: radiation could artificially induce mutations that would later be inherited according to Mendelian rules. Several Soviet geneticists were greatly enthused by this potential softening of hard heredity, which made it possible to maintain adherence to Mendelian principles without abandoning the plausibility that genes, too, can be altered, manipulated and maybe even purposefully shaped.[56]

During the 1930s, political events in both Germany and Russia left little room for such official compromises (at least at the declaratory level; what geneticists in both states actually did was a different story). The beginning of the 1930s saw the practical dissolution of eugenic societies, institutions and journals in the Soviet Union and attacks against its leading members as fascists, Menshevists or bourgeois. For a while, however, most of the protagonists managed to survive the attack by altering and accommodating their professional outlook. The mid-1930s

---

[55] Proctor, *Racial Hygiene*, 55; Fritz Lenz, in *BFL* (1927), 563–564. Fritz Kahn's quote is taken from his *Die Juden als Rasse und Kulturvolk* (Berlin: Welt Verlag, 1922), 24. Another target of Lenz' criticism was Mendel's first biographer Hugo Iltis, who was also of Jewish descent and an adherent of Lamarckian doctrines. Iltis, in turn, publicly criticized the "racial fantasies" of neo-Darwinians like Fischer and Lenz. See Veronika Lipphardt, "Das 'schwarze Schaf' der Biowissenschaftler. Ausgrenzungen und Rehabilitierungen der Rassenforschung im 20. Jahrhundert," in Lipphardt (ed.), *Pseudowissenschaft: Konzeptionen von Nicht-/Wissenschaftlichkeit in der Wissenschaftsgeschichte* (Frankfurt a. M.: Suhrkamp, 2008), 223–250, esp. 232–234.

[56] Hermann J. Muller, "Artificial Transmutation of the Gene," *Science* 66 (1927): 84–87; Adams, "Eugenics in Russia," 179.

saw the Great Terror unfold, accompanied by a much heavier blow to the Soviet genetics community. Prior name-changing practices were no longer enough, and medical genetics in its entirety was labeled as fascist, its institutions were closed down and its leaders arrested and shot. From the German point of view, Stalin seemed to have proved what the Nazis had been arguing all along: Lysenko's rise showed that Lamarckism and communism were indeed intimately related – and that both were ideologically tainted, groundless and false. For the Communists, on the other hand, Hitler's endorsement and rapid implementation of eugenic policies proved the exact reverse: namely, that Weissmannism-Mendelism-Morganism was indeed inherently fascist and anti-proletarian.

Meanwhile in the West, Nazi Germany was giving eugenics and Mendelian genetics a bad name, a fact acknowledged as a problem by eugenicists around the globe. Some (including in Britain and in the United States) looked favorably on Hitler's determination in the area of eugenics and on the Nazi Sterilization Law, and attempted to emulate it; but Nazi Germany's growing racism, antisemitism and international aggression hampered eugenics' reputation. The need to detach eugenics and genetics from racism and Nazism became ever more pressing as years went by. From the scientific perspective, undermining the racial theory of Nazi Germany – in its popular, crude, Güntherian version – was quite an easy feat. Race had been a problematic category for scientific analysis since the late nineteenth century; even at that time, German scholars considered a wholesale rejection of the term.[57] The perplexing nature of the relations between physical and cultural characteristics of human populations continued to preoccupy the European anthropological profession throughout the first half of the twentieth century; and, as is well known, the shift from physical to social and cultural anthropology, epitomized in the figure of Franz Boas, substantially predated the rise of the Nazis. In Britain, further critiques of the utility and problematics of the race concept were loudly voiced already during the 1920s, motivated both by scientific as well as political arguments.[58] Thus, by the 1930s, denouncing racial thinking – as Julian Huxley and Alfred Haddon did in their 1935 publication, *We Europeans*, or as the UNESCO statements from 1950 and 1951 would later do – represented a very small sacrifice

---

[57] Benoît Massin, "From Virchow to Fischer: Physical Anthropology and 'Modern Race Theories' in Wilhelmine Germany (1890–1914)," in George W. Stocking (ed.), *Volksgeist as Method and Ethic. Essays on Boasian Ethnography and the German Anthropological Tradition* (Madison: University of Wisconsin Press, 1996), 79–154.

[58] Nancy L. Stepan, *The Idea of Race in Science: Great Britain 1800–1960* (London: Macmillan, 1982), ch. 6; Elazar Barkan, *The Retreat of Scientific Racism* (Cambridge: Cambridge University Press, 1992), ch. 5–6.

for the global scientific community. Race as it had been rebutted was anyway an outdated framework for scientific research.

However, it was crucial to spare genetics from a similar fate. The 1939 "Geneticists Manifesto," published following the Seventh International Genetics Congress, was one of the earlier conspicuous attempts by British and U.S. geneticists to situate their discipline in stark opposition to Nazi Germany's version of it. The congress itself was originally planned for two years earlier in Moscow; the sharp turn against Western genetics in the Soviet Union eventually led to its cancellation, and it was finally convened in Edinburgh in the final week of August 1939, just before the German invasion of Poland. The Russians were missing; the German, Hungarian, Scandinavian and Swiss delegations left on the second and third days. Some of the remaining British and American delegates, headed by Hermann J. Muller, phrased the so-called manifesto as a response to a question directed at them: "How could the world's population improve most effectively genetically?"[59]

The geneticists' reply began as markedly un-genetic. First, they argued, social conditions and human attitudes should improve to provide equal opportunities for all; second, race prejudice, national antagonism, war and economic exploitation should be eliminated; third, economic conditions, medical care and education should all improve so that the bearing and rearing of children becomes easier for women and for the working classes. Only in their fourth item did the geneticists turn to birth control measures and to the raising of popular consciousness as to the "honour and a privilege, if not a duty, for a mother, married or unmarried, or for a couple, to have the best children possible, both in respect of their upbringing and of their genetic endowment, even where the latter would mean an artificial – although always voluntary – control over the process of parentage." Fifth, they continued, better understanding of the mechanisms governing heredity and selection should be popularized to foster a wider appreciation of the consequences of the lowering of selection pressures in modern society. Only on the basis of such an understanding could the public appreciate the need to implement a conscious selection policy, whose "genetic objectives, from a social point of view, are the improvement of those genetic characteristics which make (a) for health, (b) for the complex called intelligence and (c) for those temperamental qualities which favor fellow-feeling and social behavior." If all of this would be achieved, the population could improve its average qualities "within a comparatively small number of generations." To accomplish

---

[59] H. Gruenberg, "Men and Mice at Edinburgh. Reports from the Genetics Congress," *Journal of Heredity* 30, no. 9 (1939): 371–374.

this goal, however, would require more research and more scientific and international collaboration, both of which are antithetical to war.[60]

One may wonder to what extent the first three requirements laid down in the geneticists' manifesto truly emanated from their (Crew, Haldane, Hogben, Huxley, Muller and others) scientific expertise. Obviously, they felt compelled to come out against the ensuing German aggression as well as against Nazi racial-hygienic programs, to insist on the importance of social equality, and to preach genetic selection only with great care and prudence. But they also insisted on their own crucial role, as geneticists, in future planning and implementation of policies aimed at bettering the human condition. No inherent problem lay in the science of genetics itself; it was human aggression, prejudice and inequality that hindered progress.

When the war ended, it was no longer contra Nazi Germany, but contra the Soviet Union that the field of genetics needed to be defined. In 1948, Lysenkoism became the only permissible scientific policy for studying heredity (in plants) in the Soviet Union, and "Mendelism-Weismannism-Morganism" was officially banned. This juxtaposition of monikers, a favored practice in the Stalinist tradition, did not simply stand for scientific genealogy, but aimed to associate U.S. science with German-Nazi science and to present all three researchers as foreign to the native Russian "Michurinist" tradition. To the allegations that Mendelism-Morganism was inherently fascist, foreign and un-Russian, further pejoratives were added: Mendelism was formal, theoretical and therefore impractical and useless for improving the lives of the people; it was speculative, immaterial (and therefore un-Marxist) and disassociated from the living substance; it was based on pure chance, and therefore impotent.[61]

The Anglo-U.S. reaction to the Soviet Union's treatment of genetics was rife with hesitation and internal conflicts. One of the challenges troubling Western scholars was how not to conflate Cold War politics with science in a manner that would simply mimic the Soviet practice; another was how to oppose a false doctrine without jeopardizing the principle of academic freedom. In addition, practical considerations, such as existing friendly ties between U.S. and Russian scientists, led

---

[60] Ibid.

[61] See the (selective) excerpts from the 1948 Proceedings of the Lenin Academy of Agricultural Sciences of the U.S.S.R in Pamela N. Wrinch, "Science and Politics in the U.S.S.R.: The Genetics Debate," *World Politics* 3, no. 4 (1951): 486–519; William Dejong-Lambert and Nikolai Krementsov, "On Labels and Issues: The Lysenko Controversy and the Cold War," *Journal of the History of Biology* 47 (2012): 373–388.

to the former's reluctance to take any steps that could jeopardize the lives or careers of fellow Russian geneticists.[62]

Eventually, the path taken by the U.S. scientific community was a positive and not a negative one. As Audra J. Wolfe has shown, the 1950 annual meeting of the Genetics Society of America, which formally celebrated the "Mendelian Golden Jubilee," provided U.S. geneticists with an opportunity to circumvent explicit condemnation of Lysenkoism and present, instead, the practical achievements of genetics in the West. The meeting evolved into a highly publicized four-day event celebrating Mendelian genetics and, by implication, the superiority of Western over communist science.[63]

The apotheosis of Mendel himself as the father of modern genetics at the Golden Jubilee served multiple purposes. Among other things, Mendel's figure could be used to bridge scientific differences within genetics. Performing various rhetorical maneuvers, all who took part in the Golden Jubilee found a way to relate their work to Mendel's discoveries. As Wolfe details, this applied even to Boris Ephrussi, who worked on cytoplasmic inheritance. Originally, Ephrussi almost withdrew from the program, fearing that the meeting would give precedence to classical genetics. However, he eventually not only participated but, in the spirit of the entire event, also claimed that "whatever progress has been made in the study of non-Mendelian inheritance, it is ultimately due to … Gregor Mendel!"[64]

The Golden Jubilee had additional goals – and legacies. The meeting provided yet another opportunity to cleanse genetics from its ties with racism and eugenics. In the different speeches and presentations given during the event, eugenics was marginalized, while genetics was praised for its utility in agriculture, animal breeding and medicine. At the same time, in their attempt to avoid mirroring Soviet practices, the organizers of the conference took great care not to paint Mendelism itself as ideologically bound with Western science, liberalism or democracy.

---

[62] Rena Selya, "Defending Scientific Freedom and Democracy: The Genetics Society of America's Response to Lysenko," *Journal of the History of Biology* 45 (2012): 415–442. Between the end of World War II and the beginning of the Cold War there was nevertheless an interesting intermediate phase where collaboration, rather than confrontation, also characterized the Russian-U.S. relationship with regard to genetics; see Nikolai Krementsov, "A 'Second Front' in Soviet Genetics: The International Dimension of the Lysenko Controversy, 1944–1947," *Journal of the History of Biology* 29 (1996): 229–250.

[63] Audra J. Wolfe, "The Cold War Context of the Golden Jubilee, or, Why We Think of Mendel as the Father of Genetics," *Journal of the History of Biology* 45 (2012): 389–414.

[64] Cited in Wolfe, "Golden Jubilee," 402.

Mendelism was portrayed as an international, value-free, successful and useful theory, untainted by any kind of political leaning.[65]

The early 1950s also saw the publication of UNESCO's statements on race – signed, among others, by the German Mendelian researcher Hans Nachtsheim. In the educational campaign that followed, Mendel featured prominently. A huge figurative portrayal of Mendel decorated a 1953 *Life* magazine article, "How the Races of Man Developed"; Mendel's teaching was depicted not only as establishing the foundations of genetics but also refuting outdated and ideologically charged ideas on racial differences or racial superiority. In scientific writings, in popular pamphlets, in educational materials and in public speeches, Mendelian genetics was repeatedly referred to as contradicting, and not supporting, (Nazi) racism. The move from racial-anthropology to population genetics was conveyed through notions like "Mendelian populations" and through discussion of blood type differences. The latter were not only an indisputable Mendelian marker with acknowledged medical importance; they were also a proof that while human populations were different from each other in the distributions of certain markers, no clear line could be drawn between them.[66]

In many senses, this legacy of Mendelism as an a-social, anti-racist theory of heredity endures to this day. It is greatly optimistic, implying that scientific progress is antithetical to prejudice or discrimination. It is whiggish, and morally reassuring at that, entrenching the dichotomy between science and democracy, on the one hand, and superstition and tyranny, on the other. It makes us believe that the good are also the smart, and the evil are also the stupid (and vice versa). As neat as such an image may be, it is historically false. Mendel's theory was an integral part of Nazi racial thinking. This does not disqualify it as a scientific theory, nor should it inflect on our wish to mobilize it or learn from it. Like any other theory, it is its usage that determines its character. The same holds true for later developments within the field of genetics; as Elazar Barkan argued, "Science could lend itself as easily to either a racist or an anti-racist interpretation."[67] If we wish to be masters of the future, and not

---

[65] Ibid.

[66] See the reference to Mendelian teaching in L. C. Dunn, *Race and Biology* (Paris: Unesco, 1951); UNESCO, *What Is Race? Evidence from Scientists* (Paris: Unesco, 1952); Leo Kuper (ed.), *Race, Science and Society* (Paris: UNESCO Press, 1956), and the analysis offered in Jenny Bangham, "What Is Race? UNESCO, Mass Communication and Human Genetics in the 1950s," *History of the Human Sciences* 28, no. 5 (2015): 80–107.

[67] Barkan, *Retreat*, 228. Furthermore, "Conventional perceptions of the impartiality of science encourage an assumption that the growth of population genetics inherently lends support to a non-racist interpretation," but "historical records suggests otherwise."

slaves of the past, we need first to be able to discharge this past and at least temporarily detach ourselves from it. But to do so, we need to be able to look at it as it truly was. Acknowledging the true, and often dirty, paths of historical developments should not prevent us from striving toward a better world. On the contrary, it should arm us with a better understanding of the challenges facing us and motivate us in the struggles that still await us.

Bibliography

Primary Sources

Abel, Wolfgang, "Über Störungen der Papillarmuster I. Gestörte Papillarmuster in Verbindung mit einigen körperlichen und geistigen Anomalien," *Zeitschrift für Morphologie und Anthropologie* 36, no. 1 (1936): 1–38.

Adams, Eugen, "Über postembryonale Wachstumsveränderungen und Rassenmerkmale im Bereiche des menschlichen Gesichtsschädels," *Zeitschrift für Morphologie und Anthropologie* 20, no. 3 (1917): 551–628.

Agassiz, Elisabeth C., *Louis Agassiz – His Life and Correspondence*, New York, NY: Houghton, Mifflin and Company, 1885.

Albrecht, [Paul], "Gleichartige und ungleichartige Vererbung der Geisteskrankheiten," *Zeitschrift für die gesamte Neurologie und Psychiatrie* 11, no. 1 (1912): 541–580.

Alzheimer, Alois, "Ist die Einrichtung einer psychiatrischen Abteilung im Reichsgesundheitsamt erstrebenswert?" *Zeitschrift für die gesamte Neurologie und Psychiatrie* 6, no. 1 (1911): 242–246.

Ammon, Otto, "Zur Theorie der reinen Rassetypen," *Zeitschrift für Morphologie und Anthropologie* 3, no. 3 (1900): 679–685.

Ärztliches Vereinsblatt 51 (1924).

Auerbach, Elias, "Zur Plastizität des Schädels, mit Bemerkungen über den Schädelindex," *Archiv für Rassen- und Gesellschaftsbiologie* 9, no. 5 (1912): 604–611.

"Aus Rassenhygiene und Bevölkerungspolitik. Die Durchführungsbestimmungen zu den Nürnberger Gesetzen," *Volk und Rasse* 11, no. 1 (1936): 23.

Bachmaier, Fritz, "Kopfform und geistige Leistung. Eine Betrachtung an Münchner Volksschülern," *Zeitschrift für Morphologie und Anthropologie* 27, no. 1 (1928): 1–68.

Bareth, Karl & Alfred Vogel, *Erblehre und Rassenkunde für die Grund- und Hauptschule*, Bühl-Baden: Konkordia, 1937.

Bartmann, Hans, "Das Erbe der Väter: Eine kleine Erblehre," in Otto Rabes (ed.), *Mutter Natur*, 16, Berlin/Leipzig: Julius Beltz, 1934, 59–60.

Bateson, William, "An Address on Mendelian Heredity and Its Application to Man (Delivered before the Neurological Society of London, on Thursday, February 1st, 1906)," *British Medical Journal* 2376, no. 2 (1906): 61–67.

Mendel's Principles of Heredity, Cambridge: Cambridge University Press, 1909.

Baur, Erwin, "Bastardierung," in Eugen Korschelt et al. (eds.), *Handwörterbuch der Naturwissenschaften*, Jena: Gustav Fischer, 1912, 850–874.
Einführung in die experimentelle Vererbungslehre [2nd ed.: 1914, 3rd ed.: 1919, 4th ed.: 1924, 5th ed. 1930], Berlin: Gebrüder Borntraeger, 1911.
Baur, Erwin, Eugen Fischer & Fritz Lenz, *Menschliche Erblichkeitslehre und Rassenhygiene* (2nd ed.: 1923, *Grundriss der menschlichen Erblichkeitslehre und Rassenhygiene*, 3rd ed.: 1927, 4th ed.: 1931), Munich: J. F. Lehmanns.
Behr-Pinnow, Karl von, "Eine Sonderuntersuchung für Berufsvererbung," *Volksaufartung, Erbkunde, Eheberatung* 4 (1929): 59–63.
Benecke, Helmut, "Erläuterungen zu dem schematischen Grundriß einer Verwandtschaftstafel," *Zeitschrift der Zentralstelle für niedersächsische Familiengeschichte* 8, no. 10 (1926): 191–195.
Berliner Gesellschaft für Rassenhygiene, *Über den gesetzlichen Austausch von Gesundheitszeugnissen vor der Eheschließung und rassenhygienische Eheverbote*, Munich: J. F. Lehmanns, 1917.
Berliner, Max, "Blutgruppenzugehörigkeit und Rassenfragen," *Zeitschrift für Morphologie und Anthropologie* 27, no. 2 (1929): 161–170.
Bernstein, Felix, "Beiträge zur mendelistischen Anthropologie: Quantitative Rassenanalyse auf Grund von statistischen Beobachtungen über den Klangcharakter der Singstimme," *Sitzungsberichte der Preussischen Akademie der Wissenschaften, Physikalisch-mathematische Klasse* (1925): 61–70.
"Ergebnisse einer biostatischen zusammenfassenden Betrachtung über die erblichen Blutstrukturen des Menschen," *Klinische Wochenschrift* 3 (1924): 1495–1497.
Berze, Josef, "Beiträge zur psychiatrischen Erblichkeits- und Konstitutionsforschung," *Zeitschrift für die gesamte Neurologie und Psychiatrie* 87, no. 1 (1923): 94–166.
Die hereditären Beziehungen der Dementia praecox: Beitrag zur Hereditätslehre, Leipzig & Wien: F. Deuticke, 1910.
Bigler, Max, "Beitrag zur Vererbung und Klinik der sporadischen Taubstummheit," *Archiv für Ohren-, Nasen- und Kehlkopfheilkunde* 120 (1929): 81–92.
Blasbalg, Jenny, "Ausländische und deutsche Gesetze und Gesetzentwürfe über Unfruchtbarmachung," *Zeitschrift für die gesamte Strafrechtswissenschaft* 52, no. 1 (1932): 477–496.
Bleuler, Eugen, "Mendelismus bei Psychosen, speziell bei der Schizophrenie," *Schweizer Archiv für Neurologie und Psychiatrie* 1 (1916): 19–40.
Boas, Franz, *Kultur und Rasse*, Leipzig: von Veit, 1914.
"On the Variety of Lines of Descent Represented in a Population," *American Anthropologist* 18, no.1 (1916): 1–9.
"Report of an Anthropometric Investigation of the Population of the United States (1922)," in Franz Boas, *Race, Language and Culture*, New York, NY: Macmillan, 1940, 28–59.
Bonnevie, Kristine, "Was lehrt uns die Embryologie der Papillarmuster über ihre Bedeutung als Rassen- und Familiencharakter? Teil III. Zur Genetik des quantitativen Wertes der Papillarmuster," *Zeitschrift für induktive Abstammungs- und Vererbungslehre* 59 (1931): 1–60.

Bornstein, Maurycy, "Zur Frage der kombinierten Psychosen und der pathologischen Anatomie der Landryschen Paralyse," *Zeitschrift für die gesamte Neurologie und Psychiatrie* 13, no. 1 (1912): 1–16.

Boven, William, *Similarité et mendélisme dans l'hérédité de la démence précoce et de la
folie maniaque-dépressive (Thése-Univ. de Lausanne)*, Vevey: Impr. Säuberlin
& Pfeifer S.A., 1915.

Bree, [Emil], "Wie können die Kirchenbücher für die Familienforschung nutzbar
gemacht werden?" *Familiengeschichtliche Blätter* 26 no. 7/8 (1926): 199–204.

Breymann, Hans, "Genealogie und Vererbungslehre," *Familiengeschichtliche Blätter* 20, no. 9/10 (1922): 193–196.

"Über die Notwendigkeit eines Zusammengehens von Genealogen und Medizinern in der Familienforschung (Vortrag, gehalten anläßlich der Hauptversammlung der internationalen und der deutschen Gesellschaften für
Rassenhygiene zu Dresden am 6. August 1911.)," *Archiv für Rassen- und
Gesellschaftsbiologie* 9, no. 1 (1912): 18–29.

Brugger, Carl, "Die erbbiologische Stellung der Pfropfschizophrenie," *Zeitschrift
für die gesamte Neurologie und Psychiatrie* 113 (1927): 348–378.

Bryn, Halfdan, "Research into Anthropological Heredity. II. The Genetic Relation of Index Cephalicus," *Hereditas* 1 (1920): 198–212.

Burgdörfer, Friedrich, *Volk ohne Jugend*, Heidelberg/Berlin: Kurt Vowinckel, 1937.

Castle, William E., *Heredity of Coat Characters in Guinea-Pigs and Rabbits*, Washington, DC: Carnergie Institution, 1905.

"Race Mixture and Physical Disharmonies," *Science* 71 (1930): 604–605.

Cattle, Raymond B., *Psychology and Social Progress: Mankind and Destiny from the
Standpoint of a Scientist*, London: C. W. Daniel, 1933.

Chamberlain, Houston Stuart, *Die Grundlagen des neunzehnten Jahrhunderts*,
Munich: F. Bruckmann, 1899.

Chun, Carl & Wilhelm Johannsen, *Allgemeine Biologie*, Leipzig/Berlin: B. G.
Teubner, 1915.

Clauß, Ludwig Ferdinand, *Rassenseelenforschung*, Erfurt: Kurt Stenger, 1934.

Rasse und Seele, Munich: J. F. Lehmanns, 1941.

Clauß, Ludwig Ferdinand & Arthur Hoffmann, *Vorschule der Rassenkunde, auf der
Grundlage praktischer Menschenbeobachtung*, Erfurt: Kurt Stenger, 1934.

Correns, Carl, *Carl Correns gesammelte Abhandlungen zur Vererbungswissenschaft
aus periodischen Schriften 1899–1924*, ed. Fritz von Wettstein, Berlin: Julius
Springer, 1924.

"G. Mendels Regel über das Verhalten der Nachkommenschaft der Rassenbastarde," *Berichte der deutschen botanischen Gesellschaft* 18 (1900): 158–168.

"Gregor Mendel's 'Versuche über Pflanzen-Hybriden' und die Bestätigung
ihrer Ergebnisse durch die neuesten Untersuchungen," *Botanische Zeitung*
58, no. 15 (1900): 229–235.

Über Vererbungsgesetze, Berlin: Gebrüder Borntraeger, 1905.

"Ueber Levkojenbastarde," *Botanisches Centralblatt* 84 (1900): 97–113.

Crzellitzer, Arthur, "Anleitung zu biologischen Untersuchungen für Genealogen," *Familiengeschichtliche Blätter* 21, no. 4 (1923): 33–40.

"Der gegenwärtige Stand der Familienforschung," *Sexual-Probleme* 8, no. 4
(1912), 221–243.

"Methoden der Familienforschung," *Zeitschrift für Ethnologie* 41 (1909): 181–198.

"Methodik der graphischen Darstellung der Verwandtschaft mit besonderer Berücksichtigung von Familien-Karten und Familien-Stammbüchern," in Robert Sommer (ed.), *Bericht über den II. Kurs mit Kongreß für Familienforschung, Vererbungs- und Regenerationslehre in Gießen vom 9. Bis 13. April 1912*, Halle a. S.: Carl Marhold, 1912, 25–37.

Darré, Richard Walter, *Neuadel aus Blut und Boden*, Munich: J. F. Lehmanns, 1930.

Darwin, Leonard, "The Geneticists' Manifesto," *The Eugenics Review* 31, no. 4 (1940): 229–230.

Davenport, Charles B., "Determination of Dominance in Mendelian Inheritance," *Proceedings of the American Philosophical Society* 47 (1908): 59–63.

"Huntington's Chorea in Relation to Heredity and Eugenics," *Proceedings of the National Academy of Sciences of the United States of America* 1, no. 5 (1915): 283–285.

"Race Crossing in Jamaica," *Scientific Monthly* 27 (1928): 225–238.

Nomadism, or the Wandering Impulse, with Special Reference to Heredity, The Feebly Inhibited II, Washington, DC: Carnegie Institution of Washington, 1915.

Davenport, Charles B. & Morris Steggerda, *Race Crossing in Jamaica*, Washington, DC: Carnegie Institution, Publication No. 395, 1929.

Davenport, Gertrude C. & Charles B. Davenport, "Heredity of Eye Color in Man," *Science* 26 (1907): 589–592.

Deniker, Joseph, *Les races et les peuples de la terre*, Paris: Masson, 1900.

Depdolla, Philipp, "Vererbungslehre und naturwissenschaftlicher Unterricht," in Günther Just (ed.), *Vererbung und Erziehung*, Berlin: Springer, 1930, 277–303.

Devrient, Ernst, "Die Kirchenbücher und die Staatsarchive," *Mitteilungen der Zentralstelle für deutsche Personen- und Familiengeschichte* 6 (1910): 20–26.

Dinter, Artur, *Die Sünde wider das Blut*, Leipzig: Matthes und Thost, 1920.

Dresel, Kurt, "Inwiefern gelten die mendelschen Vererbungsgesetze in der menschlichen Pathologie?" *Virchows Archiv für pathologische Anatomie und Physiologie und für klinische Medizin* 224, no.3 (1917): 256–303.

Dungern, Emil von & Ludwik Hirszfeld, "Über Vererbung gruppenspezifischer Strukturen des Blutes," *Zeitschrift für Immunitätsforschung und experimentelle Therapie* 6 (1910): 284–292.

Dunn, L. C., *Race and Biology*, Paris: UNESCO, 1951.

Durham Bateson, Beatrice, *William Bateson, F. R. S., Naturalist: His Essays & Addresses*, Cambridge: Cambridge University Press, 1928.

Dürre, Konrad, *Erbstrom: Schauspiel in drei Akten*, Berlin: Bühnenverlag Ahn & Simrock, 1933.

East, Edward M., *Mankind at the Crossroads*, New York, NY: Charles Scribner's Sons, 1923.

Heredity and Human Affairs, New York, NY: Charles Scribner's Sons, 1927.

East, Edward M. & Donald F. Jones, *Inbreeding and Outbreeding: Their Genetic and Sociological Significance*, Philadelphia, PA: J. B. Lippincott, 1919.

"Eheberatungsstellen und Gesundheitszeugnisse in Preussen," *Kultur und Leben. Monatsschrift für Kulturgeschichte und biologische Familienkunde* 8 (1926): 230–236.

Eickstedt, Egon von, "Rassenelemente der Sikh. Mit einem Anhang über biometrische Methoden," *Zeitschrift für Ethnologie* 52/53, no. 4/5 (1920): 317–394.

Engelmann, Fritz & August Mayer, *Sterilität und Sterilisation. Bedeutung der Konstitution für die Frauenheilkunde*, Munich: J. F. Bergmann, 1927.

Federley, Harry, "Zur Methodik des Mendelismus in Bezug auf den Menschen," *Acta Medica Scandinavica* 56, no. 1 (1922): 393–410.

Fehlinger, Hans, "Die Giltigkeit [sic] der Mendelschen Vererbungsregeln für den Menschen," *Politisch-Anthropologische Revue* 9, no. 7 (1910): 374–379.

"Koloniale Mischehen in biologischer Beziehung," *Sexual-Probleme* 8, no. 6 (1912), 373–384.

"Kreuzung beim Menschen," *Archiv für Rassen- und Gesellschaftsbiologie* 8, no. 4 (1911), 447–457.

"Menschenarten und Menschenrassen," *Politisch-Anthropologische Revue* 9 (1910): 198–204.

Feldkampf, Hans, *Vererbungslehre, Rassenkunde, Volkspflege*, Münster: Aschendorff, 1935.

Fetscher, Rainer, "Die Sterilisierung aus eugenischen Gründen," *Zeitschrift für die gesamte Strafrechtswissenschaft* 52, no. 1 (1932): 404–424

Fischer, Eugen, "Betrachtungen über die Schädelform des Menschen," *Zeitschrift für Morphologie und Anthropologie* 24, no. 1 (1924): 37–45.

"Die Rassenmerkmale des Menschen als Domesticationserscheinungen," *Zeitschrift für Morphologie und Anthropologie* 18 (1914): 479–524.

Die Rehobother Bastards und das Bastardierungsproblem beim Menschen, Jena: Gustav Fischer, 1913.

"Ein Fall von erblicher Haararmut und die Art ihrer Vererbung," *Archiv für Rassen- und Gesellschaftsbiologie* 7, no. 1 (1910): 50–56.

Rasse und Rassenentstehung beim Menschen, Berlin: Ullstein, 1927.

"Zur Frage der 'Kreuzungen beim Menschen,'" *Archiv für Rassen- und Gesellschaftsbiologie* 9, no. 1 (1912): 8–9.

(ed.), *Deutsche Rassenkunde. Forschungen über Rassen und Stämme, Volkstum und Familien im Deutschen Volk*, vols. 1–4, Jena: Gustav Fischer, 1929–1930.

Fischer, Eugen & Hans F. K. Günther, *Deutsche Köpfe nordischer Rasse*, Munich: J. F. Lehmanns, 1927.

Fishberg, Maurice, *Die Rassenmerkmale der Juden*, Munich: Ernst Reinhardt, 1913.

Fisher, Ronald A., "XV.—The Correlation between Relatives on the Supposition of Mendelian Inheritance," *Transactions of the Royal Society of Edinburgh* 52, no. 2 (1918): 399–433.

Flügge, Ludwig, "Die Bedeutung der Genealogie für die allgemeinere Wissenschaft und für das praktische Leben," *Familiengeschichtliche Blätter* 20, no. 3 (1922): 3–8.

Förster, Adolf, "Ahnenlistenaustausch," *Familiengeschichtliche Blätter* 19, no. 2 (1921): 33.

Franceschetti, Adolf, "Die Vererbung von Augenleiden," in Franz Schieck & Arthur Brückner (eds.), *Kurzes Handbuch der Ophthalmologie*, Berlin: Julius Springer, 1930, 631–855.

Franke, Gustav, *Vererbung und Rasse: Eine Einführung in Vererbungslehre, Rassenhygiene und Rassenkunde*, Berlin: Verlag "Nationalsozialistische Erziehung," 1934.

Frets, G[errit] P[ieter], "Die Auffassungen M. W. Hauschild's über die Erblichkeit der Kopfform," *Zeitschrift für Morphologie und Anthropologie* 26, no. 2 (1927): 256–263.

Heredity of Headform in Man. Genetica, Vol. 3, The Hague: Martinus Nijhoff, 1921.

The Cephalic Index and Its Heredity, The Hague: Martinus Nijhoff, 1926.

"Über die Dominanz des brachycephalen Kopfindex," *Zeitschrift für Morphologie und Anthropologie* 29, no. 2/3 (1931): 512–517.

Friehe, Albert, *Was muß die deutsche Jugend von der Vererbung wissen?* Frankfurt a. M.: Moritz Diesterweg, 1936.

Was muß der Nationalsozialist von der Vererbung wissen? Frankfurt a. M.: Moritz Diesterweg, 1941.

Fürst, Th[eobald], "Wie kann die Tätigkeit des Schularztes der Erblichkeitsforschung und Rassenhygiene dienen?" *ARGB* 19, no. 3 (1927): 301–314.

Furuhata, Tanemoto, "A Summarized Review on the Gen-Hypothesis of Blood Groups," *American Journal of Physical Anthropology* 13, no. 1 (1929): 109–130.

Gardner, George, *Travels in the Interior of Brazil*, London: Revee Brothers, 1846.

Garrod, Archibald E., "The Incidence of Alkaptonuria: A Study in Chemical Individuality," *Lancet* 160, no. 4137 (1902): 1616–1620.

Gates, Reginald Ruggles, *Heredity and Eugenics*, London: Constable, 1923.

Gaupp, Robert, "Das Gesetz zur Verhütung erbkranken Nachwuchses und die Psychiatrie," *Klinische Wochenschrift* 13, no. 1 (1934): 1–4.

"Die Lehren Kraepelins in ihrer Bedeutung für die heutige Psychiatrie," *Zeitschrift für die gesamte Neurologie und Psychiatrie* 165, no. 1 (1939): 47–75.

Gaupp, Robert & Friedrich Mauz, "Krankheitseinheit und Mischpsychosen," *Zeitschrift für die gesamte Neurologie und Psychiatrie* 101, no. 1 (1926): 1–44.

Gelfius, Wilhelm, *Die gesetzmäßige Vererbung des Menschen*, Lübeck: Self-published, 1932.

Gerum, K., "Beitrag zur Frage der Erbbiologie der genuinen Epilepsie, der epileptoiden Erkrankungen und der epileptoiden Psychopathie," *Zeitschrift für die gesamte Neurologie und Psychiatrie* 115 (1928): 320–422.

Grant, Madison, *The Passing of the Great Race; or, The Racial Basis of European History*, New York, NY: Charles Scribner's Sons, 1916.

Grebe, Hans, "Der Nachweis der Heterozygoten bei rezessiven Erbleiden," *Der Erbarzt* 11 (1943): 1–9.

Gruber, Max von & Ernst Rüdin, *Fortpflanzung, Vererbung, Rassenhygiene: illustrierter Führer durch die Gruppe Rassenhygiene der internationalen Hygiene-Ausstellung 1911 in Dresden*, Munich: J. F. Lehmanns, 1911.

Gruenberg, H., "Men and Mice at Edinburgh. Reports from the Genetics Congress," *Journal of Heredity* 30, no. 9 (1939): 371–374.

Günther, Hans F. K., *Der nordische Gedanke unter den Deutschen*, Munich: J. F. Lehmanns, 1925.

Rassenkunde des deutschen Volkes, Munich: J. F. Lehmanns, 1923.

Gütt, Arthur, Ernst Rüdin & Falk Ruttke, *Zur Verhütung erbkranken Nachwuchses. Gesetz und Erläuterungen*, Munich: J. F. Lehmanns, 1933.

Guttenberg, Franz Carl Freiherr von, "Ein genealogisches Rätsel," *Familiengeschichtliche Blätter* 11, no. 3 (1913): 24.

Haacke, [Wilhelm], "Das Gesetz der Rassenmischung (Bericht und Notizen)," *Politisch-Anthropologische Revue* 6, no. 4 (1907): 269.

Haecker, Valentin, *Allgemeine Vererbungslehre*, Brunswick: Friedrich Vieweg & Sohn, 1911.

Die Erblichkeit im Mannesstamm und der vaterrechtliche Familienbegriff, Jena: Gustav Fischer, 1917.

"Einige Ergebnisse der Erblichkeitsforschung." *Deutsche medizinische Wochenschrift* 27 (1912): 1292–1294.

"Über die neueren Ergebnisse der Bastardlehre, ihre zellengeschichtliche Begründung und ihre Bedeutung für die praktische Tierzucht (Vortrag, gehalten im Verein für vaterländische Naturkunde am 10. März 1904)," *Archiv für Rassen- und Gesellschaftsbiologie* 1, no. 3 (1904): 321–338.

Hammerschlag, Victor, "Zur Kenntnis der hereditär-degenerativen Taubstummheit VIII. Über die hereditäre Taubheit und Gesetze ihrer Vererbung," *Zeitschrift für Ohrenheilkunde* 61 (1910): 225–253.

Hara, Sei, "Vergleichende Untersuchungen über einen planimetrischen Cranio-Facialindex," *Zeitschrift für Morphologie und Anthropologie* 30, no. 3 (1932): 571–585.

Hauschild, Max Wolfgang, "Das Mendeln des Schädels," *Zeitschrift für Ethnologie* 48, no. 1 (1916): 35–40.

"Die Göttinger Gräberschädel. Ein Beitrag zur Anthropologie Niedersachsens," *Zeitschrift für Morphologie und Anthropologie* 21, no. 3 (1921): 365–438.

"Die kleinasiatischen Völker und ihre Beziehungen zu den Juden," *Zeitschrift für Ethnologie* 52/53, no. 6 (1920): 518–528.

Grundriss der Anthropologie, Berlin: Gebrüder Borntraeger, 1926.

Heise, Hans, "Der Erbgang der Schizophrenie in der Familie D. und ihren Seitenlinien," *Zeitschrift für die gesamte Neurologie und Psychiatrie* 64, no. 1 (1921): 229–259.

Helmut, Otto, *Volk in Gefahr*, Munich: J. F. Lehmanns, 1933.

Heron, David, *Mendelism and the Problem of Mental Defect, vol. I, A Criticism of Recent American Work*, London: Dulau, 1913.

Hers, Floris, M[arie] A[nna] van Herwerden & Th. J. Boele-Mijland, "Blutgruppen-Untersuchungen in der 'Hoeksche Waard,'" *Zeitschrift für Morphologie und Anthropologie* 33, no. 1 (1934): 84–95.

Herskovits, Melville J., "A Critical Discussion of the 'Mulatto Hypothesis,'" *The Journal of Negro Education* 3, no. 3: *The Physical and Mental Abilities of the American Negro* (1934): 389–402.

Hesse, E[rich], "Die Unfruchtbarmachung aus eugenischen Gründen," *Beiheft zum Reichs-Gesundheitsblatt* 15 (April 12, 1933), 19.

Heydenreich, Eduard (ed.), *Handbuch der praktischen Genealogie*, Leipzig: Degener, 1913.

Hiemer, Ernst, *Der Jude im Sprichwort der Völker*, Nuremberg: Der Stürmer Buchverlag, 1942.

Hirschfeld, Ludwik & Hanka Hirschfeld, "Serological Differences between the Blood of Different Races," *The Lancet* 194, no. 5016 (1919): 675–679.

Hiss, Wilhelm, "Beschreibung einiger Schädel altschweizerischer Bevölkerung nebst Bemerkungen über die Aufstellung von Schädeltypen," *Archiv für Anthropologie* 1, no. 1 (1866): 61–74.

Hitler, Adolf, *Mein Kampf*, Munich: Verlag Franz Eher Nachf., 1943.

Hoche, Alfred E., "Die Bedeutung der Symptomenkomplexe in der Psychiatrie," *History of Psychiatry* 2, no. 7 (1991): 334–343.

Hoffmann, Hermann, *Die individuelle Entwicklungskurve des Menschen*, Berlin: Springer, 1922.

Studien über Vererbung und Entstehung geistiger Störungen, Berlin: Julius Springer, 1921.

Vererbung und Seelenleben, Berlin: Julius Springer, 1922.

Huxley, Julian, *Evolution: The Modern Synthesis*, London: George Allen & Unwin, 1942.

Iltis, Hugo, *Gregor Johann Mendel. Leben, Werk und Wirkung*, Berlin: Springer, 1924.

Life of Mendel, translated by Eden and Cedar Paul, London: George Allen & Unwin, 1932.

Isenburg, Wilhelm Karl Prinz von, "Aus der Werkstatt eines Ahnentafelforschers," *Familiengeschichtliche Blätter* 23 (1925), 7–14.

Jennings, Herbert. S., "Public Health Progress and Race Progress – Are They Incompatible?" *Science* 66, no. 1698 (1927): 45–50.

The Biological Basis of Human Nature, New York, NY: W. W. Norton, 1930.

Johannsen, Wilhelm, *Elemente der exakten Erblichkeitslehre*, Jena: Gustav Fischer, 1909.

Jörns, Emil, *Meine Sippe. Ein Arbeitsheft für rassebewusste deutsche Jugend*, Görlitz: Verlag für Sippenforschung und Wappenkunde C. A. Starke, 1934.

Kaestner, Sandor, *Was muss der Familiengeschichtsforscher von der Vererbungswissenschaft wissen? Praktikum für Familienforscher 5*, Leipzig: Degener & Co., 1924.

Kahn, Eugen, "Konstitution, Erbbiologie und Psychiatrie," *Zeitschrift für die gesamte Neurologie und Psychiatrie* 57, no. 1 (1920): 280–311

Schizoid und Schizophrenie im Erbgang, Studien über Vererbung und Entstehung geistiger Störungen IV, Berlin: Julius Springer, 1923.

Kahn, Fritz, *Die Juden als Rasse und Kulturvolk*, Berlin: Welt Verlag, 1922.

Kaul, Ingo, *Das Wunder des Lebens: Ausstellung Berlin 1935, 25. März bis 5. Mai*, Berlin: Meisenbach, 1935.

Kaznelson, Paul, "Kritische Besprechungen und Referate," *Archiv für Rassen- und Gesellschaftsbiologie* 10, no. 5 (1913): 685–686.

"Kritische Besprechungen und Referate," *Archiv für Rassen- und Gesellschaftsbiologie* 10, no. 6 (1913): 796–802.

"Über einige 'Rassenmerkmale' des jüdischen Volkes," *Archiv für Rassen- und Gesellschaftsbiologie* 10, no. 5 (1913): 484–502.

Kehrer, Ferdinand Adalbert & Ernst Kretschmer, *Die Veranlagung zu seelischen Störungen*, Berlin: Julius Springer, 1924.

Klein, W. & H. Osthoff, "Haemagglutinine, Rasse- und anthropologische Merkmale," *Archiv für Rassen- und Gesellschaftsbiologie* 17, no. 4 (1926): 371–378.

Klocke, Friedrich von, "Deutsche Ahnentafeln," *Familiengeschichtliche Blätter* 19, no. 9 (1921): 257–262.

Die Entwicklung der Genealogie vom Ende des 19. bis zur Mitte des 20. Jahrhunderts, Prolegomena zu einem Lehrbuch der Genealogie, Schellenberg bei Berchtesgaden: Degener & Co., 1950.

"Vom Begriff Genealogie und den Verdeutschungen des Wortes," *Familiengeschichtliche Blätter* 17, no. 12 (1919): 217–228.

Köhler, W[olfgang], "'Referate,'" *Psychologische Forschung* 6, no. 1 (1925): 208.

Kohls, Erna, "Über die Sterilisation zur Verhütung geistig minderwertiger Nachkommen," *Archiv für Psychiatrie* 77 (1926): 285–302.

Kraemer, Rudolf, *Kritik der Eugenik, vom Standpunkt des Betroffenen*, Berlin: Reichsdeutscher Blindenverband, 1933.

"Neujahrs-Zauber, 2. Bild: Das Blindenpferd, der Triumph der angewandten Vererbungswissenschaft," *Die Blindenwelt* 21, no. 1 (1933): 7–11.

Kraepelin, Emil, *Psychiatrie: Ein Lehrbuch für Studierende und Aerzte*, vol. VI, Leipzig: Johann Ambrosius Barth, 1899.

Kraus, F., "Blutsverwandtschaft in der Ehe und deren Folgen für die Nachkommenschaft," in Hermann Senator & Siegfried Kaminer (eds.), *Krankheiten und Ehe: Darstellung der Beziehungen zwischen Gesundheitsstörungen und Ehegemeinschaft*, Munich: J. F. Lehmanns, 1904, 56–88.

Kronacher, Carl, *Grundzüge der Züchtungbiologie*, Berlin: Paul Parey, 1912.

Krueger, Hermann, "Zur Frage nach einer vererbbaren Disposition zu Geisteskrankheiten und ihren Gesetzen," *Zeitschrift für die gesamte Neurologie und Psychiatrie* 25, no. 1 (1914): 113–182.

Kruse, W[alther], "Ueber Blutzusammensetzung und Rasse," *Archiv für Rassen- und Gesellschaftsbiologie* 19, no. 1 (1927): 20–33.

Kuper, Leo (ed.), *Race, Science and Society*, Paris: UNESCO Press, 1956.

Lasas, "Ueber die Blutgruppen der Litauer, Letten und Ostpreußen," *Archiv für Rassen- und Gesellschaftsbiologie* 22, no. 3 (1930): 270–274.

Laughlin, Harry H., *Bulletin No. 10B: Report of the Committee to Study and to Report on the Best Practical Means of Cutting Off the Defective Germ-Plasm in the American Population. II. The Legal, Legislative and Administrative Aspects of Sterilization*, New York, NY: Eugenics Record Office, 1914.

Eugenical Sterilization in the United States, Chicago, IL: Psychopathic Laboratory of the Municipal Court of Chicago, 1922.

Lehmann, Ernst, "Zur Terminologie und Begriffsbildung in der Vererbungslehre," *Induktive Abstammungs- und Vererbungslehre* 22 (1919): 237–260.

Lenz, Fritz, "Die Bedeutung der statistisch ermittelten Belastung mit Blutsverwandtschaft der Eltern," *Münchener medizinische Wochenschrift* 66 (1919): 1340–1342.

"Gregor Mendel," *Münchener medizinische Wochenschrift* 69 (1922): 1349–1350.

"Kritische Besprechungen und Referate," *Archiv für Rassen- und Gesellschaftsbiologie* 11, no. 6 (1914): 811.

"Kritische Besprechungen und Referate," *Archiv für Rassen- und Gesellschafts-biologie* 21, no. 2 (1929): 194–204.

"Über die idioplasmatischen Ursachen der physiologischen und patholo-gischen Sexualcharaktere des Menschen," *Archiv für Rassen- und Gesellschaftsbiologie* 9, no. 5 (1912): 545–603.

Leveringhaus, Herbert, "Die Bedeutung der menschlichen Isohämagglutination für Rassenbiologie und Klinik," *Archiv für Rassen- und Gesellschaftsbiologie* 19, no. 2 (1927): 1–17.

Lösener, Bernhard & Friedrich August Knost, *Die Nürnberger Gesetze über das Reichsbürgerrecht und den Schutz des deutschen Blutes und der deutschen Ehre*, Berlin: Vahlen, 1936.

Lorenz, Ottokar, *Lehrbuch der gesammten wissenschaftlichen Genealogie*, Berlin: Wilhelm Hertz, 1898.

Lundborg, Herman, *Medizinisch-biologische Familienforschungen innerhalb eines 2232-köpfigen Bauerngeschlechtes in Schweden*, Jena: Gustav Fischer, 1913.

"Über die Erblichkeitsverhältnisse der konstitutionellen (hereditären) Taub-stummheit und einige Worte über die Bedeutung der Erblichkeitsforschung für die Krankheitslehre," *Archiv für Rassen- und Gesellschaftsbiologie* 9, no. 2 (1912): 133–149.

Luschan, Felix von, "Afrika," in Georg Buschan (ed.), *Illustrierte Völkerkunde*, Stuttgart: Strecker & Schröder, 1909, 357–380.

Völker, Rassen, Sprachen, Berlin: Welt-Verlag, 1922.

Luschan, Felix von & Hermann Struck, *Kriegsgefangene. Ein Beitrag zur Völk-erkunde im Weltkriege*, Berlin: D. Reimer, 1917.

Luther, A. "Erblichkeitsbeziehungen der Psychosen," *Zeitschrift für die gesamte Neurologie und Psychiatrie* 24, no. 1 (1914): 12–81.

Luxenburger, Hans, "Die wichtigsten Ergebnisse der psychiatrischen Erb-forschung und ihre Bedeutung für die eugenische Praxis (Vortrag, gehalten in der Sitzung der Münchner gynäkologischen Gesellschaft, am 23. 1. 1930)," *Archiv für Gynäkologie* 141 (1930): 237–254.

"Der Begriff der Belastung in der Eheberatungstätigkeit des Arztes," *Der Erbarzt* 1 (1935): 12–15.

Magdeburg, Paul, *Rassenkunde und Rassenpolitik. Zahlen, Gesetze und Verordnun-gen*, Bildung und Nation: Schiftenreihe zur nationalpolitischen Erziehung, Leipzig: Eichblatt-Verlag (Max Zedler), 1933.

Marcuse, Max, "Inzucht und Verwandtenehe," in Max Marcuse (ed.), *Hand-wörterbuch der Sexualwissenschaft*, 2nd ed., Bonn: A. Marcus & E. Webers Verlag, 1926, 311–314.

"Über die christlich-jüdische Mischehe," *Sexual-Probleme* 8, no. 10 (1912): 691–749.

Martin, Rudolf, *Lehrbuch der Anthropologie*, Jena: Gustav Fischer, 1928.

Medow, W., "Zur Erblichkeitsfrage in der Psychiatrie," *Zeitschrift für die gesamte Neurologie und Psychiatrie* 26, no. 1 (1914): 493–545.

Meggendorfer, Friedrich, "Ueber die hereditaere Disposition zur Dementia seni-lis," *Zeitschrift für die gesamte Neurologie und Psychiatrie* 101 (1925): 387–405.

"Erblichkeitsforschung und Psychiatrie," *Zeitschrift der Zentralstelle für nieder-sächsische Familiengeschichte* 7, no. 10 (1925): 225–229.

Michelsson, Gustav, "Kritische Betrachtungen über Methoden und Aufgaben der Anthropologie," *Zeitschrift für Morphologie und Anthropologie* 23, no. 2 (1923): 263–294.

Mjøen, Jon Alfred, "Harmonische und unharmonische Kreuzungen (Vortrag, ausserordentliche Sitzung vom 31. Mai 1921)," *Zeitschrift für Ethnologie* 53, no. 4/5 (1921): 470–479.

Moszkowski, Max, "Klima, Rasse und Nationalität in ihrer Bedeutung für die Ehe," in C[arl] Noorden & Siegfried Kaminer (eds.), *Krankheiten und Ehe: Darstellung der Beziehungen zwischen Gesundheitsstörungen und Ehegemeinschaft*, 2nd ed., Munich: J. F. Lehmanns, 1916, 100–156.

Muckermann, Hermann, *Die Erblichkeitsforschung und die Wiedergeburt von Familie und Volk*, Freiburg i. B.: Herdersche Verlagshandlung, 1919.

Grundriss der Biologie. Erster Teil: Allgemeine Biologie, Freiburg i. B.: Herdersche Verlagshandlung, 1909.

Muller, Hermann J., "Artificial Transmutation of the Gene," *Science* 66 (1927): 84–87

Nachtsheim, Hans, "Die Frage der Sterilisation vom Standpunkt der Erbbiologen," *Berliner Gesundheitsblatt* 1, no. 24 (1950): 603–604

Für und wider die Sterilisierung aus eugenischer Indikation, Stuttgart: Georg Thieme, 1952.

"Mendelismums und Tierzucht," *Die Naturwissenschaften* 29 (1922): 635–640.

"Sterilisation aus eugenischer Indikation," *Rheinisches Ärzteblatt* 10, no. 10 (1956): 215–219.

Naturwissenschaftliche Wochenschrift, Neue Folge 12, no. 14 (1913): 222–223.

Nettleship, Edward, "A History of Congenital Stationary Night Blindness in Nine Consecutive Generations," *Transactions of the Ophthalmological Societies of the United Kingdom* 27 (1907): 269–293.

"On Heredity in the Various Forms of Cataract," *The Royal London Ophthalmic Hospital Reports* 16 (1905): 1.

The Bowman Lecture on Some Hereditary Diseases of the Eye, London: Adlard, 1909.

Neubauer, Gabriele, "Experimentelle Untersuchungen über die Beeinflussung der Schädelform," *Zeitschrift für Morphologie und Anthropologie* 23, no. 3 (1925): 411–442.

Orel, Herbert, "Die Verwandtenehen in der Erzdiözese Wien," *Archiv für Rassen- und Gesellschaftsbiologie* 26, no. 3 (1932): 249–278.

Ottenberg, Reuben, "A Classification of Human Races Based on Geographic Distribution of the Blood Groups," *Journal of the American Medical Association* 84, no. 19 (1925): 1393.

Otto, Hermann & Werner Stachowitz, *Abriß der Vererbungslehre und Rassenkunde einschließlich der Familienkunde, Rassenhygiene und Bevölkerungspolitik*, Frankfurt a. M.: Moritz Diesterweg, 1934.

Park, Robert Ezra, "Human Migration and the Marginal Man," *American Journal of Sociology* 33, no. 6 (1928): 881–893.

Paulsen, Jens, "Über Die Erblichkeit von Thoraxanomalien mit besonderer Berücksichtigung der Tuberkulose," *Archiv für Rassen- und Gesellschaftsbiologie* 13, no. 1 (1921): 10–31.

Peters, W[ilhelm], *Die Vererbung geistiger Eigenschaften und die psychische Konstitution*, Jena: Gustav Fischer, 1925.

Plate, Ludwig, "Kritische Besprechungen und Referate," *Archiv für Rassen- und Gesellschaftsbiologie* 2, no. 5–6 (1905): 852.

"Kritische Besprechungen und Referate," *Archiv für Rassen- und Gesellschaftsbiologie* 6, no. 1 (1909): 101–102.

"Über Vererbung und die Notwendigkeit der Gründung einer Versuchsanstalt für Vererbungs- und Züchtungskunde (Vortrag, gehalten am 24. Oktober in der deutschen Gesellschaft für Züchtungskunde)," *Archiv für Rassen- und Gesellschaftsbiologie* 2 no. 5/6 (1905): 777–796.

Vererbungslehre: Mit besonderer Beruecksichtigung des Menschen, für Studierende, Ärzte und Züchter, Leipzig: Wilhelm Engelmann, 1913.

Vererbungslehre mit besonderer Berücksichtigung der Abstammungslehre und des Menschen, Jena: Gustav Fischer, 1932.

Prell, Heinrich, "Die Grundtypen der gesetzmäßigen Vererbung," *Naturwissenschaftliche Wochenschrift, Neue Folge* 20 (1921): 289–297.

Punnett, R[eginald] C., "Mendelism in Relation to Disease," *Proceedings, Royal Society of Medicine* 1 (1908): 135–168.

"Rassenkunde in der Volksschule," *Neues Volk* 7 (1934): 7–11.

Reche, Otto, "Der Wert des erbbiologischen Abstammungsnachweises für die richterliche Praxis," *Deutsches Recht* 9, no. 28 (1939): 1606–1612.

Regge, Jürgen & Werner Schubert (eds.), *Quellen zur Reform des Straf- und Strafprozeßrechts*, vol. 2.2 (Berlin: de Gruyter, 1989), 223–348.

Reibmayr, Albert, *Inzucht und Vermischung beim Menschen*, Leipzig/Vienna: Franz Deuticke, 1897.

Reinstorf, Ernst, *Wie erforsche und schreibe ich als Bürgerlicher meine Familiengeschichte? Eine kurze Anleitung dazu* [2nd. ed.: 1924, 3rd ed.: 1926, 4th ed. 1935] Hamburg: Zentralstelle für niedersächsische Familiengeschichte, 1919.

Reutlinger, Wilhelm, "Über die Häufigkeit der Verwandtenehen bei den Juden in Hohenzollern und über Untersuchungen bei Deszendenten aus jüdischen Verwandtenehen," *Archiv für Rassen- und Gesellschaftsbiologie* 14, no. 3 (1922): 301–305.

Richter, Brigitte, *Burkhards und Kaulstoß: Zwei oberhessische Dörfer*, Deutsche Rassenkunde, vol. 14, Jena: Gustav Fischer, 1936.

Ripley, William Z., *The Races of Europe: A Sociological Study*, New York, NY: D. Appleton and Co., 1899.

Rodenwaldt, Ernst, "Vom Seelenkonflikt des Mischlings," *Zeitschrift für Morphologie und Anthropologie* 34 (1934): 364–375.

Roemer, Hans, "Eine Einteilung der Psychosen und Psychopathien, für die Zwecke der Statistik vereinbart zwischen der psychiatrischen Klinik Heidelberg und den Heil- und Pflegeanstalten Illenau und Wiesloch," *Zeitschrift für die gesamte Neurologie und Psychiatrie* 11, no. 1 (1912): 69–90.

"Über psychiatrische Erblichkeitsforschung," *Archiv für Rassen- und Gesellschaftsbiologie* 9, no. 3 (1912): 292–329.

Römpp, Hermann, "Mendelismus," *Der praktische Schulmann* 10 (1934): 45.

Röse, Carl, "Beiträge zur europäischen Rassenkunde," *Archiv für Rassen- und Gesellschaftsbiologie* 2, no. 5–6 (1905): 689–798.

Roesler, Gottfried, "Genealogie als Grundlage der Familienpolitik," *Volksaufartung, Erbkunde, Eheberatung* 5, no. 5 (1930): 101–107.

Rüdin, Ernst, "Einige Wege und Ziele der Familienforschung, mit Rücksicht auf die Psychiatrie," *Zeitschrift für die gesamte Neurologie und Psychiatrie* 7, no. 1 (1911): 487–585.

"Empirische Erbprognose," *Archiv für Rassen- und Gesellschaftsbiologie* 27, no. 3 (1933): 271–283.

"Kritische Besprechungen und Referate," *Archiv für Rassen- und Gesellschaftsbiologie* 3, no. 5 (1906): 750.

"Kritische Besprechungen und Referate," *Archiv für Rassen- und Gesellschaftsbiologie* 5, no. 1 (1908): 133–135.

"Kritische Besprechungen und Referate," *Archiv für Rassen- und Gesellschaftsbiologie* 5, no. 2 (1908): 272–275.

"Ueber die Vorhersage von Geistesstörung in der Nachkommenschaft (Vortrag im Februar 1928 in der Gesellschaft für Rassen-Hygiene in München)," *Archiv für Rassen- und Gesellschaftsbiologie* 20, no. 4 (1928): 394–407.

Zur Vererbung und Neuentstehung der Dementia praecox (Studien über Vererbung und Entstehung geistiger Störungen I.), Berlin: Julius Springer, 1916.

Ruttmann, W[ilhelm] J[ulius], *Erblichkeitslehre und Pädagogik*, Leipzig: Schulwissenschaftlicher Verlag A. Haase, 1917.

Salaman, Redcliffe N., "Heredity and the Jew," *Journal of Genetics* 1, no. 3 (1911): 273–292.

Saller, Karl, "Ein Meßkoffer für anthropologische Reisen," *Zeitschrift für Morphologie und Anthropologie* 27, no. 3 (1930): 492–496.

Einführung in die menschliche Erblichkeitslehre und Eugenik, Berlin: Julius Springer, 1932.

Scheidt, Walter, *Allgemeine Rassenkunde als Einführung in das Studium der Menschenrassen*, Munich: J. F. Lehmanns, 1925.

"Erbbiologische und bevölkerungsbiologische Aufgaben der Familienforschung," *Archiv für Sippenforschung und alle verwandten Gebiete* 9 (1928): 289–315.

Rassenunterschiede des Blutes, Leipzig: Georg Thieme, 1927.

"Zur Theorie der Auslese," *Zeitschrift für induktive Abstammungs- und Vererbungslehre* 46, no. 1 (1928): 318–332.

Schlaginhaufen, Otto, "Die Rassenmerkmale der Juden by Maurice Fishberg," *Zeitschrift für Morphologie und Anthropologie* 19, no. 1 (1915): 265–269.

Schopohl, Heinrich (eds.), *Die Eugenik im Dienste der Volkswohlfahrt. Bericht über die Verhandlungen eines zusammengesetzten Ausschusses des preussischen Landesgesundheitsrats vom 2. Juli 1932. Veröffentlichungen aus dem Gebiete der Medizinalverwaltung*, vol. 38, no. 5, Berlin: Richard Schoetz, 1932.

Schulz, Bruno, "Zum Problem der Erbprognose-Bestimmung. Die Erkrankungsaussichten der Neffen und Nichten von Schizophrenen," *Zeitschrift für die gesamte Neurologie und Psychiatrie* 102 (1925): 1–37.

"Zur Erbpathologie der Schizophrenie," *Zeitschrift für die gesamte Neurologie und Psychiatrie* 143 (1932): 175–298.

"Zur Frage einer Belastungsstatistik der Durchschnittsbevölkerung. Geschwisterschaften und Elternschaften von 100 hirnarterioskleritiker-Ehegatten," *Zeitschrift für die gesamte Neurologie und Psychiatrie* 109, no. 1 (1927): 15–48.

Schuppius, "Über Erblichkeitsbeziehungen in des Psychiatrie," *Zeitschrift für die gesamte Neurologie und Psychiatrie* 13, no. 1 (1912): 218–284.

Schwarzburg, Walter, "Statistische Untersuchungen über den menschlichen Scheitelwirbel und seine Vererbung," *Zeitschrift für Morphologie und Anthropologie* 26, no. 2 (1927): 195–224.

Schwidetzky, Ilse, "Merkmalszählung oder Rassenforschung?" *Zeitschrift für Rassenkunde und die gesamte Forschung am Menschen* 13 (1942): 177–182.

Semigothaisches genealogisches Taschenbuch ari(st)okratisch-jüdischer Heiraten, Munich: F. Bruckmann, 1914.

Siemens, Hermann Werner, "Bastard," in Max Marcuse (ed.), *Handwörterbuch der Sexualwissenschaft*, Bonn: A. Marcus & E. Webers Verlag, 1926 [reprint 2011], 45–46.

"Bedeutung und Methodik der Ahnentafelforschung," *Archiv für Rassen- und Gesellschaftsbiologie* 24 (1930): 185–197.

Grundzüge der Vererbungslehre, Rassenhygiene und Bevölkerungspolitik, Munich: J. F. Lehmanns, 1930.

Snyder, Laurence H., "Human Blood Groups: Their Inheritance and Racial Significance," *American Journal of Physical Anthropology* 9, no. 2 (1926): 233–263.

Sofer, Leo, "Auf den Spuren der mendelschen Gesetze," *Politisch-Anthropologische Revue* 7, no. 7 (1908): 345–351.

"Über die Plastizität der menschlichen Rassen," *Archiv für Rassen- und Gesellschaftsbiologie* 5, no. 5/6 (1908): 660–668.

Sommer, Robert, *Bericht über den II. Kurs mit Kongress für Familienforschung, Vererbungs- und Regenerationslehre in Giessen von 9. Bis 13. April 1912*, Halle a. S.: Carl Marhold, 1912.

Familienforschung und Vererbungslehre, Leipzig: Johann Ambrosius Barth, 1907.

Sprel, August, "Wie wir zur Genealogie gekommen," *Familiengeschichtliche Blätter* 22, no. 1/2 (1924): 5–12.

Steche, Otto, *Leitfaden der Rassenkunde und Vererbungslehre, der Erbgesundheitspflege und Familienkunde für die Mittelstufe*, Leipzig: Quelle & Meyer, 1934.

Steiger, Adolf, "Über die Bedeutung von Augenuntersuchungen für die Vererbungsforschung," *Archiv für Rassen- und Gesellschaftsbiologie* 5, no. 5/6 (1908): 623–634.

Steffan, Paul, "Die Bedeutung der Blutuntersuchung für die Bluttransfusion und die Rassenforschung," *Archiv für Rassen- und Gesellschaftsbiologie* 15, no. 2 (1924): 137–150.

Stenberg, Sven, "Zur Frage der kombinierten Psychosen," *Zeitschrift für die gesamte Neurologie und Psychiatrie* 129 (1930): 724–738.

Stoll, Adolf, "Über Familienforschung und Vererbung," *Zeitschrift für Kulturgeschichte und biologische Familienkunde* 1, no. 2 (1924): 62–67.

Stonequist, Everett V., *The Marginal Man: A Study in Personality and Cultural Conflict*, New York, NY: Scribner, 1937.

"The Problem of the Marginal Man," *The American Journal of Sociology* 41, no. 1 (1935): 1–12.

Stradonitz, Stephan Kekulé von, "Die Genealogie auf der internationalen Hygiene-Ausstellung zu Dresden," *Familiengeschichtliche Blätter* 10 (1912): 3–4, 19–20, 39–40.

Strohmayer, Wilhelm, "Die Ahnentafel des Könige Ludwig II. und Otto I. von Bayern," *Archiv für Rassen- und Gesellschaftsbiologie* 7, no. 1 (1910): 65–92.

"Die Vererbungs des Habsburger Familientypus," *Archiv für Rassen- und Gesellschaftsbiologie* 8, no. 6 (1911): 775–785.

"Die Vererbung des Habsburger Familientypus (Zweite Mitteilung)," *Archiv für Rassen- und Gesellschaftsbiologie* 9, no. 2 (1912): 150–164.

Psychiatrisch-genealogische Untersuchung der Abstammung König Ludwigs II. und Ottos I. von Bayern, Wiesbaden: J. F. Bergmann, 1912.

Stuckart, Wilhelm & Hans Globke, *Kommentare zur deutschen Rassengesetzgebung*, Munich: C. H. Beck, 1936.

Teichman, Ernst, *Handwörterbuch der Naturwissenschaften*, Jena: Gustav Fischer, 1912.

Tornow, Karl & Herbert Weinert, *Erbe und Schicksal: Von geschädigten Menschen, Erbkrankheiten und deren Bekämpfung*, Berlin: Alfred Metzner, 1942.

Török, Aurel von, "Inwiefern kann das Gesichtsprofil als Ausdruck der Intelligenz gelten? Ein Beitrag zur Kritik der heutigen physischen Anthropologie," *Zeitschrift für Morphologie und Anthropologie* 3, no. 3 (1901): 351–484.

Tschermak, Erich, "Die Mendelsche Lehre und die Galtonsche Theorie vom Ahnenerbe," *Archiv für Rassen- und Gesellschaftsbiologie* 2, no. 5–6 (1905): 663–672.

UNESCO, *What Is Race? Evidence from Scientists*, Paris: UNESCO, 1952.

Velden, Adolf von den, "Wert und Pflege der Ahnentafel (Vortrag, gehalten in der ersten Hauptversammlung am 21. November 1904)," *Mitteilungen der Zentralstelle für deutsche Personen- und Familiengeschichte* 1 (1905): 17–22.

Velden, [Reinhard] Freiherr von den, "Gelten die Mendelschen Regeln für die Vererbung menschlicher Krankheiten?" *Politisch-Anthropologische Revue* 9, no. 2 (1910): 91–97.

Verschuer, Ottmar Freiherr von, *Blindheit und Eugenik*, Berlin: Reichsdeutscher Blindenverband, 1933.

"Rassenbiologie der Juden," in *Forschungen zur Judenfrage*, vol. 3, Hamburg: Hanseatische Verlagsanstalt, 1938, 137–151.

"Vom Umfang der erblichen Belastung im deutschen Volke," *Archiv für Rassen- und Gesellschaftsbiologie* 24 (1930): 238–268.

Vogt, Alfred & Richard Klainguti, "Weitere Untersuchungen über die Entstehung der Rotgrünblindheit beim Weibe," *Archiv für Rassen- und Gesellschaftsbiologie* 14, no. 2 (1922): 129–140.

Vries, Hugo de, "Das Spaltungsgesetz der Bastarde," *Berichte der deutschen botanischen Gesellschaft* 18, no. 3 (1900): 83–90.

"Sur la loi de disjonction des hybrides," *Comptes Rendus de l'Académie des sciences, Paris* 130 (1900): 845–847.

Wagenseil, Ferdinand, "Beiträge zur physischen Anthropologie der spaniolischen Juden und zur jüdischen Rassenfrage," *Zeitschrift für Morphologie und Anthropologie* 23, no. 1 (1923): 33–150.

Wecken, Friedrich, *Die Ahnentafel als Nachweis deutscher Abstammung*, Leipzig: Degener & Co., 1934.

Taschenbuch für Familiengeschichtsforschung, [3rd ed.: 1924; 4th ed.: 1930] Leipzig: Degener & Co., 1919.

Weinberg, Wilhelm, "Aufgabe und Methode der Familienstatistik bei medizinisch-biologischen Problemen (Vortrag, gehalten in der 8. Sektion des XIV. Kongresses für Hygiene und Demographie in Berlin)," *Zeitschrift für Soziale Medizin, Medizinalstatistik, Arbeitsversicherung, soziale Hygiene und die Grenzfragen der Medizin und Volkswirtschaft* 3 (1908): 4–26.

"Auslesewirkungen bei biologisch-statistischen Problemen," *Archiv für Rassen- und Gesellschaftsbiologie* 10, no. 4 (1913): 557–581.

"Die württembergischen Familienregister und ihre Bedeutung als Quelle wissenschaftlicher Untersuchungen," *Württembergische Jahrbücher für Statistik und Landeskunde* 1 (1907): 174–198.

"Über neuere psychiatrische Vererbungsstatistik," *Archiv für Rassen- und Gesellschaftsbiologie* 10, no. 3 (1913): 303–312.

"Weitere Beiträge zur Theorie der Vererbung. 4. Ueber Methode und Fehlerquellen der Untersuchung auf Mendelsche Zahlen beim Menschen," *Archiv für Rassen- und Gesellschaftsbiologie* 9 (1912): 165–174.

Weinert, Hans, *Biologische Grundlagen für Rassenkunde und Rassenhygiene*, Stuttgart: Ferdinand Enke, 1943.

Weitz, Wilhelm, "Über die Häufigkeit des Vorkommens des gleichen Leidens bei den Verwandten eines an einem einfach rezessiven Leiden Erkrankten," *Archiv für Rassen- und Gesellschaftsbiologie* 27, no. 1 (1933): 12–24.

Whitney, Leon F. *The Case for Sterilization*, New York, NY: Frederick A. Stockes, 1934.

Wilmanns, Karl, "Die Schizophrenie," *Zeitschrift für die gesamte Neurologie und Psychiatrie* 78, no. 1 (1922): 325–372.

Wittermann, Ernst, "Psychiatrische Familienforschungen," *Zeitschrift für die gesamte Neurologie und Psychiatrie* 20, no. 1 (1913): 153–278.

Wulz, Gustav, "Ein Beitrag zur Statistik der Verwandtenehen," *Archiv für Rassen- und Gesellschaftsbiologie* 17, no. 1 (1927): 82–94.

Zehlendorf, A. W., "Sterilisation Erbkranker," *Berliner Gesundheitsblatt* 1, no. 19 (1950), 486.

Ziegler, Heinrich E., "Die Chromosomen-Theorie der Vererbung in ihrer Anwendung auf den Menschen," *Archiv für Rassen- und Gesellschaftsbiologie* 3 no. 6 (1906): 797–812.

Die Vererbungslehre in der Biologie und in der Soziologie, Jena: Gustav Fischer, 1918.

Zoller, Erwin, "Zur Erblichkeitsforschung bei Dementia praecox," *Zeitschrift für die gesamte Neurologie und Psychiatrie* 55, no. 1 (1920): 275–293.

Secondary Literature

Adams, Mark B., "Eugenics in Russia 1900–1940," in Mark B. Adams (ed.), *The Wellborn Science: Eugenics in Germany, France, Brazil, and Russia*, New York, NY: Oxford University Press, 1990, 153–216.

"Toward a Comparative History of Eugenics," in Mark B. Adams (ed.), *The Wellborn Science. Eugenics in Germany, France, Brazil and Russia*, New York, NY: Oxford University Press, 1990, 217–231.

Allen, Garland E., "Eugenics as an International Movement," in James D. Wright (ed.), *International Encyclopedia of the Social & Behavioral Sciences*, 2nd ed., vol. 8, Oxford: Elsevier, 2015, 224–232.

Aly, Götz, *Cleansing the Fatherland: Nazi Medicine and Racial Hygiene*, Baltimore, MD: Johns Hopkins University Press, 1994.

Aly, Götz & Susanne Heim, *Architects of Annihilation: Auschwitz and the Logic of Destruction*, translated by A. G. Blunden, London: Weidenfeld and Nicolson, 2002.

Améry, Jean, "On the Necessity and Impossibility of Being a Jew," *New German Critique* 20, special issue: Germans and Jews (1980): 15–29.

Angst, Jules, "Historical Aspects of the Dichotomy between Manic-Depressive Disorders and Schizophrenia," *Schizophrenia Research* 57, no. 1 (2002): 5–13.

Annas, George J. & Michael A. Gordin (eds.), *The Nazi Doctors and the Nuremberg Code*, New York, NY: Oxford University Press, 1992.

Arbeitsgruppe Pädagogisches Museum (ed.), *Heil Hitler, Herr Lehrer: Volksschule 1933–1945, Das Beispiel Berlin*, Hamburg: Rowohlt, 1983.

Arendt, Hanna, *Elements of Totalitarianism*, New York, NY: Harcourt Brace Jovanovich, 1951.

Ash, Mitchell G., *Gestalt Psychology in German Culture, 1890–1967: Holism and the Quest for Objectivity*, Cambridge: Cambridge University Press, 1995.

"Wissenschaft und Politik als Ressourcen füreinander," in Rüdiger vom Bruch & Brigitte Kaderas (eds.), *Wissenschaften und Wissenschaftspolitik: Bestandaufnahmen zu Formationen, Brüchen und Kontinuitäten im Deutschland des 20. Jahrhunderts*, Stuttgart: Franz Steiner, 2002, 32–51.

Avraham, Doron, "The 'Racialization' of Jewish Self-Identity: The Response to Exclusion in Nazi Germany, 1933–1938," *Nationalism and Ethnic Politics* 19 (2013): 354–374.

Bachrach, Susan & Dieter Kuntz (eds.), *Deadly Medicine: Creating the Master Race*, Washington, DC: United States Holocaust Memorial Museum, 2004.

Bangham, Jenny, "What Is Race? UNESCO, Mass Communication and Human Genetics in the 1950s," *History of the Human Sciences* 28, no. 5 (2015): 80–107.

Barkan, Elazar, *The Retreat of Scientific Racism*, Cambridge: Cambridge University Press, 1992.

Bashford, Alison, "Internationalism, Cosmopolitanism, and Eugenics," in Alison Bashford and Philippa Levine (eds.), *The Oxford Handbook of the History of Eugenics*, New York, NY: Oxford University Press, 2010.

Bashford, Alison & Philippa Levine (eds.), *The Oxford Handbook of the History of Eugenics*, New York, NY: Oxford University Press, 2010.

Bassie, Ashley, "The Metropolis and Modernity," in Ashley Bassie, *Expressionism*, New York, NY: Parkstone International, 2012, 71–86.

Bauman, Zygmunt, *Modernity and Ambivalence*, Ithaca, NY: Cornell University Press, 1991.

Modernity and the Holocaust, Ithaca, NY: Cornell University Press, 1989.

Baumann, Stefanie Michaela & Andreas Scheulen, "Zur Rechtslage und Rechtsentwicklung des Erbgesundheitsgesetzes 1934," in Margret Hamm (ed.), *Lebensunwert – zerstörtes Leben. Zwangssterilisation und "Euthanasie,"* Frankfurt a. M.: VAS, 2005, 212–219

Benzenhöfer, Udo, *Zur Genese des Gesetzes zur Verhütung erbkranken Nachwuchses*, Münster: Klemm & Oelschläger, 2006.

Benzenhöfer, Udo & Hanns Ackermann, *Die Zahl der Verfahren und der Sterilisationen nach dem Gesetz zur Verhütung erbkranken Nachwuchses*, Münster: Kontur-Verlag, 2015.

Berger, Silvia, *Bakterien in Krieg und Frieden. Eine Geschichte der medizinischen Bakteriologie in Deutschland 1890–1933*, Göttingen: Wallstein, 2009.

Bergman, Jerry, *Hitler and the Nazi Darwinian Worldview: How the Nazi Eugenic Crusade for a Superior Race Caused the Greatest Holocaust in World History*, Kitchener, ON: Joshua Press, 2012.

Berrios, German E. & Dominic Beer, "The Notion of Unitary Psychosis: A Conceptual History," *History of Psychiatry* 5, no. 17 (1994): 13–36.

Beyerchen, Alan, "What We Now Know About Nazism and Science," *Social Research* 59 (1992): 615–641.

Black, Edwin, *War against the Weak: Eugenics and America's Campaign to Create a Master Race*, New York, NY: Four Walls Eight Windows, 2003.

Bland, Lucy & Lesley A. Hall, "Eugenics in Britain: The View from the Metropole," in Alison Bashford & Philippa Levine (eds.), *The Oxford Handbook of the History of Eugenics*, New York, NY: Oxford University Press, 2010, 213–227.

Boaz, Rachel E., *In Search of "Aryan Blood": Serology in Interwar and National Socialist Germany*, Budapest: CEU Press, 2012.

Bock, Gisela, "Nazi Sterilization and Reproductive Policies," in Susan Bachrach & Dieter Kuntz (eds.), *Deadly Medicine: Creating the Master Race*, Washington, DC: United States Holocaust Memorial Museum, 2004, 61–87.

Zwangssterilisation im Nationalsozialismus, Opladen: Westdeutscher Verlag, 1986.

Bonneuil, Christophe, "Pure Lines as Industrial Simulacra: A Cultural History of Genetics from Darwin to Johannsen," in Staffan Müller-Wille & Christina Brandt (eds.), *Heredity Explored: Between Public Domain and Experimental Science, 1850–1930*, Cambridge, MA: MIT Press, 2016, 213–242.

Bowler, Peter J., *The Mendelian Revolution: The Emergence of Hereditarian Concepts in Modern Science and Society*, London: Athlone Press, 1989.

Boyd, Byron A., *Rudolf Virchow: The Scientist as Citizen*, New York, NY: Garland, 1991.

Brannigan, Augustine, "The Reification of Mendel," *Social Studies of Science* 9, no. 4 (1979): 423–454.

Braund, James & Douglas G. Sutton, "The Case of Heinrich Wilhelm Poll (1877–1939): A German-Jewish Geneticist, Eugenicist, Twin Researcher, and Victim of the Nazis," *Journal of the History of Biology* 41, no. 1 (2008): 1–35.

Bruinius, Harry, *Better for All the World: The Secret History of Forced Sterilization and America's Quest for Racial Purity*, New York, NY: Vintage Books, 2007.

Buck-Zerchin, Dorothea S., "Seventy Years of Coercion in Psychiatric Institutions, Experienced and Witnessed," in Thomas W. Kallert, Juan E. Mezzich & John Monahan (eds.), *Coercive Treatment in Psychiatry: Clinical, Legal and Ethical Aspects*, Chichester: John Wiley & Sons, 2011, 235–243.

Bulmer, Michael, "Galton's Law of Ancestral Heredity," *Heredity* 81 (1998): 579–585.

Burian, Richard M., Jean Gayon & Doris Zallen, "The Singular Fate of Genetics in the History of French Biology, 1900–1940," *Journal of the History of Biology* 21, no. 3 (1988): 357–402.

Burleigh, Michael, *Death and Deliverance: "Euthanasia" in Germany, 1900–1945*, Cambridge: Cambridge University Press, 1994.

Burleigh, Michael & Wolfgang Wippermann, *The Racial State: Germany 1933–1945*, Cambridge: Cambridge University Press, 1991.

Cassata, Francesco, *Building the New Man: Eugenics, Racial Science and Genetics in Twentieth-Century Italy*, Budapest/New York, NY: CEU Press, 2011.

Chamberlain, J. Edward & Sander L. Gilman (eds.), *Degeneration: The Dark Side of Progress*, New York, NY: Columbia University Press, 1985.

Confino, Alon, *A World without Jews: The Nazi Imagination from Persecution to Genocide*, New Haven, CT: Yale University Press, 2014.

Cooter, Roger, "Pseudo-science and Quackery," in John L. Heilborn (ed.), *The Oxford Companion to the History of Modern Science*, New York, NY: Oxford University Press, 2003, 683–684.

Corcos, Alain F. & Floyd V. Monaghan, "The Real Objective of Mendel's Paper," *Biology and Philosophy* 5 (1990): 267–292.

Corni, Gustavo, *Hitler and the Peasants. Agrarian Policy of the Third Reich, 1930–1939*, translated by David Kerr, New York, NY/Oxford/Munich: Berg, 1990.

Cottebrune, Anne, "The Emergence of Genetic Counselling in the Federal Republic of Germany: Continuity and Change in the Narratives of Human Geneticists, c. 1968–80," in Bernd Gausemeier, Staffan Müller-Wille & Edmund Ramsden (eds.), *Human Heredity in the Twentieth Century*, London: Pickering and Chatto, 2013, 193–204.

"Eugenische Konzepte in der westdeutschen Humangenetik, 1945–1980," *Journal of Modern European History* 10, no. 4, special issue: Eugenics after 1945 (2012): 500–518.

"Zwischen Theorie und Deutung der Vererbung psychischer Störungen," *NTM Zeitschrift für Geschichte der Wissenschaften, Technik und Medizin* 17, no. 1 (2009): 35–54.

Crow, James F., "Hardy, Weinberg and Language Impediments," *Genetics* 152 (1999): 821–825.

Deichmann, Ute, *Biologists under Hitler*, Cambridge, MA: Harvard University Press, 1999.

Dejong-Lambert, William & Nikolai Krementsov, "On Labels and Issues: The Lysenko Controversy and the Cold War," *Journal of the History of Biology* 47 (2012): 373–388.

Dietz, Bernhard, "Countryside-versus-City in European Thought: German and British Anti-Urbanism between the Wars," *European Legacy* 1, no. 7 (2008): 801–814.

Dörner, Klaus, *Gestern minderwertig – heute gleichwertig?* Gütersloh: Jakob van Hoddis, 1986.

Dorr, Gregory Michael, *Segregation's Science: Eugenics and Society in Virginia*, Charlottesville: University of Virginia Press, 2008.

Dowbiggin, Ian, "Degeneration and Hereditarianism in French Mental Medicine 1840–90," in William. F. Bynum, Roy Porter & Michael Shepherd (eds.), *The Anatomy of Madness: Essays in the History of Psychiatry*, vol. 1., London/New York, NY: Routledge, 1987, 188–232.

Efron, John, *Defenders of the Race: Jewish Doctors and Race Science in Fin-de-Siecle Europe*, New Haven, CT: Yale University Press, 1994.

Ehrenreich, Eric, *The Nazi Ancestral Proof*, Bloomington: Indiana University Press, 2007.

El-Tayeb, Fatima, *Schwarze Deutsche. "Rasse" und nationale Identität 1890–1933*, Frankfurt a. M./New York, NY: Campus, 2001.

Engstrom, Eric J., *Clinical Psychiatry in Imperial Germany*, Ithaca, NY/London: Cornell University Press, 2003.

"Emil Kraepelin. Leben und Werk des Psychiaters im Spannungsfeld zwischen positivistischer Wissenschaft und Irrationalitaet," MA thesis, University of Munich, 1990. www.engstrom.de/KRAEPELINBIOGRAPHY.pdf

"'On the Question of Degeneration' by Emil Kraepelin (1908)," *History of Psychiatry* 18, no. 3 (2007): 389–404.

Engstrom, Eric J. & Matthias M. Weber, "Making Kraepelin History: A Great Instauration?" *History of Psychiatry* 18, no. 3 (2007): 267–273.

Essner, Cornelia, *Die "Nürnberger Gesetze" oder die Verwaltung des Rassenwahns 1933–1945*, Paderborn: Ferdinand Schöningh, 2002.

"Nazi Anti-Semitism and the Question of 'Jewish Blood,'" in Christopher H. Johnson, Bernhard Jussen, David Warren Sabean & Simon Teuscher (eds.), *Blood and Kinship: Matter for Metaphor from Ancient Rome to the Present*, New York, NY: Berghahn, 2013, 227–243.

Evans, Richard J., "In Search of German Social Darwinism. The History and Historiography of a Concept," in Manfred Berg & Geoffrey Cocks (eds.), *Medicine and Modernity: Public Health and Medical Care in 19th- and 20th-Century Germany*, Washington, DC: German Historical Institute, 1997, 55–79.

The Third Reich at War, London/New York, NY: Allen Lane, 2008.

Falk, Raphael, "Commentary: A Century of Mendelism: On Johannsen's Genotype Conception," *International Journal of Epidemiology* 43, no. 4 (2014): 1002–1007.

Falk, Raphael & Sahotra Sarkar, "The Real Objective of Mendel's Paper: A Response to Monaghan and Corcos," *Biology and Philosophy* 6 (1991): 447–451.

Fangerau, Heiner, *Etablierung eines rassenhygienischen Standardwerkes 1921–1941. Der Baur-Fischer-Lenz im Spiegel der zeitgenössischen Rezensionsliteratur*, Frankfurt a. M.: Lang, 2001.

Flesch-Thebesius, Marlies, "'Wir saßen zwischen allen Stühlen.' Als Mischling zweiten Grades in Frankfurt am Main," in Monica Kingreen (ed.), *"Nach der Kristallnacht." Jüdisches Leben und antijüdische Politik in Frankfurt am Main 1938–1945*, Schriftenreihe des Fritz Bauer Instituts, vol. 17, Frankfurt: Campus, 1999, 415–434.

Fogarty, Richard S. & Michael A. Osborne, "Eugenics in France and the Colonies," in Alison Bashford & Philippa Levine (eds.), *The Oxford Handbook of the History of Eugenics*, New York, NY: Oxford University Press, 2010, 332–346.

Friedlander, Henry, *The Origins of the Nazi Genocide: From Euthanasia to the Final Solution*, Chapel Hill: University of North Carolina Press, 1995.

Friedländer, Saul, *Nazi Germany and the Jews, vol. 1, The Years of Persecution 1933–1939*, New York, NY: Perennial, 1998.

Nazi Germany and the Jews, vol. 2, The Years of Extermination, New York, NY: HarperCollins, 2007.

Ganssmüller, Christian, *Die Erbgesundheitspolitik des Dritten Reiches: Planung, Durchführung und Durchsetzung*, Cologne: Böhlau, 1987.

Gasman, Daniel, *The Scientific Origins of National Socialism*, London: Macdonald & Co., 1971.

Gaudillière, Jean-Paul, "Mendelism and Medicine: Controlling Human Inheritance in Local Contexts, 1920–1960," *Comptes Rendus de l'Académie des Sciences – Series III – Sciences de la Vie* 323, no. 12 (2000): 1117–1126.

Gausemeier, Bernd, "Auf der 'Brücke zwischen Natur- und Geschichtswissenschaft': Ottokar Lorenz und die Neuerfindung der Genealogie um 1900," in Florence Vienne & Christina Brandt (eds.), *Wissensobjekt Mensch: Humanwissenschafliche Praktiken im 20. Jahrhundert*, Berlin: Kulturverlag Kadmos, 2008, 137–164.

"Borderlands of Heredity: The Debate about Hereditary Susceptibility to Tuberculosis, 1882–1945," in Bernd Gausemeier, Staffan Müller-Wille & Edmund Ramsden (eds.), *Human Heredity in the Twentieth Century*, London: Pickering and Chatto, 2013, 13–26.

"From Pedigree to Database. Genealogy and Human Heredity in Germany, 1890–1914," in *Conference: A Cultural History of Heredity III: 19th and Early 20th Centuries*, Max-Planck Institute Preprint Series 294 (2006): 179–192.

"Genetics as a Modernization Program: Biological Research at the Kaiser Wilhelm Institutes and the Political Economy of the Nazi State," *Historical Studies in the Natural Sciences* 40, no. 4 (2010): 429–456.

"In Search of the Ideal Population: The Study of Human Heredity before and after the Mendelian Break," in Staffan Müller-Wille & Christina Brandt (eds.), *Heredity Explored: Between Public Domain and Experimental Science, 1850–1930*, Cambridge, MA: MIT Press, 2016, 337–364.

Natürliche Ordnungen und politische Allianzen: Biologische und biochemische Forschung an Kaiser-Wilhelm-Instituten, 1933–1945, Göttingen: Wallstein, 2005.

"Pedigree vs. Mendelism. Concepts of Heredity in Psychiatry before and after 1900," in *Conference: A Cultural History of Heredity IV: Heredity in the Century of the Gene*. Max-Planck Institute Preprint Series 343 (2008): 149–162.

"Pedigrees of Madness: The Study of Heredity in Nineteenth and Early Twentieth Century Psychiatry," *History and Philosophy of the Life Sciences* 36, no. 4 (2015): 467–483.

Gayon, Jean & Richard M. Burian, "France in the Era of Mendelism (1900–1930)", *Comptes Rendus de l'Académie des Sciences – Series III – Sciences de la Vie* 324, no. 12 (2000): 1097–1106.

Gerhard, Gesine, "Das Ende der deutschen Bauernfrage – Ländliche Gesellschaft im Umbruch," in Daniela Münkel (ed.), *Der lange Abschied vom Agrarland. Agrarpolitik, Landwirtschaft und ländliche Gesellschaft zwischen Weimar und Bonn*, Göttingen: Wallstein, 2000, 124–142.

Gilman, Sander L. *Difference and Pathology: Stereotypes of Sexuality, Race, and Madness*, Ithaca, NY: Cornell University Press, 1985.

The Jew's Body, New York, NY/London: Routledge, 1991.

Glass, Bentley, "Geneticists Embattled: Their Stand against Rampant Eugenics and Racism in America during the 1920s and 1930s," *Proceedings of the American Philosophical Society* 130, no. 1 (1986): 130–154.

Goeschel, Christian, "Suicides of German Jews in the Third Reich," *German History* 25 no. 1 (2007): 22–45.

Gordin, Michael D., *The Pseudo-Science Wars: Immanuel Velikovsky and the Birth of the Modern Fringe*, Chicago, IL/London: University of Chicago Press, 2012.

Goschler, Constantin, *Rudolf Virchow: Mediziner, Anthropologe, Politiker*, Cologne: Boehlau, 2002.

Gould, Stephen Jay, *The Mismeasure of Man*, London/New York, NY: W. W. Norton, 1981.

Gradmann, Christoph, "Invisible Enemies: Bacteriology and the Language of Politics in Imperial Germany," *Science in Context* 13, no. 1 (2000): 9–30.

Hagner, Michael, "Bye-Bye Science, Welcome Pseudoscience? Reflexionen über einen beschädigten Status," in Dirk Rupnow et al. (eds.), *Pseudowissenschaft: Konzeptionen von Nichtwissenschaftlichkeit in der Wissenschaftsgeschichte*, Frankfurt a. M.: Suhrkamp, 2008, 21–50.

Hahn, Daphne, *Modernisierung und Biopolitik. Sterilisation und Schwangerschaftsabbruch in Deutschland nach 1945*, Frankfurt a. M.: Campus, 2000.

Hahn Rafter, Nicole (ed.), *White Trash: The Eugenic Family Studies, 1877–1919*, Boston, MA: Northeastern University Press, 1988.

Hänseler, Marianne, *Metaphern unter dem Mikroskop*, Zurich: Chronos, 2009.

Hansen, Bent Sigurd "Something Rotten in the State of Denmark: Eugenics and the Ascent of the Welfare State," in Gunnar Broberg & Nils Roll-Hansen (eds.), *Eugenics and the Welfare State. Sterilization Policy in Denmark, Sweden, Norway and Finland*, East Lansing: Michigan State University Press, 1996, 9–76.

Harrington, Anne, *Reenchanted Science: Holism in German Culture from Wilhelm II to Hitler*, Princeton, NJ: Princeton University Press, 1996.

Hart, Mitchell B., "Racial Science, Social Science, and the Politics of Jewish Assimilation," *Isis* 90, no. 2 (1999): 268–97.

Social Science and the Politics of Modern Jewish Identity, Stanford, CA: Stanford University Press, 2000.

The Healthy Jew: The Symbiosis of Judaism and Modern Medicine, Cambridge: Cambridge University Press, 2007.

Harten, Hans-Christian, Uwe Neirich & Matthias Schwerendt, *Rassenhygiene als Erziehungsideologie des Dritten Reichs*, Berlin: De Gruyter, 2006.

Harvey, Rosemary D., "Pioneers of Genetics: A Comparison of the Attitudes of William Bateson and Erwin Baur to Eugenics," *Notes and Records of the Royal Society of London* 49, no. 1 (1995): 105–117.

Harwood, Jonathan, *Styles of Scientific Thought: The German Genetics Community, 1900–1933*, Chicago, IL/London: University of Chicago Press, 1993.

Harwood, Jonathan & Michael Banton, *The Race Concept*, Newton Abbot: David & Charles, 1975.

Heiber, Helmut (ed.), *Reichsführer!* ... *Briefe an und von Himmler*, Stuttgart: Deutsche Verlags-Anstalt, 1968.

Heinemann, Isabel, "Ambivalente Sozialingenieure? Die Rasseexperten der SS," in Gerhard Hirschfeld & Tobias Jersak (eds.), *Karrieren im Nationalsozialismus. Funktionseliten zwischen Mitwirkung und Distanz*, Frankfurt a. M.: Campus, 2004, 73–95.

Herbert, Ulrich, *Geschichte Deutschlands im 20. Jahrhundert*, Munich: C. H. Beck, 2014.

Herf, Jeffrey, *The Jewish Enemy: Nazi Propaganda during World War II and the Holocaust*, Cambridge, MA: Harvard University Press, 2006.

Hilberg, Raul, *The Destruction of the European Jews*, 3rd ed., New Haven, CT: Yale University Press, 2003.

Hirsch, Dafna, "Zionist Eugenics, Mixed Marriage, and the Creation of a 'New Jewish Type,'" *Journal of the Royal Anthropological Institute (JRAI)* 15 (2009): 592–609.

Hochman, Gilberto, Nísia Trinidade Lima & Marcos Chor Maio, "The Path of Eugenics in Brazil: Dilemmas of Miscegenation," in Alison Bashford & Philippa Levine (eds.), *The Oxford Handbook of the History of Eugenics*, New York, NY: Oxford University Press, 2010, 463–510.

Hoff, Paul, "Kraepelin and Degeneration Theory," *European Archives of Psychiatry and Clinical Neuroscience* 258, no. 2 (Suppl.) (2008): 12–17.

Holtkamp, Martin, *Werner Villinger (1887–1961). Die Kontinuität des Minderwertigkeitsgedankens in der Jugend- und Sozialpsychiatrie*, Husum: Matthiesen, 2002.

Hossfeld, Uwe, *Geschichte der biologischen Anthropologie in Deutschland*, Stuttgart: Franz Steiner, 2005.

Hufenreuter, Gregor, "Der 'Semi-Gotha' (1912–1919). Entstehung und Geschichte eines antisemitischen Adelshandbuches," *Herold-Jahrbuch, Neue Folge* 9 (2004): 71–88.

Hutton, Christopher M., *Race and the Third Reich: Linguistics, Racial Anthropology and Genetics in the Dialectic of Volk*, Cambridge: Polity Press, 2005.

Kallmann, Franz. J., "In Memoriam Bruno Schulz, 1890–1958," *Archiv für Psychiatrie und Nervenkrankheiten* 197, no. 2 (1958): 121–123.

Kater, Michael H., *Doctors under Hitler*, Chapel Hill: University of North Carolina Press, 1989.

Kiefer, Annegret, *Das Problem einer "jüdischen Rasse." Eine Diskussion zwischen Wissenschaft und Ideologie (1870–1930)*, Frankfurt a. M.: Peter Lang, 1991.

Klee, Ernst, *Deutsche Medizin im Dritten Reich: Karrieren vor und nach 1945*, Frankfurt a. M.: S. Fischer, 2001.

"Euthanasie" im NS-Staat: Die "Vernichtung Lebensunwerten Lebens," 2nd ed. Frankfurt a. M.: S. Fischer, 1983.

Kochavi, Arieh J., *Prelude to Nuremberg: Allied War Crimes Policy and the Question of Punishment*, Chapel Hill: University of North Carolina Press, 1998.

Krementsov, Nikolai, "A 'Second Front' in Soviet Genetics: The International Dimension of the Lysenko Controversy, 1944–1947," *Journal of the History of Biology* 29 (1996): 229–250.

"Eugenics in Russia and the Soviet Union," in Alison Bashford & Philippa Levine (eds.), *The Oxford Handbook of the History of Eugenics*, New York, NY: Oxford University Press, 2010, 413–429.

"From 'Beastly Philosophy' to Medical Genetics: Eugenics in Russia and the Soviet Union," *Annals of Science* 68, no. 1 (2011): 61–92.

Kühl, Stefan, *For the Betterment of the Race: The Rise and Fall of the International Movement for Eugenics and Racial Hygiene*, New York, NY: Palgrave Macmillan, 2013.

Kundrus, Birthe, "Von Windhoek nach Nürnberg? Koloniale 'Mischehenverbote' und die nationalsozialistische Rassengesetzgebung," in Birthe Kundrus (ed.), *Phantasiereiche. Der deutsche Kolonialismus aus kulturgeschichtlicher Perspektive*, Frankfurt a. M.: Campus, 2003, 110–131.

Kyllingstad, Jon Røyne, *Measuring the Master Race: Physical Anthropology in Norway, 1890–1945*, Cambridge: Open Book Publishers, 2016.

Larson, Edward J., *Sex, Race and Science. Eugenics in the Deep South*, Baltimore, MD: Johns Hopkins University Press, 1995.

Lifton, Robert J., *The Nazi Doctors: Medical Killing and the Psychology of Genocide*, New York, NY: Basic Books, 1986.

Lilienthal, Georg, "Arier oder Jude? Die Geschichte des Erb- und Rassenkundlichen Abstammungsgutachtens," in Peter Propping & Heinz Schott (eds.), *Wissenschaft auf Irrwegen: Biologismus – Rassenhygiene – Eugenik*, Bonn: Bouvier, 1992, 66–84.

"Die jüdischen 'Rassenmerkmale': zur Geschichte der Anthropologie der Juden," *Medizinhistorisches Journal* 28, no. 2/3 (1993): 173–198.

Lipphardt, Veronika, *Biologie der Juden. Jüdische Wissenschaftler über "Rasse" und Vererbung 1900–1935*, Göttingen: Vandenhoeck & Ruprecht, 2008.

"Das 'schwarze Schaf' der Biowissenschaftler. Ausgrenzungen und Rehabilitierungen der Rassenforschung im 20. Jahrhundert," in Veronika Lipphardt (ed.), *Pseudowissenschaft: Konzeptionen von Nicht-/Wissenschaftlichkeit in der Wissenschaftsgeschichte*, Frankfurt a. M.: Suhrkamp, 2008, 223–250.

"Isolates and Crosses in Human Population Genetics; or, A Contextualization of German Race Science," *Current Anthropology* 53, no. S5, special issue: "The Biological Anthropology of Living Human Populations: World Histories, National Styles, and International Networks," (2012): S69–S82.

Lloyd, Jill, *German Expressionism. Primitivism and Modernity*, New Haven, CT/ London: Yale University Press, 1991.

Longerich, Peter, *Holocaust: The Nazi Persecution and Murder of the Jews*, Oxford: Oxford University Press, 2010.

Lösch, Niels, *Rasse als Konstrukt: Leben und Wirken Eugen Fischers*, Frankfurt a. M.: Peter Lang, 1997.

Lowenstein, Steven M., Paul Mendes-Flohr, Peter Pulzer & Monika Richarz (eds.), *Deutsch-jüdische Geschichte, vol. 3, Umstrittene Integration 1871–1918*, Munich: C. H. Beck, 2000.

Lüthi, Barbara, "Germs of Anarchy, Crime, Disease and Degeneracy: Jewish Migration to the United States and the Medicalization of European Borders around 1900," in Tobias Brinkmann (ed.), *Points of Passage: Jewish Migrants from Eastern Europe in Scandinavia, Germany and Britain 1800–1914*, New York, NY: Berghan Books, 2013, 27–46.

Mackenzie, Donald, "Statistical Theory and Social Interests: A Case-Study," *Social Studies of Science* 8, no. 1, theme issue: *Sociology of Mathematics* (1978): 35–83.

Madley, Benjamin, "From Africa to Auschwitz: How German South West Africa Incubated Ideas and Methods Adopted and Developed by the Nazis in Eastern Europe," *European History Quaterly* 35, no. 3 (2005): 429–464.

Makowski, Christine C., *Eugenik, Sterilisationspolitik, "Euthanasie" und Bevölkerungspolitik in der nationalsozialistischen Parteipresse*, Husum: Matthiesen, 1996.

Markel, Howard, *Quarantine! East European Jewish Immigrants and the New York City Epidemics of 1892*, Baltimore, MD/London: Johns Hopkins University Press, 1997.

Marks, Jonathan, "The Construction of Mendel's Laws," *Evolutionary Anthropology: Issues, News, and Reviews* 17, no. 6 (2008): 250–253.

"The Legacy of Serological Studies in American Physical Anthropology," *History and Philosophy of the Life Sciences* 18 (1996): 345–362.

Massin, Benoît, "From Virchow to Fischer: Physical Anthropology and 'Modern Race Theories' in Wilhelmine Germany (1890–1914)," in George W. Stocking (ed.), *Volksgeist as Method and Ethic. Essays on Boasian Ethnography and the German Anthropological Tradition*. Madison: University of Wisconsin Press, 1996, 79–154.

Masterman, Margaret, "The Nature of a Paradigm," in Imre Lakatos & Alan Musgrave (eds.), *Criticism and the Growth of Knowledge: Proceedings of the 1965 International Colloquium in the Philosophy of Science*, vol. 4, 3rd ed., Cambridge: Cambridge University Press, 1965, 59–90.

Mayr, Ernst, *The Growth of Biological Thought*, Cambridge, MA: Belknap Press of Harvard University Press, 1982.

Mayr, Ernst & William B. Provine (eds.), *The Evolutionary Synthesis: Perspectives on the Unification of Biology*, Cambridge, MA: Harvard University Press, 1998.

Mazumdar, Pauline, "Blood and Soil: The Serology of the Aryan Racial State," *Bulletin of the History of Medicine* 64, no. 2 (1990): 187–219.

Eugenics, Human Genetics and Human Failings: The Eugenics Society, Its Sources and Its Critics in Britain, London: Routledge, 1992.

"Two Models for Human Genetics: Blood Grouping and Psychiatry in Germany between the World Wars," *Bulletin of the History of Medicine* 70 (1996): 609–657.

McMahon, Richard, *The Races of Europe: Anthropological Race Classification of Europeans, 1839–1939*, London: Palgrave Macmillan, 2016.

McNeely, Ian Farell, *Medicine on a Grand Scale: Rudolf Virchow, Health Politics, and Liberal Social Reform in Nineteenth-Century Germany*, London: The Welcome Trust Centre for the History of Medicine at University College London, 2002.

Meloni, Maurizio, *Political Biology*, Basingstoke: Palgrave Macmillan, 2016.

Mendelsohn, Andrew "Message in a Bottle: Vaccines and the Nature of Heredity after 1880," in Staffan Müller-Wille & Christina Brandt (eds.), *Heredity*

Explored: Between Public Domain and Experimental Science, 1850–1930, Cambridge, MA: MIT Press, 2016, 243–264.

Mendelsohn, John (ed.), *The Holocaust: Selected Documents in Eighteen Volumes*, vol. 2, New York, NY: Garland, 1982.

Miller Lane, Barbara, *Architecture and Politics in Germany 1918–1945*, Cambridge, MA/London: Harvard University Press, 1985.

Monaghan, Floyd V. & Alain F. Corcos, "The Real Objective of Mendel's Paper: A Response to Falk and Sarkar's Criticism," *Biology and Philosophy* 8 (1993): 95–98.

Moore Jr., Barrington, *Moral Purity and Persecution in History*, Princeton, NJ: Princeton University Press, 2000.

Morabia, Alfredo & Regina Guthold, "Wilhelm Weinberg's 1913 Large Retrospective Cohort Study: A Rediscovery," *American Journal of Epidemiology* 165, no. 7 (2007): 727–733.

Morris-Reich, Amos, *Race and Photography. Racial Evidence as Scientific Evidence, 1876–1980*, Chicago, IL: University of Chicago Press, 2016.

"Race, Ideas, and Ideals: A Comparison of Franz Boas and Hans F. K. Günther," *History of European Ideas* 32, no. 3 (2006): 313–332.

Müller-Hill, Benno, *Tödliche Wissenschaft: Die Aussonderung von Juden, Zigeunern und Geisteskranken, 1933–1945*, Reinbeck bei Hamburg: Rowohlt, 1984. [In English: *Murderous Science: Elimination by Scientific Selection of Jews, Gypsies, and Others in Germany, 1933–1945*, translated by George R. Fraser, Oxford: Oxford University Press, 1988.]

Müller-Wille, Staffan, "Hybrids, Pure Cultures, and Pure Lines: From Nineteenth-Century Biology to Twentieth-Century Genetics," *Studies in History and Philosophy of Science Part C: Studies in History and Philosophy of Biological and Biomedical Sciences* 38, no. 4 (2007): 796–806.

"Leaving Inheritance Behind: Wilhelm Johannsen and the Politics of Mendelism," *Conference: A Cultural History of Heredity IV: Heredity in the Century of the Gene*, Max-Planck Institute Preprint Series 343 (2008): 7–18.

Müller-Wille, Staffan & Christina Brandt (eds.), *Heredity Explored: Between Public Domain and Experimental Science, 1850–1930*, Cambridge, MA/London: MIT Press, 2016.

Müller-Wille, Staffan & Hans-Jörg Rheinberger, *A Cultural History of Heredity*, Chicago, IL: University of Chicago Press, 2012.

Nanjundiah, Vidyanand, "Why Are Most Mutations Recessive?" *Journal of Genetics* 72, no. 2/3 (1993): 85–97.

Nitschke, Asmus, *Die "Erbpolizei" im Nationalsozialismus: Zur Alltagsgeschichte der Gesundheitsämter im Dritten Reich*, Opladen/Wiesbaden: Westdeutscher Verlag, 1999.

Noakes, Jeremy, "The Development of Nazi Policy towards the German-Jewish 'Mischlinge' 1933–1945," *Leo Baeck Institute Yearbook* 34 (1989): 291–354.

Norton, Bernard, "Fisher's Entrance into Evolutionary Science: The Role of Eugenics," in Marjorie Grene (ed.), *Dimensions of Darwinism. Themes and Counterthemes in Twentieth-Century Evolutionary Theory*, Cambridge/Paris: Cambridge University Press, 1983, 19–30.

Olby, Robert, "Mendel, Mendelism and Genetics," Self-published, 1997. www.mendelweb.org/MWolby.intro.html.

"Mendel no Mendelian?" *History of Science* 17 (1979): 53–72.

Orel, Vítězslav, *Gregor Mendel: The First Geneticist*, Oxford: Oxford University Press, 1996.

"The Spectre of Inbreeding in the Early Investigation of Heredity," *History and Philosophy of the Life Sciences* 19, no. 3 (1997): 315–330.

Outlaw, Lucius, "Toward a Critical Theory of 'Race,'" in Bernard Boxill (ed.), *Race and Racism*, Oxford: Oxford University Press, 2001, 58–82.

Paul, Diane B., "Darwin, Social Darwinism and Eugenics," in Jonathan Hodge & Gregory Radick (eds.), *The Cambridge Companion to Darwin*, 2nd ed., Cambridge: Cambridge University Press, 2009, 219–245.

"Genes and Contagious Disease: The Rise and Fall of a Metaphor," in Diane B. Paul, *The Politics of Heredity: Essays on Eugenics, Biomedicine, and the Nature-Nurture Debate*, Albany, NY: SUNY Press, 1988, 157–171.

Pemberton, Stephen, *The Bleeding Disease: Hemophilia and the Unintended Consequences of Medical Progress*, Baltimore, MD: Johns Hopkins University Press, 2011.

Penny, H. Glenn & Matti Bunzl (eds.), *Worldly Provincialism: German Anthropology in the Age of Empire*, Ann Arbor: University of Michigan Press, 2003.

Pendas, Devin O., Mark Roseman & Richard F. Wetzell (eds.), *Beyond the Racial State: Rethinking Nazi Germany*, Cambridge: Cambridge University Press, 2017.

Pergher, Roberta, Mark Roseman, Jürgen Zimmerer, Shelley Baranowski, Doris L. Bergen & Zygmunt Bauman, "The Holocaust: A Colonial Genocide? A Scholars' Forum," *Dapim: Studies on the Holocaust* 27, no. 1 (2013): 40–73.

Pick, Daniel, *Faces of Degeneration: A European Disorder, c.1848–c.1918*, Cambridge: Cambridge University Press, 1989.

Picker, Henry, *Hitler's Tischgespräche im Führerhauptquartier, 1941–1942*, Munich: Deutscher Taschenbuch Verlag, 1968.

Portin, Petter & Adam S. Wilkins, "The Evolving Definition of the Term 'Gene,'" *Genetics* 205, no. 4 (2017): 1353–1364.

Poore, Carole, *Disability in Twentieth-Century German Culture*, Ann Arbor: University of Michigan Press, 2007.

Porter, Theodore M., *Genetics in the Madhouse. The Unknown History of Human Heredity*, Princeton, NJ: Princeton University Press, 2018.

Preuss, Dirk, Uwe Hossfeld & Olaf Breidbach, *Anthropologie nach Haeckel*, Munich: Franz Steiner, 2006.

Pringle, Heather, *The Master Plan: Himmler's Scholars and the Holocaust*, New York, NY: Hyperion, 2006.

Proctor, Robert, "From 'Anthropologie' to 'Rassenkunde' in the German Anthropological Tradition," in George W. Stocking (ed.), *Bones, Bodies, Behavior. Essays on Biological Anthropology*, Madison: University of Wisconsin Press, 1988, 138–179.

Racial Hygiene: Medicine under the Nazis, Cambridge, MA: Harvard University Press, 1988.

Pross, Christian, *Paying for the Past: The Struggle over Reparations for Surviving Victims of the Nazi Terror*, Baltimore, MD: Johns Hopkins University Press, 1998.

Provine, William B., "Geneticists and the Biology of Race Crossing," *Science* 182, no. 4114 (1973): 790–796.

The Origins of Theoretical Population Genetics, Chicago, IL: University of Chicago Press, 1971.

Quine, Maria Sophia, "The First-Wave Eugenic Revolution in Southern Europe: Science sans Frontières," in Alison Bashford & Philippa Levine (eds.), *The Oxford Handbook of the History of Eugenics*, New York, NY: Oxford University Press, 2010, 377–397.

Reschke, Horst A., "Arthur Czellitzer," in *Encyclopedia of Jewish Genealogy*, Salt Lake City, UT: Center of Jewish History, 1987. http://digital.cjh.org:1801/webclient/DeliveryManager?pid=523964

Resta, Robert G., "The Crane's Foot: The Rise of the Pedigree in Human Genetics," *Journal of Genetic Counseling* 2, no. 4 (1993): 235–260.

Rheinberger, Hans-Jörg & Staffan Müller-Wille, *The Gene: From Genetics to Post-Genomics*, Chicago, IL: University of Chicago Press, 2017.

Richards, Robert J., *Was Hitler a Darwinian? Disputed Questions in the History of Evolutionary Theory*, Chicago, IL: University of Chicago Press, 2013.

Risch, N[eil], "Estimating Morbidity Risks with Variable Age of Onset: Review of Methods and a Maximum Likelihood Approach," *Biometrics* 39, no. 4 (1983): 929–939.

Röder, Thomas, Volker Kubillus & Anthony Burwell, *Psychiatrists: The Men behind Hitler.* Los Angeles, CA: Freedom Publishing, 1995.

Roelcke, Volker, "Die Entwicklung der Psychiatrie zwischen 1880 und 1932. Theoriebildung, Institutionen, Interaktionen mit zeitgenössischer Wissenschafts- und Sozialpolitik," in Rüdiger vom Bruch & Brigitte Kaderas (eds.), *Wissenschaften und Wissenschaftspolitik: Bestandsaufnahmen zu Formationen, Brüchen und Kontinuitäten im Deutschland des 20. Jahrhunderts*, Stuttgart: Franz Steiner, 2002, 109–124.

"Die Etablierung der psychiatrischen Genetik in Deutschland, Großbritannien und den USA, ca. 1910–1960. Zur untrennbaren Geschichte von Eugenik und Humangenetik," *Acta Historica Leopoldina* 48 (2007): 173–190.

"Programm und Praxis der psychiatrischen Genetik an der deutschen Forschungsanstalt für Psychiatrie unter Ernst Rüdin: Zum Verhältnis von Wissenschaft, Politik und Rasse-Begriff vor und nach 1933," *Medizinhistorisches Journal* 37 (2002): 21–55.

Roll-Hansen, Nils, "Commentary: Wilhelm Johannsen and the Problem of Heredity at the Turn of the 19th Century," *International Journal of Epidemiology* 43, no. 4 (2014): 1007–1013.

"Norwegian Eugenics: Sterilization as Social Reform," in Gunnar Broberg & Nils Roll-Hansen (eds.), *Eugenics and the Welfare State. Sterilization Policy in Denmark, Sweden, Norway and Finland*, East Lansing, MI: Michigan State University Press, 1996, 151–194.

Roth, Karl Heinz, "Schöner neuer Mensch. Das Paradigmenwechsel der klassischen Genetik und seine Auswirkungen auf die Bevölkerungsbiologie des

'Dritten Reichs,'" in Heidrun Kaupen-Haas (ed.), *Der Griff nach der Bevölkerung*, Nördlingen: F. Greno, 1986, 11–63.

Roth, Karl Heinz & Götz Aly, *Die restlose Erfassung. Volkszählen, Identifizieren, Aussondern im Nationalsozialismus*, Berlin: Rothbuch Verlag, 1984. [In English: Götz Aly & Karl Heinz Roth, *The Nazi Census*, Philadelphia, PA: Temple University Press, 2004.]

Rotzoll, Maike & Gerrit Hohendorf, "Murdering the Sick in the Name of Progress? The Heidelberg Psychiatrist Carl Schneider as a Brain Researcher and 'Therapeutic Idealist,'" in Paul Weindling (ed.), *From Clinic to Concentration Camp. Reassessing Nazi Medical and Racial Research, 1933–1945*, London: Routledge, 2017, 163–182.

Rotzoll, Maike, Gerrit Hohendorf, Petra Fuchs, Paul Richter, Wolfgang U. Eckart & Christoph Mundt (eds.), *Die nationalsozialistische "Euthanasie" – Aktion "T4" und ihre Opfer: Geschichte und ethische Konsequenzen für die Gegenwart*, Paderborn: Schöningh, 2010.

Rotzoll, Maike, Petra Fuchs, Paul Richter & Gerrit Hohendorf, "Die nationalsozialistische 'Euthanasie-Aktion T4.' Historische Forschung, individuelle Lebensgeschichten und Erinnerungskultur," *Nervenarzt* 81 (2010): 1326–1332.

Sabean, David Warren & Simon Teuscher, "Kinship in Europe: A New Approach to Long-Term Development," in David Warren Sabean, Simon Teuscher & Jon Mathieu (eds.), *Kinship in Europe: Approaches to Long-Term Development (1300–1900)*, New York, NY: Oxford University Press, 2007, 1–32.

Schaffer, Gavin, *Racial Science and British Society, 1930–62*, Basingstoke: Palgrave Macmillan, 2008.

Schell, Patience A., "Eugenics Policy and Practice in Cuba, Puerto Rico, and Mexico," in Alison Bashford & Philippa Levine (eds.), *The Oxford Handbook of the History of Eugenics*, New York, NY: Oxford University Press, 2010, 477–492.

Schleunes, Karl A., *The Twisted Road to Auschwitz*, Urbana: University of Illinois Press, 1970.

Schmidt, Ulf, *Medical Films, Ethics, and Euthanasia* in *Nazi Germany: The History of Medical Research and Teaching Films of the Reich Office for Educational Films/ Reich Institute for Films in Science and Education, 1933–1945*, Husum: Matthiesen, 2002.

Schmuhl, Hans-Walter, *Grenzüberschreitungen: Das Kaiser-Wilhelm-Institut für Anthropologie, menschliche Erblehre und Eugenik 1927–1945*, Göttingen: Wallstein, 2005. [In English: *The Kaiser Wilhelm Institute for Anthropology, Human Heredity, and Eugenics, 1927–1945: Crossing Boundaries*, Boston Studies in the Philosophy of Science, vol. 259, Dordrecht: Springer, 2008.]

Rassenhygiene, Nationalsozialismus, Euthanasie, Göttingen: Vandenhoeck & Ruprecht, 1987.

"Zwischen vorauseilendem Gehorsam und halbherziger Verweigerung. Werner Villinger und die nationalsozialistischen Medizinverbrechen," *Nervenarzt* 73, no. 11 (2002): 1058–1063.

Schneider, William H., *Quality and Quantity: The Quest for Biological Regeneration in Twentieth-Century France*, Cambridge: Cambridge University Press, 1990.

Schulle, Diana, *Das Reichssippenamt: Eine Institution nationalsozialistischer Rassen-politik*, Berlin: Logos, 2001.

Schwerin, Alexander von, *Experimentalisierung des Menschen. Der Genetiker Hans Nachtsheim und die Vergleichende Erbpathologie 1920–1945*, Göttingen: Wall-stein, 2004.

Schwidetzky, Ilse, "History of Biological Anthropology in Germany," *International Association of Human Biologists: Occasional Papers* 3, no. 4 (1992).

Selya, Rena, "Defending Scientific Freedom and Democracy: The Genetics Society of America's Response to Lysneko," *Journal of the History of Biology* 45 (2012): 415–442.

Smith, Woodruff D., *Politics and the Sciences of Culture in Germany, 1840–1920*, Oxford: Oxford University Press, 1991.

Spencer, Frank (ed.), *History of Physical Anthropology: An Encyclopedia*, New York, NY/London: Garland, 1997.

Spiro, Jonathan P., *Defending the Master Race: Conservation, Eugenics, and the Legacy of Madison Grant*, Burlington: University of Vermont Press, 2009.

Steinweiss, Alan E., *Kristallnacht 1938*, Cambridge, MA: Belknap Press of Harvard University Press, 2009.

Stachura, Peter D., *The German Youth Movement 1900–1945. An Interpretative and Documentary History*, London/Basingstoke: Macmillan, 1981.

Stepan, Nancy L. *The Hour of Eugenics: Race, Gender, and Nation in Latin America*, Ithaca, NY/London: Cornell University Press, 1991.

The Idea of Race in Science: Great Britain 1800–1960, London: Macmillan, 1982.

Stern, Curt & Eva R. Sherwood, *The Origins of Genetics*, San Francisco/London: W. H. Freeman, 1966.

Stone, Dan, *Breeding Superman: Nietzsche, Race and Eugenics in Edwardian and Interwar Britain*, Liverpool: Liverpool University Press, 2002.

"Race in British Eugenics," *European History Quarterly* 31 (2000): 397–425.

Surmann, Rolf, "Rehabilitation and Indemnification for the Victims of Forced Sterilization and 'Euthanasie.' The West German Policies of 'Compensation' ('Wiedergutmachung')," in Volker Roelcke, Sascha Topp & Etienne Lepicard (eds.), *Silence, Scapegoats, Self-reflection: The Shadow of Nazi Medical Crimes on Medicine and Bioethics*, Göttingen: V & R Unipress, 2014, 113–127.

Taylor, Peter & Richard Lewontin, "The Genotype/Phenotype Distinction," in Edward N. Zalta (ed.), *The [Online] Stanford Encyclopedia of Philosophy*, 2017. https://plato.stanford.edu/archives/sum2017/entries/genotype-phenotype/.

Teicher, Amir, "'Ahnenforschung macht frei': On the Correlation between Research Strategies and Socio-Political Bias in German Genealogy, 1898–1935," *Historische Anthropologie* 22, no. 1 (2014): 67–90.

"Why Did the Nazis Sterilize the Blind? Genetics and the Shaping of the Sterilization Law of 1933," How Developments within Genetics Shaped the Nazi Sterilization Law," *Central European History* 52, no. 2 (2019): 239–309.

Teo, Thomas, "The Historical Problematization of Mixed Race in Psychological and Human-Scientific Discourse," in Andrew S. Winston (ed.), *Defining*

Difference: Race and Racism in the History of Psychology, Washington, DC: American Psychological Association, 2004, 79–108.

Tucker, William H. "'Inharmoniously Adapted to Each Other': Science and Racial Crosses," in Andrew S. Winston (ed.), *Defining Difference: Race and Racism in the History of Psychology*, Washington, DC: American Psychological Association, 2004, 109–133.

Turda, Marius, *Modernism and Eugenics*, New York, NY: Palgrave Macmillan, 2010.

Tydén, Mattias, "The Scandinavian States: Reformed Eugenics Applied," in Alison Bashford & Philippa Levine (eds.), *The Oxford Handbook of the History of Eugenics*, New York, NY: Oxford University Press, 2010, 363–376.

Wailoo, Keith & Stephen Pemberton, *The Troubled Dream of Genetic Medicine: Ethnicity and Innovation in Tay-Sachs, Cystic Fibrosis, and Sickle Cell Disease*, Baltimore, MD: Johns Hopkins University Press, 2006.

Watkins, W[inifred] M., "The ABO Blood Group System: Historical Background," *Transfusion Medicine* 11, no. 4 (2001): 243–265.

Weber, Matthias M., *Ernst Rüdin. Eine kritische Biographie*, Berlin: Springer, 1993.

Weber, Matthias M., Wolfgang Burgmair & Eric J. Engstrom, "Emil Kraepelin (1856-1926): Zwischen klinischen Krankheitsbildern und 'psychischer Volkshygiene,'" *Deutsches Ärzteblatt* 103, no. 41 (2006): 2685–2690.

Weber, Matthias M. & Eric J. Engstrom, "Kraepelin's 'Diagnostic Cards': The Confluence of Clinical Research and Preconceived Categories," *History of Psychiatry* 8, no. 31 (1997): 375–385.

Wegner, Gregory Paul, *Anti-Semitism and Schooling under the Third Reich*, New York, NY: Routledge, 2002.

Wehler, Hans-Ulrich, *Deutsche Gesellschaftsgeschichte, vol. 4, Vom Beginn des Ersten Weltkrieges bis zur Gründung der beiden deutschen Staaten 1914–1949*, Munich: C. H. Beck, 2003.

Weikart, Richard, *From Darwin to Hitler: Evolutionary Ethics, Eugenics, and Racism in Germany*, New York, NY: Palgrave Macmillan, 2004.

Weindling, Paul, *Epidemics and Genocide in Eastern Europe, 1890–1945*, Oxford/New York, NY: Oxford University Press, 2000.

Weindling, Paul, (ed.), *From Clinic to Concentration Camp: Reassessing Nazi Medical and Racial Research, 1933–1945*, London/New York, NY: Routledge, 2017.

Health, Race, and German Politics between National Unification and Nazism: 1870–1945, Cambridge: Cambridge University Press, 1989.

Nazi Medicine and the Nuremberg Trials: From Medical War Crimes to Informed Consent, New York, NY: Palgrave Macmillan, 2004.

Weinreich, Max, *Hitler's Professors: The Part of Scholarship in Germany's Crimes against the Jewish People*, New York, NY: Yiddish Scientific Institute, 1946.

Weingart, Peter, Jürgen Kroll & Kurt Bayertz, *Rasse, Blut und Gene. Geschichte der Eugenik und Rassenhygiene in Deutschland*, Frankfurt a. M.: Suhrkamp, 1988.

Weiss, Sheila F., "The Loyal Genetic Doctor. Otmar Freiherr von Verschuer, and the Institut für Erbbiologie und Rassenhygiene: Origins, Controversy, and Racial Political Practice," *Central European History* 45 no. 4 (2012): 631–668.

The Nazi Symbiosis, Chicago, IL: University of Chicago Press, 2010.

"Pedagogy, Professionalism and Politics: Biology Instruction during the Third Reich," in Monika Renneberg & Mark Walker (eds.), *Science, Technology and National Socialism*, Cambridge: Cambridge University Press, 1994, 184–196.

Westermann, Stefanie, "'Die Gemeinschaft hat ein Interesse daran, dass sie nicht mit Erbkranken verseucht wird' – Der Umgang mit den nationalsozialistischen Zwangssterilisationen und die Diskussion über eugenische (Zwangs-) Maßnahmen in der Bundesrepublik," in Stefanie Westermann, Richard Kühl & Dominik Groß (eds.), *Medizin im Dienst der "Erbgesundheit." Beitrag zur Geschichte der Eugenik und "Rassenhygiene,"* Berlin: LIT, 2009, 169–199.

Verschwiegenes Leid. Der Umgang mit den NS-Zwangssterilisationen in der Bundesrepublik Deutschland, Cologne: Böhlau, 2010.

Wetzell, Richard F., *Inventing the Criminal: A History of German Criminology, 1880–1945*, Chapel Hill, NC: University of North Carolina Press, 2000.

Wexler, Alice, *The Woman Who Walked into the Sea: Huntington's and the Making of a Genetic Disease*, New Haven, CT: Yale University Press, 2008.

Wolfe, Audra J., "The Cold War Context of the Golden Jubilee, or, Why We Think of Mendel as the Father of Genetics," *Journal of the History of Biology* 45 (2012): 389–414.

Wolff, Tamsen, *Mendel's Theatre: Heredity, Eugenics, and Early Twentieth-Century American Drama*, Basingstoke: Palgrave Macmillan, 2009.

Worboys, Michael, "From Heredity to Infection? Tuberculosis, 1870-1890," in Jean-Paul Gaudillière & Ilana Löwy (eds.), *Heredity and Infection. The History of Disease Transmission*, London/New York, NY: Routledge, 2001, 81–100.

Wrinch, Pamela N., "Science and Politics in the U.S.S.R.: The Genetics Debate," *World Politics* 3, no. 4 (1951): 486–519.

Zielke, Roland, *Sterilisation per Gesetz. Die Gesetzesinitiativen zur Unfruchtbarmachung in den Akten der Bundesministerialverwaltung (1949–1976)*, Berlin: Buchmacherei, 2006.

Zimmerer, Jürgen, "The Birth of the *Ostland* out of the Spirit of Colonialism: A Postcolonial Perspective on the Nazi Policy of Conquest and Extermination," *Patterns of Prejudice* 39, no. 2 (2005): 197–219.

Zimmerman, Andrew, *Anthropology and Antihumanism in Imperial Germany*, Chicago, IL/London: University of Chicago Press, 2001.

Index

Agassiz, Louis 220
Ahnentafel 44, 47, 49, 120, 177
  as genealogical tool 47–50, 58–63, 87
  as proof of German descent 165–166
alcohol 160
  alcoholism 106, 131, 134–135, 141–142, 154
  effects on heredity 43
Alpine (race) 24, 75, 83–84, 121
amaurotic idiocy (Tay sachs) 208, 213
Améry, Jean 167
Ammon, Otto 79, 94
Aryan (race) 24, 114, 165, 180–181, 189–192
Ast, Friedrich (Fritz) 138
Astel, Karl 175
Auerbach, Elias 75

Bareth, Karl 188
Bartenstein, Max 37
Bastards/Hybrids/*Mischlinge* 28–32, 90, 185–186, 189–192. *See Rehoboth Bastards*
  the effect of hybridization 8–11, 38, 90, 92, 97–100, 111
  as inferior/dangerous 24, 92–95, 98, 100–101, 167–169, 173–176, 188, 196–203
Bateson, William 32–33, 36, 86, 91, 111, 137
  presence-and-absence model 95
  views on eugenics 217–219
Bauman, Zygmunt 103
Baur, Erwin 33, 83, 85–86, 143, 172
Behr-Pinnow, Karl von 55
Bernstein, Felix 79, 82, 84
Berze, Josef 66, 71–72
blending inheritance 86
Bleuler, Eugen 65–67, 71
blindness
  color-blindness 137, 154
  as a Mendelian condition 136, 138

  as target of eugenic legislation 131–135, 142–143
blondness 75, 184–187
  among Jews 110, 185, 201
  as recessive trait 103, 121, 185
blood type 79–84, 87, 195, 230
blue eyes
  among Jews 110, 201
  among Nordic and Alpine races 75
  as example of Mendelian heredity 21, 126, 152, 184–187
Boas, Franz 79, 92, 226
Boeters, Gustav 132–133
Bormann, Martin 200–201
brachycephaly (short/broad skull) 75–76, 79, 110
Breymann, Hans 56, 59–62

Castle, William E. 16, 172
Central Office for German Personal and Family History in Leipzig 27, 49, 56, 62
cephalic index *See* skull
Chamberlain, Houston Stewart 24, 83
chromosomes
  in racial mixture 99
  study of – 6, 33, 86
  as taught in schools 147
Clauss, Ludwig Ferdinand 123, 184
consanguinity 55, 105–107, 111–112, 115–116, 119–120, 160, 220
Correns, Carl 28, 31–33, 83, 86, 146–147
criminality
  gene for – 17, 81, 98, 106
  in German sterilization law 141
  as impediment to German citizenship 174
  as target of eugenics 131, 133, 139, 142, 154, 221
Crzellitzer, Arthur 50–53, 56, 59–60, 118